AF760069

Introduction to Mathematical Techniques Used in GIS

GIS technology and applications have advanced as fast as computing technology to enhance business analytics, predictive modeling, virtual reality, and artificial intelligence. The third edition addresses these new topics of interest to students and practitioners who are using geographic information systems but have a limited mathematical background. Thoroughly updated and reorganized to focus more on applications and problem-solving by mathematical techniques, this book explains the basic architecture of computing as it relates to GIS, includes new application examples of selected mathematical methods, and introduces 3D modeling, machine learning, and more.

Features

- Explains the basic mathematics that underpins the manipulation of spatially related data and adds new technology direction, such as machine learning.
- Builds logically step-by-step from simple basic assumptions to real-world GIS applications to illustrate mathematical techniques covered in each chapter.
- Explains computing fundamentals, including databases, and modeling techniques such as network modeling and topology overlay.
- Includes two new chapters focused on how computing relates to mathematics and new popular applications of GIS, which connect with data science and artificial intelligence.
- Prepares today's GIS students who do not have STEM backgrounds to follow the thought processes behind the practice of GIS.

This textbook is written for those who use global information systems and applications but have a limited mathematical background. It explains the mathematics behind the applications, making it an accessible book for both undergraduate and graduate students in GIS, Computer Science, Geography, and Geomatics.

Introduction to Mathematical Techniques Used in GIS

Third Edition

Gordon Ye and Peter Dale

CRC Press
Taylor & Francis Group
Boca Raton London New York

CRC Press is an imprint of the
Taylor & Francis Group, an **informa** business

Designed cover image: Getty Images

Third edition published 2026
by CRC Press
2385 NW Executive Center Drive, Suite 320, Boca Raton FL 33431

and by CRC Press
4 Park Square, Milton Park, Abingdon, Oxon, OX14 4RN

CRC Press is an imprint of Taylor & Francis Group, LLC

Second Edition published by CRC Press 2014

First Edition published by CRC Press 2004

ISBN: 978-1-032-73472-9 (hbk)
ISBN: 978-1-032-73475-0 (pbk)
ISBN: 978-1-003-46440-2 (ebk)

DOI: 10.1201/9781003464402

Typeset in Times
by KnowledgeWorks Global Ltd.

Access the Support Material: www.Routledge.com/9781032734729

Contents

List of Illustrations

List of Tables

List of Boxes

List of Examples

Preface to Third Edition

New developments in GIS technology and applications in the industry have occurred at a fast pace to be in step with the continued advancement of computing technology and the use of such technology to enhance business analytics, predictive modeling, virtual reality, and artificial intelligence. Data Science has emerged as an increasingly important and powerful toolset in applied sciences disciplines. As with previous editions, the book's objective is to prepare today's GIS students who do not have STEM backgrounds to follow the thought processes behind the practice of GIS or spatial computing. The book intends to enable students to develop their own tool bag of mathematical and analytical techniques to meet today's demands for producing geographic information, deriving meaning, identifying patterns, and making predictions from such information. Crucial considerations of data integrity and computational efficiency and understanding the limitations and potential pitfalls of widely used analytical methods are also explained. In the Third Edition, we have updated the book's contents to include the following:

1. Added material needed by today's GIS practitioners to understand the basic architecture of computing as it relates to GIS, as well as growing application fields of GIS including 3D modeling and computer graphics, modeling networks, machine learning, and artificial intelligence.
2. Organized the existing chapters of the book into five higher-level "units" to give focus to what types of applications or problems can be solved by the mathematical technique.
3. Added application examples of selected mathematical methods taught in the book, with consideration of using the right level of rigor that satisfies the needs of the application.

The tool bag of mathematical techniques covered in the revised/updated book will enable students to understand how to approach challenges or demands encountered in today's GIS practice and identify a path to meeting them.

Preface to Second Edition

Many people wishing to make use of geographic information systems (GIS) start from a limited mathematical background. In *Mathematical Techniques in GIS*, Second Edition, the text as before focuses on those who are unfamiliar with mathematics and need to understand the principles behind the manipulation of spatial data. The new text adds additional material. The first nine chapters lay out the basic foundations and introduce the reader to the relevant techniques and shorthand notations that frequently occur in mathematics; the remaining five chapters build on the earlier material. These later chapters place particular emphasis on the use of the techniques in GIS and geomatics. Throughout the text, there are a number of examples shown in boxes, and there is a summary at the end of each chapter listing all the important ideas that have been introduced.

Preface to First Edition

This book has been written for nonmathematicians who wish to understand some of the assumptions that underlie the manipulation and display of geographic information. It assumes very little basic knowledge of mathematics but moves rapidly through a wide range of data transformations, outlining the techniques involved. Many of these are precise, building logically from certain underlying assumptions; others are based on statistical analysis and the pursuit of the optimum rather than the perfect and definite solution.

Mathematics has its own form of shorthand that often gets in the way of understanding what is going on. For those who are unfamiliar with mathematical notation, this can be daunting, but it cannot be avoided. It can, in many cases, be kept to a minimum and in what follows, the derivation of some of the formulae is placed in boxes that can be digested at leisure without interrupting the narrative. But at the end of the day, a compromise has had to be made, and as the text progresses, there is an increasing use of symbols.

This spirit of compromise is most apparent in the selection of topics discussed. Many things have had to be left out—indeed, every chapter could be expanded to a full book, and most would require several volumes in order to do their subject justice. *Introduction to Mathematical Techniques Used in GIS* is therefore a book that allows the reader to get started and then to turn to the many more informative texts that are available.

The text begins with an introduction to geographic data but soon focuses on the "where" rather than the "what." It assumes that the data have been measured and refrains from discussing the techniques of measurement science, other than to recognize that measurement is prone to error. Pure mathematics, even when dealing with vague concepts, provides precise answers that can be verified by anyone. Even statistical analysis uses processes that can be programmed into a computer to give a consistent answer, even when the underlying assumptions are not met or the hypothesis has been incorrectly formulated. The apparent exactness of an answer does not mean that it is correct. To understand the output from, for example, a geographic information system, one needs to understand the quality of the data that are entered into the system, the algorithms behind the data processing, and the limitations of the graphic displays.

This book deals with only part of the bigger picture. It focuses on the basic mathematical techniques, building the whole of mathematics in a series of steps that are the foundations for a deeper understanding. It seeks to lay the foundations for the more complex forms of manipulation that arise in the handling of spatially related data.

The technology behind geographic information systems (GIS) allows such data to be gathered, processed, and displayed. The power and appeal of such systems often lie in their graphical output, the maps that they create. Users of GIS need to understand the quality of that output so that they can advise others on the integrity of their results. The issue is not a matter of which buttons to push but rather of the quality of the information that has been produced. Quality means "fitness for purpose" and "safety in use."

This book, therefore, looks at some of the fundamentals and provides an introduction to spatial data manipulation through which users of GIS may come to understand whether what they do results in what can genuinely be described as a "quality product." It has been copy-edited for an American market, hence the spelling of words such as "meter" for the English "metre" and "center" for "centre."

The Authors

Gordon Ye majored in engineering physics University of California Berkeley and soon branched out specializing in Geographic Information Systems at the UC Berkeley City Regional Planning Department. He was an instructor at the City College of San Francisco over a span of 21 years and developed and taught a well-regarded GIS certificate program. During his over 20-year tenure working for San Francisco-based Pacific Gas and Electric Co., he developed GIS mapping systems for the management of both gas and electric infrastructure. He also worked briefly for a defense contractor as a software engineer and developed software tools to build 3D virtual reality models of war zones for military training games. After the devastating gas pipeline explosion in San Bruno, California, in 2010, he has been focused on modeling gas pipeline integrity risks such as corrosion and earthquakes and their potential impacts on people and the environment. He would like to thank his long-time friend and colleague Vladimir for inspiration and feedback for this project. Gordon Ye can be contacted via e-mail at: gordonye@yahoo.com.

Peter Dale trained as a land surveyor and worked for seven years in Uganda before entering the academic world. He ultimately became a professor in land information management at the University College London. He is an Honorary President of the International Federation of Surveyors and was awarded an OBE in recognition of his services to surveying. He is now retired and lives in a remote area of Scotland.

Unit I

Data, Information, Numbers

Today's geographic information system practitioners are increasingly accessing high-powered computing systems to capture, store, and analyze information about the world around us. Handling large quantities of information has become far easier than a few decades ago, and the phrase "inundated with data" accurately describes the predicament of today's practitioners. Unit I in the book is about understanding the most fundamental element of knowledge, that is, data, which are measured by various sensing tools much like human visual, audial, tactile, and other sensory systems. Any piece of data must be stored before it can be analyzed, and it also must be interpreted in order to derive the right meaning and logic behind it. This leads to the concept of numbers: number is a logical creation of the human brain in order to correctly categorize each piece of data or observation. Modern-day computers are designed to help humans handle massive amounts of data, perform logical analysis, and communicate insights derived from these analyses to others, much like how humans created ideas and solutions before computers were invented.

The process of human logical analysis is based on mathematics, and this is the foundation that modern computing is built on, through manipulating numbers and performing mathematical tasks or calculations. In Chapter 1, we introduce the concept of numbers and how information is derived from numbers. In Chapter 2, we introduce arithmetic methods that are utilized to solve simple everyday problems. In Chapter 3, we introduce how data are stored inside computers and how computer databases work to streamline the handling of massive amounts of data and information.

DOI: 10.1201/9781003464402-1

1 Characteristics of Geographic Information

1.1 GEOGRAPHIC INFORMATION AND DATA

It used to be said that geography was about "maps," as distinct from "chaps." Without doubt, today it is about both, and a lot more besides. Ultimately, geography is about making sense of the world around us, and this is done by observing, measuring, and processing data about the environment and then presenting the information either as text or pictorially. In particular, it is concerned with why things are, where they are.

In recent years, much has been said and written about geographic information systems or GIS, which are tools that can help the process of understanding. Although there are various interpretations of what is meant by the acronym "GIS," the majority of people would accept that it includes a computer system of hardware and software that can be used to record, manage, integrate, manipulate, analyze, and display data that are spatially referenced to the Earth. The term *spatially referenced* means that their location can be described by measured quantities. *Data* are basic facts that can somehow or other be measured and turned into information.

Information is the commodity that is used by people when they make decisions. Too many facts can be confusing—there may be many different possible routes from one's home to the nearest shopping mall, each of which has its own quality of road surface, slopes, twists, turns and intersections, street lamps, drain covers, and so forth. All these facts about each route can be measured and recorded, but all that the average user really wants to know is which is the shortest route. This is a piece of information that can be extracted from the basic data.

The term *shortest* is ambiguous since it could mean shortest in terms of time or shortest in terms of distance; these are not necessarily the same thing. The types of data that need to be collected depend on the use to which the information is to be put. The required output determines the required input and the manner in which the data may need to be processed. One can of course start with a set of data and see what sense can be made of all the facts and figures. Frequently, the most effective way to do this is through pictures, especially maps and graphs. Advocates of the use of GIS often quote the 19th-century case in London where the location of cases of cholera were plotted on a map, which then showed clearly that they formed a cluster around an infected well whose water had become contaminated.

When processing data, two golden rules always apply.

1. Bad data plus good processing gives rise to unreliable information.
2. Good data plus bad processing also gives rise to unreliable information.

If data are to be converted into good quality information, then both the data and the way in which they are processed must be of good quality; that is, they must be "fit for purpose" or "safe in use." In the discussions that follow, we will focus on the basic mathematical principles underlying how data are processed and not on the technical aspects of measurement or how the data are acquired.

1.2 CATEGORIES OF DATA

Data come essentially in two forms—*categorical* and *numerical*. As their name suggests, categorical data are those that are placed in a category or codified according to a classification system. Such data are sometimes referred to as *nominal data* and have no numerical value as such. Whether a

DOI: 10.1201/9781003464402-2

piece of fruit is an apple or a pear or something else depends on the object itself and the way in which fruits have been classified. For many objects, there are internationally and scientifically recognized standards for classification, though even then there is the occasional dispute over whether some new discovery belongs as a subset of an existing class or whether it represents a totally new species.

With some data, categorization is less scientific, for instance, when designating the type of land use at a particular location. Although within each country there may be a national land use classification system, it does not mean that all those who record land use abide by it and it certainly does not follow that every country uses the same system. A building may be used in several different ways with, for example, the basement as a gymnasium, the ground floor as a shop, the next floor as commercial offices, and the top floor as a residential accommodation. In spite of national guidelines, investigators may still disagree as to how the use of the building should be categorized. It is, however, not the aim of this book to analyze the problems of data classification but rather to note that it is an issue that intimately affects the quality of data.

Once data have been categorized, they can be subjected to comparison without being quantified. Thus, the data can be placed in a rank so that a is said to be more than b, which is more than c, and so forth. Such data are described as *ordinal*, an example of which is a list of preferences (area a is a nicer place to live than area b, etc.). Various statistical tests exist to process and analyze the differences between ranks or sequences of ordinal data, but these too will not be discussed here.

Once the data have been categorized, it is often necessary to indicate their magnitude. This may be done through the use of *discrete* or *continuous* variables. A discrete variable is one that can only take distinct values, while a continuous variable is one that changes only gradually, allowing any intermediate values. Some data can only be measured in terms of whole numbers (called *integers*—such as the number of children in a family), while other items can be measured on a continuous scale (such as the height of each child). One can, of course, talk about the average family size being in the decimal system (see Chapter 2), 2.54 children, even though it is impossible to have 54 of 100 parts of a child. Such a figure is useful for some practical purposes, especially when associated with an estimate of its reliability, as discussed in Chapter 12.

Discrete variables are precise and are often expressed as whole numbers or integers (0, 1, 2, 3, etc.). More particularly, they can take a succession of distinct values at set intervals along a scale for which there are no intermediate values. Such data are often referred to as *interval data* (see Figure 1.1).

The data may be positive or negative but such items can only be compared quantitatively on the basis of the differences between them. Only when the values are *absolute* can valid conclusions be drawn about their relative sizes. One can say that a family with four children has twice as many youngsters as a family with two children because "zero children" is an absolute point of reference. One should not, however, say that a temperature of 16°C is twice as hot as a temperature of 8°C, as zero on the centigrade scale is an arbitrarily chosen point.

The highest level of measurement is the *ratio scale*, which differs from the interval scale in that it relates to absolute zero (in the case of temperature, this is approximately –273°C). Absolute temperature, length, and breadth are examples of measures on a ratio scale. They are *continuous variables* in that they are not restricted to integer forms but can take any value whatsoever from zero

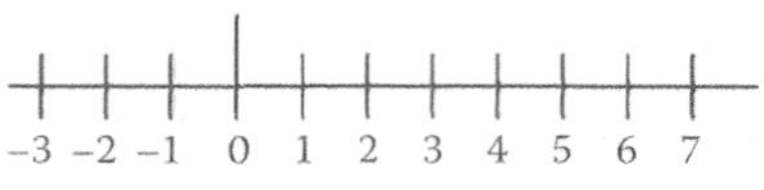

FIGURE 1.1 A scale bar.

TABLE 1.1
Points, Lines, and Areas on Maps

Feature	Points	Lines	Areas
Physical objects	Corner of building	Road network	Planning zone
Statistical values	Sampling point	Isoline	Layer tints
Areas	Central point	Boundary line	Polygon
Surfaces	Height point	Contour	Hill shading
Text	House numbers	Street names	District names

upward. The numerical quantity used to express the measurement of a continuous variable, such as the length of a line or the area of a field, presupposes a standard unit of measure. The numerical value represents the ratio between the quantity measured and the unit of measurement (e.g., the meter or "metre").

Geographical data have one particular characteristic that distinguishes them from all other forms of data, namely, location. Graphical data can be plotted on a map and be represented by points, lines, and areas. From a theoretical perspective, a point on its own has no dimension, a line has one dimension (length), and an area has two and a volume three. In practice, a point on a map is a blob or very small area, while a line has thickness and also direction. Each has a category (the "what") representing some attribute or attributes associated with it, and each has a location ("where"). Examples of how the "what" may be categorized as points, lines, and areas when used by cartographers are shown in Table 1.1.

To define the location of any point, there must be some reference to which the point can be related. The most common reference system uses a rectangular grid made up of squares of a standard size. For absolute position (as distinct from relative position), the grid must have a point of *origin* from which measurements are taken. Points may then be located so far east (or to the right) of the origin and so far north (or up the page), using a standard unit of measure. The two distances are called the *coordinates* or more particularly the *rectangular Cartesian coordinates* (named after the French mathematician Descartes).

In Figure 1.2, the rectangular grid coordinates of P are (x, y) relative to the origin and are shown here as (6, 5). The idea can be extended to three dimensions by adding the height above the origin. Although in some countries the x-direction is taken as being to the north or upward, here we will follow the convention that the direction across the page is the x-direction while up the page is described as the y-direction (x, y, or "in the door and up the stairs," which is the opposite of many

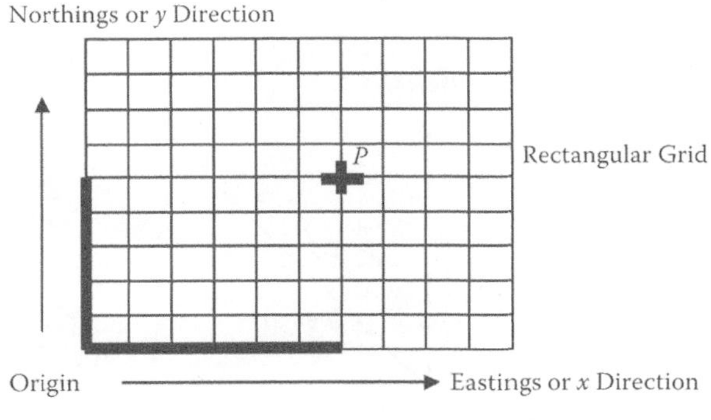

FIGURE 1.2 Rectangular Cartesian coordinates.

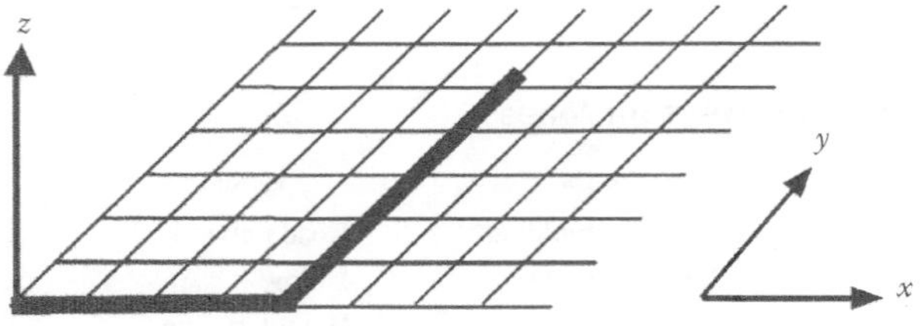

FIGURE 1.3 Nonrectangular or skewed grid.

computer software graphics packages that measure the position of points from the top of the screen downward). Height is then in the *z*-direction. For any point on a three-dimensional object, the coordinates would be (x, y, z). For two-dimensional ("flat Earth") displays, $z = 0$.

Simple mathematical techniques can be used to analyze the locations of points that have been referenced to a rectangular grid. Sometimes, it is useful to use a non-rectangular or skewed grid, for example, when trying to show three dimensions on a flat piece of paper (Figure 1.3). Data manipulation is slightly more complicated in these circumstances, although the underlying principles are the same.

An alternative way of measuring the location of a point is through the use of *polar coordinates* (Figure 1.4). These describe points by their distance from an origin and their direction relative to some reference line. The direction is known as the *bearing* and is normally measured clockwise from the north (or up the page). In Figure 1.4, *P* has polar coordinates (θ, d) relative to the origin, *d* being the distance, and θ (the Greek letter "theta") being the direction or bearing from the north.

Angles and distances are examples of measures on the ratio scale. Distances are normally expressed as a ratio, for instance, between the amount of space between two points in comparison with a standard length. Under the *Systeme Internationał* (SI), the standard unit of length is known as the *metre* and was once defined as the distance between two marks on a bar of platinum kept at constant temperature in Paris. It is now defined by stipulating that the speed of light is 299,792,458 meters (or "metres") per second. As we will show later, trigonometrical formulae allow polar coordinates (θ, d) to be converted into Cartesians (eastings and northings or *x* and *y*) and vice versa.

Angles are a ratio between the amount of turning and a complete turn. They may be expressed as a proportion of either 360 degrees—written as 360° with each degree being subdivided into 60 minutes (60′) and each minute into 60 seconds (60″); or 400 grads (where 100 grads equate with a quarter turn, with submeasurements being expressed as decimals) or 2π (two pi) radians where "pi" is the ratio between the diameter of a circle and its circumference.

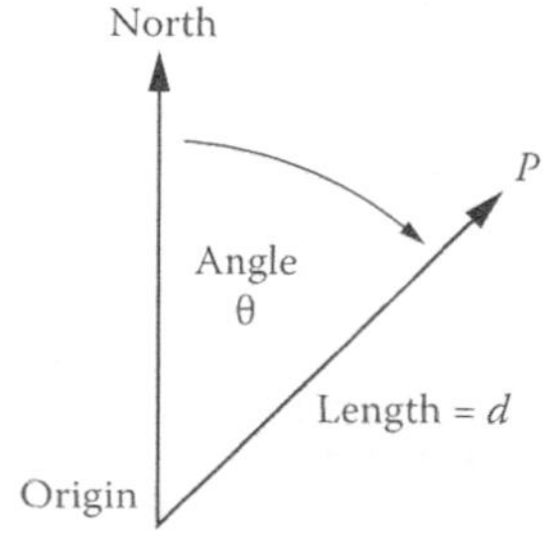

FIGURE 1.4 Polar coordinates.

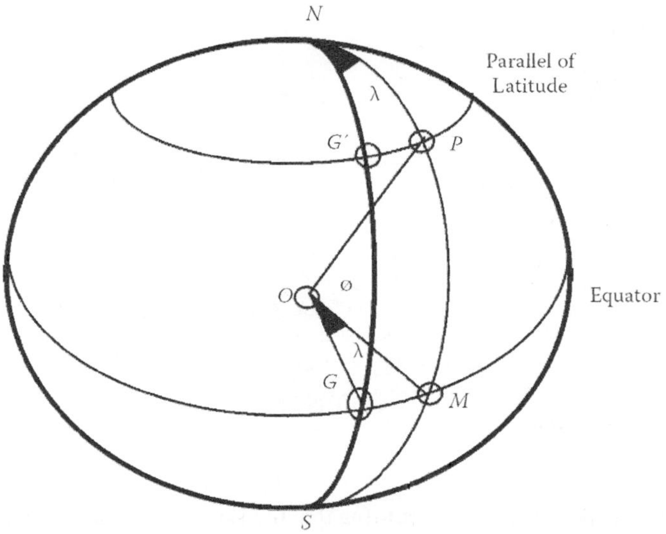

GG′ is the Greenwich Meridian of Origin (Zero Longitude)
GM is the Equator (Zero Latitude)
O is the center of the spherical Earth
The angle *GOM* is the *longitude* of *P* = λ
The angle *MOP* is the *latitude* of *P* = ø

FIGURE 1.5 Latitude and longitude.

Angular measures are important in surveying where positions may be expressed as if the Earth were a sphere using what are known as *spherical coordinates*. The *latitude* of a point is its angular distance north or south of the equator and is often represented by the Greek letter "phi" or ø. The *longitude* of a point is an angular measure east or west of the Greenwich standard meridian: it is normally represented by the Greek letter "lambda" or λ (see Figure 1.5).

The altitude or height of any point is measured as a distance above a reference level or surface, such as a mathematical shape that best approximates the size and shape of the Earth. It is not normally given a Greek letter, and hence the coordinates of points are either expressed as (ø, λ) or as (ø, λ, *H*). The use of Greek letters is common in mathematics. The full alphabet is given in Table 1.2.

TABLE 1.2
The Greek Alphabet

Letter	Name	Letter	Name	Letter	Name
Α α	**alpha**	Ι ι	**iota**	Ρ ρ	**rho**
Β β	**beta**	Κ κ	**kappa**	Σ σ	**sigma**
Γ γ	**gamma**	Λ λ	**lambda**	Τ τ	**tau**
Δ δ	**delta**	Μ μ	**mu**	Υ υ	**upsilon**
Ε ε	**epsilon**	Ν ν	**nu**	Φ ø	**phi**
Ζ ζ	**zeta**	Ξ ξ	**xi**	Χ χ	**chi**
Η η	**eta**	Ο ο	**omicron**	Ψ ψ	**psi**
Θ θ	**theta**	Π π	**pi**	Ω ω	**omega**

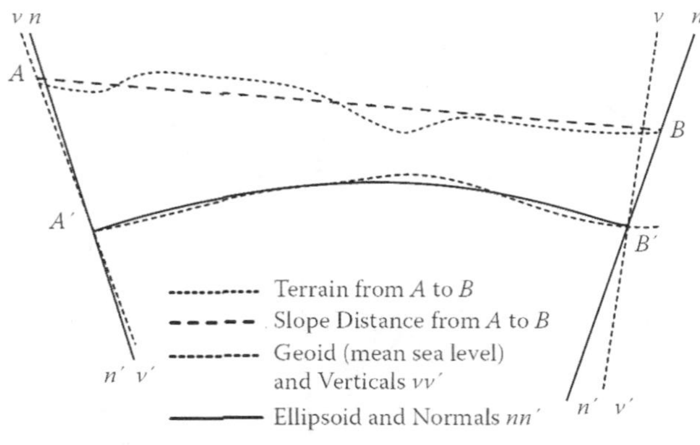

FIGURE 1.6 Straight-line distances.

For more accurate work, rather than assuming that the Earth is a perfect sphere, its shape is taken to be an *ellipsoid* (an ellipse rotated on its shorter axis creating a squashed sphere) as discussed in Chapter 4. However, for many practical purposes, the Earth can be regarded as a sphere. The word *accurate* as used here relates to nearness to the truth. The word *precision* will be used to refer to the exactness with which a value is expressed, whether the value is right or wrong. The word *precision* is often used to describe the number of decimal places that represent the quantity—for example, a distance expressed as 2.105 m (where "m" means meters) is more precise than 2 m, though the latter may be nearer to the truth and hence more accurate.

On a flat surface, a straight line represents the shortest distance between two points. On a curved surface such as a sphere, a so-called straight line bends with the surface, while a line of sight is even more bent because the light is refracted in the atmosphere. We will not deal with the consequences of the latter effect. In Chapter 11, we will consider how to transform measurements on curved surfaces onto a plane through map projections. Before doing so, measurements taken on the surface of the Earth need to be adjusted so that they fit on a mathematical surface known as the *spheroid* or *ellipsoid* of reference that is the best approximation to mean sea level. The surface defined by the mean level of the seas, assuming that they extend under the mountains, is known as the *geoid.* The difference in length between what is measured and what is calculated is important in *geodesy,* which is the scientific study of the size and shape of the Earth. Figure 1.6 illustrates how the vertical as shown by a plumb line (here the lines *vv′*) may not in practice point at the center of the Earth (the lines *nn′* are perpendicular to the mathematical surface *A′B′*) due, for example, to it being pulled aside by nearby mountains. We will discuss some of these matters later but for now we will focus on a flat Earth and two-dimensional representations.

GIS handle lines either as vectors or rasters. A *vector* is a quantity that represents both direction and distance. A polar coordinate as described above is an example of a vector quantity. Every straight line has a direction and length (such as *A* to *B* in Figure 1.7), while a curved line may be considered as being made up of a series of short lines or vectors. In Chapter 8, there is further discussion on vectors.

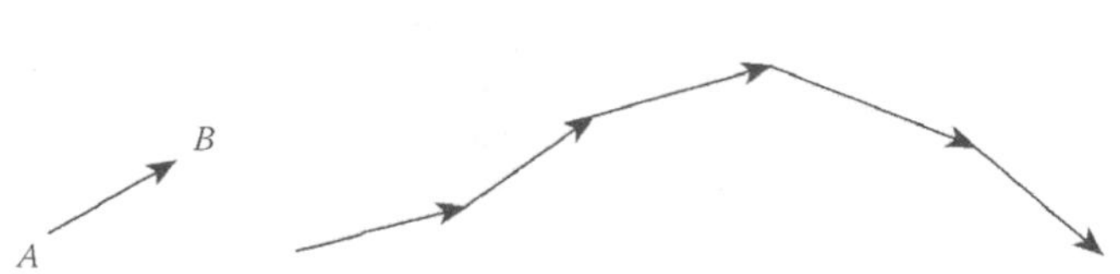

FIGURE 1.7 Lines as vectors.

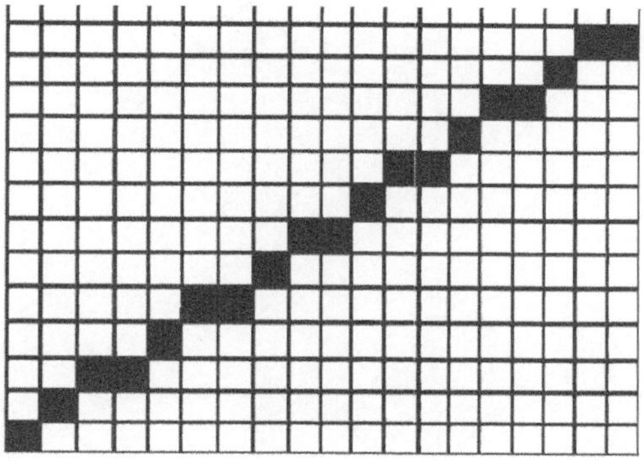

FIGURE 1.8 A line as a raster image.

On the other hand, a line can be considered as a series of points adjacent to each other, each point being of small but finite size (Figure 1.8). A television screen or a dot matrix printer produces what to the eye may appear like smooth curved lines but which in practice are a series of points on a grid. Such a representation is called a *raster image*, and each small square is known as a *pixel*.

Raster imaging can be used to show lines or areas. Raster data are simple for a computer to handle but require relatively large amounts of computer storage. Given an area 20 cm by 20 cm and a grid cell size or resolution of 100 dots to the centimeter (i.e., one-tenth of a millimeter), there will need to be a storage capacity for four million bits of information. For each vector, it is necessary to record only the coordinates of the start and end points of straight-line sections.

One of several advantages of pixelation is that each pixel can be allocated a number or set of numbers that indicate the characteristics of the point concerned. Those numbers may represent the color of the point, such as the proportion of red, green, and blue (referred to as *RGB*) to be used on a color television; or they may, for instance, represent the various wavelengths in the electromagnetic spectrum that have been picked up by a scanner, for example, in remote sensing. The values can either be analyzed through "number crunching" on a computer, or they can be displayed visually as an image that the human brain can analyze and interpret.

A line on a flat surface may be regarded as an item in its own right; alternatively, it may be regarded as the division between two areas: one on the left and the other to the right of the line. *Topology* involves the study of *adjacency*, that is, what lies beside a given area, *containment* (what is contained within it), and *connectivity* (how lines or areas are connected to other lines or areas).

Thus, in Figure 1.9, the area *B* is adjacent to the area *A*, while the area *C* is contained within the area *A*. Point *P* is connected to point *Q*. The fact that the line *PQ* may be straight or curved is of no significance; what is important is that there is a division and areas *A* and *B* lie on opposite sides

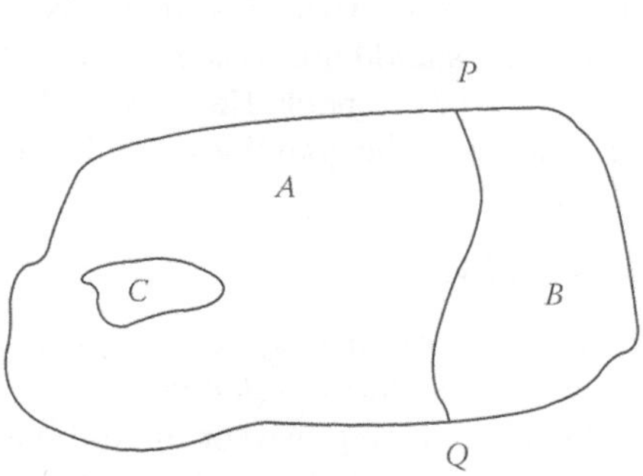

FIGURE 1.9 Topology.

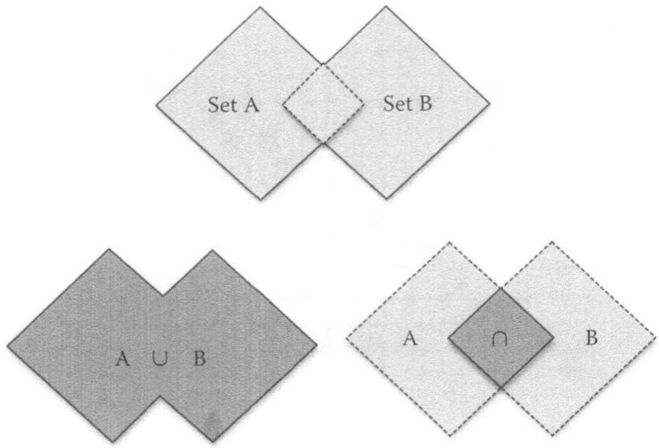

FIGURE 1.10 Union and intersection of data sets A and B.

of it. Whether something belongs to one group or another is often an important consideration in mathematics, and a specific form of symbolism or shorthand has been developed to analyze this, known as *set theory*.

As shown in Table 1.1, points, lines, and areas (and volumes) have size, shape, location, and also a classification. They each form part of a set of information. A set, also called a *class*, is a collection of related objects that can be treated as an entity in its own right. A set may be finite in size, such as the letters in an alphabet, or infinite, such as the set of integers that has no limit. A set is often identified by a pair of curved brackets such as (all human beings), or if they are in a specific sequence, angled brackets <letters in the Roman alphabet>. The symbol "∈" is used to indicate that an object is a member of a specified set {"a" ∈ < Roman alphabet >} while ∉ indicates that the object is not a member {"6" ∉ < Roman alphabet >}.

"A ⊂ B" shows that set A is a subset of set B (and "A ⊄ B" that A is not a subset), while "A ⊃ B" means that set A includes set B. "A ∪ B" is the *union* of sets A and B; that is, it is the combination of set A and set B, the symbol "∪" sometimes being referred to as *cup* while "∩," sometimes referred to as *cap*, is the *intersection* of A and B. See Figure 1.10.

Ideas of topology will be developed in more detail in Chapter 7. We will not develop the processes of classifying data, although we will be concerned with the outcomes of such classification when, in Chapters 14 and 15, we touch on elementary statistical techniques commonly used in geomatics and GIS. Throughout this book, we will focus on two and three dimensions (length, breadth, and height, or latitude, longitude, and altitude) and how relevant data may be manipulated. Time is, of course, a further dimension, and strictly speaking, we should consider not only the (x, y) or (x, y, z) coordinates of a point but also the (x, y, z, t), where t is time. Also, from a mathematical perspective, there is no inherent reason why we should not consider a world in which there are many more dimensions, but this is outside the scope of this book. Here, we will concentrate on the manipulation of numbers that represent either location or the quantities found at a location.

1.3 LINEAR REFERENCE SYSTEMS

After describing Cartesian and angular coordinate systems in the previous section, now we discuss a type of spatial coordinate system called *linear reference system*. A linear reference system is used to greatly simplify spatial data by reducing two- or three-dimensional coordinates to a single coordinate plus a reference object. The rationale for setting up linear reference coordinate systems is because the spatial location data for a specific application are constrained to a specific range of possible values, or, in other words, the location data for this application have fewer *degrees of freedom* than unconstrained location data.

For example, addresses of residences or commercial establishments in many countries are based on a linear reference system. Using the USA as an example, the physical addresses of buildings and dwelling units are constrained to be along the edges of streets that exist in each city. As a result, there is no need to specify a set of two- or three-dimensional coordinates in order to identify a building; one only needs to count the number of buildings along the street in the correct direction from the correct starting point. This is the original idea for the invention of physical address systems.

When new buildings are built in between existing buildings, the method of creating address numbers by counting houses no longer works. We will need a more absolute way to nail down the location of a building, which is, obviously, using a distance-based coordinate. By measuring the distance from the starting point of a street in the correct direction rather than counting the number of buildings, the distance measured using a length unit (such as meters, kilometers, or miles) becomes the address, and it will not change as long as said building is not physically moved to a different location (highly difficult to do so in most cases).

In real-life applications, linear reference systems are usually set up to measure distance or time along fixed paths or *space curves*. This is because the objects of interest are constrained to be located along such curves. When collecting location data of objects along such a path, if these are stationary objects, then only the spatial location is needed. If they are moving objects, then both spatial location and time may be of interest. Essentially, the coordinate system is reduced from two- or three-dimensional to one-dimensional, which greatly simplifies things. In GIS, this technique is called Linear Referencing. Figure 1.11 illustrates how accident locations and speed limits on a stretch of highway can be stored using linear referencing.

Using linear referencing terminology, the space curves that constrain the location of objects are called *Routes*. The one-dimensional coordinate values along the space curves are called *Route Measures*. Space curves are linear shapes created from location data of fixed infrastructures such as roads, pipeline networks, transmission lines, rivers, borders of land areas, coastlines, etc. Intangible infrastructure such as bus routes or shipping lanes can also be used as space curves. When creating route measures along shipping lanes or bus routes, both distance and elapsed time from the beginning point of the path can be used.

Route measure values are usually obtained from an instrument that measures the distance traveled, the time it took to travel the distance or by measuring distances on a map. Different measurement methods will yield slightly different values. This is because a physical instrument would measure distance in three-dimensional space, while a value measured from a map comes from a two-dimensional space, plus the object measured is a path drawn on the map that is commonly a simplified curve compared to the actual space curve.

How does one convert location data between a linear reference system and a Cartesian coordinate system? Using a GIS drawing of the space curve, it is necessary to record the one-dimensional

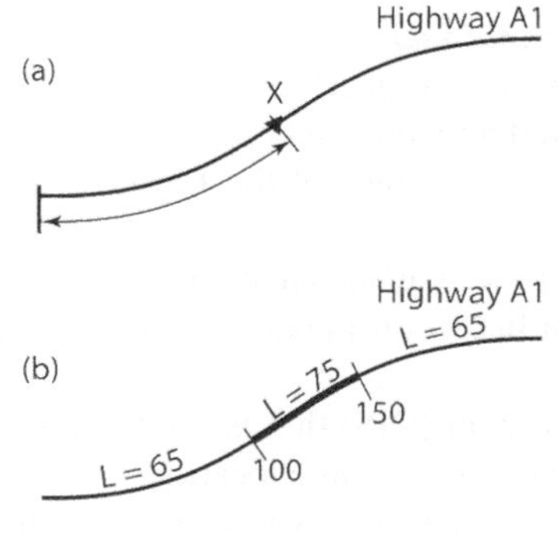

FIGURE 1.11 Linear referencing on a highway. X: accident; L: speed limit.

coordinate—the Route Measure—at selected locations along the curve. Once these data are available, one can *interpolate* via two-dimensional map distances to obtain the Route Measure value of any Cartesian coordinates of point locations on the curve or vice versa. Detailed explanations of interpolation are found in Chapter 4. Typically, a GIS software will add the Route Measure values to the Cartesian coordinates of each vertex of the space curve, resulting in a data structure in Cartesian coordinates to include four values: X, Y, Z, and M (M stands for "measure"). Having the *M coordinate* stored in the space curves enables coordinate conversions between Cartesian and Linear Referencing coordinate systems.

GLOSSARY

Absolute value The value of a quantity for which there is an absolute zero and hence the quantity can be measured irrespective of its relation to other values. It is also used in mathematics as the magnitude of a number without regard to whether it is positive or negative.

Accuracy The nearest to truth or correctness of a quantity. Distinguished from *precision*, which relates to the exactness of a quantity regardless of whether it is nearer to the truth.

Adjacency A term used in *topology* for areas that are side by side.

Bearing The direction of an object, normally quoted as an angle measured clockwise relative to the direction of north.

Cartesian coordinates Numbers that indicate the distance from an origin and in directions parallel to two or three fixed lines known as the axes.

Categorical data Data that are identified by their category or class.

Connectivity A term used in *topology* to describe which lines or areas are connected to each other.

Containment A term used in *topology* to describe objects within a given area or volume.

Continuous variable A quantity that can take any intermediate value and is not restricted, for instance, to being an *integer.*

Data Raw facts that have not been processed into *information.*

Decimal A number that relates to a tenth part (or hundredth or thousandth, etc.), part of a whole number or *integer.*

Decimal places The number of digits to the right of a *decimal point.*

Decimal point A symbol after an integer that indicates the start of a series of decimal numbers. Here, a full stop (.) or period is used while in some countries a comma (,) is adopted. Thus, one thousand will be shown as 1,000.00 rather than 1.000,00.

Degree of freedom An independent parameter that define a state of an object. Usually multiple independent parameters are required to fully describe the state, called multiple *degrees of freedom.*

Discrete variable A quantity that can only take one of a distinct set of values such as an *integer.*

Distance The amount of space between two points, often though not necessarily measured in a straight line.

Eastings The distance in a *Cartesian coordinate system* east from an *origin.*

Ellipsoid A mathematical shape that is, in effect, a squashed sphere and is a better approximation to the shape of the Earth than a pure sphere.

Equator An imaginary line on the surface of the Earth that is equidistant from the North and South Poles.

Geodesy The scientific study of the size and shape of the Earth.

Geoid The shape of the Earth based on mean sea level and its imagined extension under or over land.

GIS Geographic Information System (GIS) that is used to record, analyze, manipulate, and display information, which relates to some location.

Great circle A line on the surface of a sphere where the plane that contains the line passes through the center of the sphere.

Information Data that have been processed and presented in a form that permits decisions to be made.

Integer A whole number such as 1, 2, 3, 4, and so on.

Interpolation A method to estimate new data values based on the known data values in nearby locations.

Intersection Those elements of two or more data sets that overlap.

Interval data A set of *discrete variables* that take values at set intervals.

Latitude The angular distance of a point north or south of the *equator.*

Linear reference system A method for storing data about linear features, like roads and rivers, using relative positions along a space curve.

Longitude The angular distance of a point on the Earth's surface that is east or west of a *meridian* of origin, usually taken as the meridian through Greenwich in England.

M coordinate *Route measure* value of a point along a linear feature (called the *route*, like a road, river, or pipeline) that is stored together with Cartesian coordinates in a GIS system.

Meridian of longitude A *great circle* on the Earth's surface that passes through the North and South Poles and a given point.

Nominal data Data that are identified only by their class or category, such as apples or pears.

Northings The distance in a *Cartesian coordinate system* north from an *origin.*

Numerical data Data that are expressed in terms of numbers.

Ordinal data Data that can be arranged in a sequence, such as preferences.

Origin A fixed point that is the start point for a coordinate system from which *eastings* and *northings* can be measured.

Parallel of latitude A *small circle* on the Earth's surface along the line of which all points have the same *latitude.*

Pixel The smallest area on a display screen that can be uniquely identified.

Polar coordinates A coordinate system based on the measurements of *bearing* and *distance*, the latter being either the straight-line distance or, on a sphere, the distance along a *great circle.*

Precision The exactness of a quantity, for instance, the number of *decimal places* to which the quantity is expressed.

Raster A rectangular pattern of parallel lines that divides a screen up into a series of *pixels.*

Ratio scale data Data that are *continuous* and *absolute.*

Rectangular coordinates *Cartesian coordinates* in which the axes are at right angles.

Route A specific linear feature (like a road, river, or pipeline) with a defined measurement system along its length.

Route measure A numerical value that indicates a specific location along a linear feature (called the *route*, like a road, river, or pipeline) based on its distance from a known starting point on that route. The distance measure can be spatial or temporal.

Small circle A line on the surface of a sphere where the plane that contains the line does not pass through the center of the sphere.

Space curve Describes the path of a continuously moving point in three-dimensional space.

Spherical coordinates The location of points based on the angular measurements of *latitude* and *longitude.*

Topology The study of geometric properties and relationships that are not related to their size or shape.

Union The combination of two data sets.

Vector A quantity that has direction as well as magnitude. (The term is also used for a matrix that has only one row or one column—see Chapter 8.)

2 Numbers and Numerical Analysis

2.1 THE RULES OF ARITHMETIC

The mathematics that underpins all geographical analysis involves the application of rules, most of which are straightforward. Mathematics makes use of symbols, the most basic of which are shown in Table 2.1. We will add and explain other symbols later. In this chapter, we will consider arithmetic, which is concerned with numerical calculations such as adding, subtracting, multiplying, and dividing.

The whole of arithmetic is based essentially on seven axioms, as shown in Box 2.1. Outside arithmetic, these axioms may not apply, for instance, when two raindrops running down a windowpane come together to make one raindrop so that in symbolic form: $1 + 1 = 1$. Furthermore, computer programmers often write "N = N + 1," meaning "Take the number in the box labeled N, add one to that number and put it back in the box labeled N"; although partially an arithmetic operation, the use of the "=" sign has a different meaning from that which we are considering here.

Axiom 1 in Box 2.1 states that adding a to b has the same result as adding b to a or in symbolic form, $a + b = b + a$, for instance, $2 + 5 = 5 + 2$. This applies to basic arithmetic but elsewhere the sequence of operations is important. If you rotate a dice forward and then sideways it will finish in a different position than if you rotate it sideways and then forward (see Figure 2.1). This illustrates how in some circumstances the sequence of operations can be important and we will discuss this further in Chapter 7 in the context of what are called matrices.

In this chapter, we will deal with simple arithmetic for which the axioms in Box 2.1 are fundamental. They are all necessary but they are not quite sufficient. Consider the calculation $2 + 3 * 4$. A pocket calculator will show $2 + 3$ equals 5; then enter 4 and multiply to give the answer 20. On the other hand, $3 * 4$ equals 12. Adding 2 gives 14. The same sum done in a different order gives a different answer.

Hence, we must have rules of priority. The simplest way to handle this is to place brackets around the groups that are together. Thus, in the first case, we have $(2 + 3) * 4$, while in the second case, we have calculated $2 + (3 * 4)$. We must distinguish between these two cases. This leads to Rule 2.1 in Box 2.2.

TABLE 2.1
Standard Symbols

Add	+	Less than	<	Equal	=
Subtract	–	More than	>	Nearly equal	≈
Multiply	*	Less or equal	≤	Not equal	≠
Divide	/	More or equal	≥		

DOI: 10.1201/9781003464402-3

BOX 2.1 AXIOMS OF ARITHMETIC

AXIOM 1: THE COMMUTATIVE LAW

For any numbers ***a*** and ***b***,

a plus ***b*** has the same value as ***b*** plus ***a***.

Also,

a multiplied by ***b*** has the same value ***b*** multiplied by ***a***.

Put into symbols,

$$a + b = b + a$$
$$a * b = b * a$$

The latter may also be written as

$$ab = ba$$

AXIOM 2: THE ASSOCIATIVE LAW

For any three numbers ***a***, ***b***, ***c***,

$$(a + b) + c = a + (b + c)$$

Also,

$$(ab)c = a(bc)$$

AXIOM 3: THE DISTRIBUTIVE LAW

For any numbers ***a***, ***b***, ***c***, then

$$a * (b + c) = a * b + a * c$$

or

$$a(b + c) = ab + ac$$

AXIOM 4

There is a number called **zero** (0) that for any ***a***

$$a + 0 = a$$

(Historically, there was a debate as to whether 0 was actually a number. The Romans, for example, used the symbols I, V, X, L, C, D, and M to represent 1, 5, 10, 50, 100, 500, and 1000 in the decimal system but had no zero.)

AXIOM 5

There is also a number 1 such that

$$a * 1 = a$$

AXIOM 6

For every value of ***a*** there is a number ***d*** such that

$$a + d = 0$$

(This then introduces the whole range of negative numbers.)

AXIOM 7

Provided c does *not* equal zero ($c \neq 0$), then

$$\text{if } c * a = c * b, \text{ then } a = b$$

(Similarly, if $a + c = b + c$, then $a = b$ although this also applies if $c = 0$ but not when c is infinitely large.)

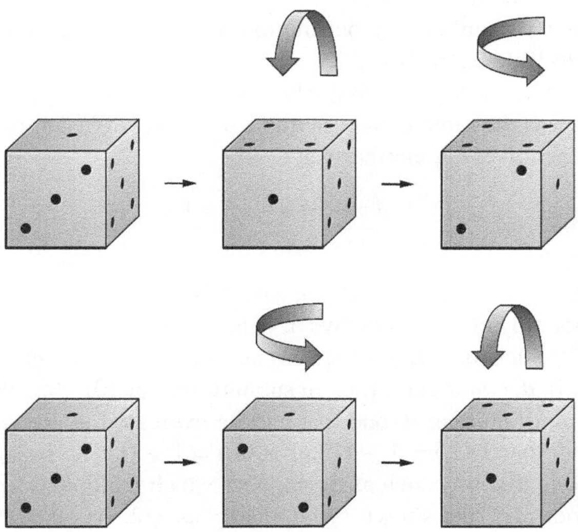

FIGURE 2.1 Rotating a dice (forward + sideways or sideways + forward).

BOX 2.2 RULES OF ARITHMETIC

RULE 2.1

Place items that are together in brackets and deal with what is inside the brackets first.

RULE 2.2

If there are no brackets or when what is inside the brackets has been evaluated, then deal with multiplication or division before addition or subtraction.

RULE 2.3

$$(\text{positive}) * (\text{positive}) = + = (\text{positive}) / (\text{positive})$$
$$(\text{positive}) * (\text{negative}) = - = (\text{positive}) / (\text{negative})$$
$$(\text{negative}) * (\text{positive}) = - = (\text{negative}) / (\text{positive})$$
$$(\text{negative}) * (\text{negative}) = + = (\text{negative}) / (\text{negative})$$

RULE 2.4

To add or subtract fractions they must share a common denominator.

Where there are no brackets, then we have to decide which comes first—addition (+), subtraction (–), multiplication (*), or division (/). In fact, it does not normally matter whether we multiply and then divide in that order, or divide first and then multiply. Thus,

$$3 * 4 / 2 = (3 * 4 / 2) = 12 / 2 = 6$$

while 3 * (4 / 2) = 3 * 2, which also equals 6.

The same happens with addition and subtraction. Thus,

$$2 + 3 - 4 = (2 + 3) - 4 = 5 - 4 = 1$$

while 2 + (3 – 4) = 2 – 1, which also equals 1.

When there is doubt, we should carry out the multiplication or division before the addition or subtraction. This leads to Rule 2.2 in Box 2.2.

Numbers may be positive or negative and when handling these, simple rules also apply. Thus, adding a negative number is the same as subtracting a positive, while subtracting a negative number is the same as adding a positive. Put another way:

$$4 + (-3) = 4 - 3 = 1$$

while 4 – (–3) = 4 + 3 = 7.

Multiplication and division follow Rule 2.3 in Box 2.2.

Ordinary numbers not only come as positive or negative but also as integers or real numbers. An *integer* is a whole number such as 1, 2, 3, or 4. *Real* numbers are basically what happens in between and are often expressed in *decimal* parts (i.e., in submultiples of 10). Real numbers are either rational or irrational. A *rational* number is one that can be expressed as the ratio of two integers, for example, the decimal number 1.125 = 1 + (125 / 1000) = 1 + (1 / 8). *Irrational* numbers are ones that cannot be expressed in this way such as the number which multiplied by itself gives the value 2, known as the square root of 2 and shown symbolically as $\sqrt{2}$. Another example of an irrational number is π, which is the ratio between the circumference of a circle and its diameter. π has been calculated to millions of decimal places with the sequence of numbers never being repeated.

A *fraction* is a number that is expressed in the form of two integers such as (*I* / *J*) where *I* and *J* are whole numbers; for example, $\frac{1}{3} = 1/3$; $\frac{5}{8} = 5/8$; and $\frac{13}{25} = 13/25$ are all fractions. *I* is known as the *numerator* (the number that indicates how many there are) and *J* is the *denominator* (giving the denomination or category of the fraction). Fractions can be converted into decimal parts simply by dividing the denominator into the numerator; thus, 5 and 8 are integers, while 5 / 8 is a fraction that can be expressed as the decimal number 0.625 (= 625 / 1000).

Numbers can also be *complex* or *imaginary* in that they have two parts, the first of which is real and the second a real number multiplied by the square root of (–1), often referred to as *i*, for instance, (2 + 3*i*). Complex numbers obey special rules that contradict some of the rules listed above, notably that when multiplied by themselves they can produce a negative answer, contrary to Rule 2.3 in Box 2.2. This cannot happen in normal arithmetic. Complex numbers are particularly useful when handling vectors (which will be discussed in Chapter 8) but will not be dealt with here. For now, we will treat all numbers as integers or as real numbers.

Many integers are the product of smaller integers each of which is a *factor* of the bigger number. Thus, the factors of 14 are two and seven since (2 * 7) = 14. Positive numbers that cannot be expressed as the product of at least two smaller integers (apart from 1) are said to be *prime* numbers. Thus, 1, 2, 3, 5, 7, 11, 13, 17, 19, 23, 29, 31, 37, 41, 43, and 47 are all the prime numbers under 50. None of these numbers can be "factorized," that is, they have no factors other than 1.

Factorizing is important when dealing with fractions as it permits simplification. By removing all the common factors from the numerator and the denominator, we can, for example, simplify $\frac{561}{2431}$ to $\frac{187*3}{187*13}$ which reduces to $\frac{3}{13}$ Finding the *highest common factor* (here 187) can greatly simplify subsequent data processing.

Conversely, we may wish to combine two factors such as adding $\frac{7}{16}$ to $\frac{13}{40}$. Here, we need to find the *lowest common denominator*, which is the smallest number into which the two denominators can divide. In this case, it would be 80 since that is the smallest integer number that can be divided by both 16 and 40. There are, of course, higher numbers such as (16 * 40 or 640) but 80 is the smallest number that is divisible by both. Hence, by expressing the first fraction as $\frac{5*7}{5*16}$ or $\frac{35}{80}$ and the second as $\frac{2*13}{2*40}$ or $\frac{26}{80}$, the two fractions can be added; $\frac{7}{16}+\frac{13}{40}=\frac{35}{80}+\frac{26}{80}=\frac{61}{80}$. This leads to Rule 2.4 in Box 2.2.

Alternatively, fractions can be expressed as decimal numbers and then processed accordingly. A decimal number consists of an integer, a decimal point, and a series of fractions of the number 10. In some countries, notably in parts of mainland Europe, a comma is used instead of a decimal point, while a full stop (or period) is used to separate out thousands. Thus, one million plus twenty-three hundredths would be written as 1.000.000,23 but in this text, we will use the comma to separate the thousands and a full stop as the decimal separator so that our number will be shown as 1,000,000.23.

Decimals are easier to add or subtract than fractions. Thus, in the example above, $\frac{35}{80}=0.4375$ while $\frac{26}{80}=0.3250$ giving a sum of 0.7625, which is the same as $\frac{61}{80}$ expressed as a decimal.

Decimal numbers use a sequence of numerals that were derived from the Arabic system, with numbers represented as the symbols 0, 1, 2, 3, 4, 5, 6, 7, 8, 9. In Arabic the actual shapes of the symbols used for these numbers are different (٠ ١٢٣٤٥٦٧٨٩) but they are based on the number 10. In the decimal system, the number 8642 represents

$$8*(10*10*10)+6*(10*10)+4*(10)+2*1.$$

By using indices as discussed later, we can write 10 * 10 * 10 as 10^3, 10 * 10 as 10^2, 10 as 10^1 where the "3," "2," and "1" are called the *indices*. 1 can be written as 10^0 (in fact, any number whose index is 0 will equal 1). This means that

$$8642=8*10^3+6*10^2+4*10^1+2*10^0$$

Continuing the sequence downward, (1 / 10) = 10^{-1}, (1 / 100) = 10^{-2}, and the like, so that 0.325, for example, represents 3 * (1 / 10) + 2 * (1 / 100) + 5 * (1 / 1000), which may be written as:

$$0.325=3*10^{-1}+2*10^{-2}+5*10^{-3}$$

2.2 THE BINARY SYSTEM

Although the decimal numbering system is convenient for human beings, there are useful alternatives. The most common is the binary system, because it is appropriate for computers. The binary system works only with the numbers 0 and 1 (*binary digits or bits*) and increases by multiples of 2 rather than 10 (Example 2.1).

Early computers used to store numbers in 8 bits, a "bit" being simply "0 or 1," "off or on," or "no charge or with charge." Each storage box within the computer system could hold 8 of these bits at a time, giving any number from 0 and 255, since 256 = 2^8. This is known as a *byte*. Now many computers use at least 64 bits. 2^{64}, which is "2" multiplied by "2" 64 times, represents a very large number.

Example 2.1 The Binary System

2 in decimals = **10** = **1** * 2^1 + **0** * 2^0 in binary (since 2^0 = 1)
3 in decimals = **11** = **1** * 2^1 + **1** * 2^0 in binary
4 in decimals = **100** = **1** * 2^2 + **0** * 2^1 + **0** * 2^0
8 in decimals = **1000** = **1** * 2^3 + **0** * 2^2 + **0** * 2^1 + **0** * 2^0
12 in decimals = **1100** = **1** * 2^3 + **1** * 2^2 + **0** * 2^1 + **0** * 2^0
14 in decimals = **1110** = **1** * 2^3 + **1** * 2^2 + **1** * 2^1 + **0** * 2^0
16 in decimals = **10000** = **1** * 2^4 + **0**

31 in decimals = **11111** = $\mathbf{1} * 2^4 + \mathbf{1} * 2^3 + \mathbf{1} * 2^2 + \mathbf{1} * 2^1 + \mathbf{1} * 2^0$
32 in decimals = **100000** = $\mathbf{1} * 2^5 + \mathbf{0}$, and so on.
255 in decimals = **11111111** = $\mathbf{1}* 2^7 + \mathbf{1}* 2^6 + \mathbf{1}* 2^5 + \mathbf{1}* 2^4 + \mathbf{1}* 2^3 + \mathbf{1}* 2^2 + \mathbf{1}* 2^1 + \mathbf{1}* 2^0$
$= 128 + 64 + 32 + 16 + 8 + 4 + 2 + 1$ in decimals

and consists of 8 bits or 1 byte.

In the binary system, the rules for addition, subtraction, multiplication, and division are similar to those for decimal numbers. For example, when adding two decimal numbers, work from the right to the left and "carry one" if over the base number 10 (Example 2.2).

The same procedure is followed with the binary system where increments go from 1 to 2 (= 10) to 4 (= 100) to 8 (= 1000) to 16 (= 10000), and so on, doubling each time and adding a zero, as shown in Example 2.3.

Similar procedures are followed for subtraction, multiplication, and division. A pocket calculator, for example, turns the decimal number that is typed in on the keypad into a binary number, processes it in that form and then returns the answer in decimals (which is why sometimes answers such as 79.99999999999999 appear instead of 80). Although numbers such as 11010101001010101 may be ideal for computers, they are tedious for human beings so in what follows we will stick to the decimal system.

Example 2.2 Decimal (Base 10) Addition

To add 83 + 649: 649 = $\mathbf{6} * 10^2 + \mathbf{4} * 10^1 + \mathbf{9} * 1$. Then, to add

$$+83 = \mathbf{0} * 10^2 + \mathbf{8} * 10^1 + \mathbf{3} * 1$$

Start on the right (singles) column. Add

$$(\mathbf{3} + \mathbf{9} = \mathbf{12} = \mathbf{1} * 10^1 + \mathbf{2} * 10^0).$$

Keep **2** and carry one over to the tens column.
Add the tens (= **4** + **8** + **1** carried over = **13** = $\mathbf{1} * 10^2 + \mathbf{3} * 10^1$).
Keep the **3** and carry one over to the hundreds column.

$$(\mathbf{6} + \mathbf{0} + \mathbf{1} \text{ carried over} = \mathbf{7}) \text{ finishing with}$$

$$\text{Total} = \mathbf{7} * 10^2 + \mathbf{3} * 10^1 + \mathbf{2} * 10^0 = 732$$

Example 2.3 Binary (Base 2) Addition

To add the binary numbers **111** + **101** (Using **bold** for binary numbers):
Given:

$$\mathbf{111} \text{ (in decimals } \mathbf{1} * 2^2 + \mathbf{1} * 2^1 + \mathbf{1} * 2^0 = 7)$$

Add:

$$\mathbf{101} \text{ (in decimals } \mathbf{1} * 2^2 + \mathbf{0} * 2^1 + \mathbf{1} * 2^0 = 5)$$

Start on the right (singles or 2^0) column

$$\mathbf{1} + \mathbf{1} = \mathbf{10} (\mathbf{1} \text{ lot of } 2^1 \text{ plus } \mathbf{0} \text{ lot of } 2^0)$$

Giving $+\mathbf{0} * 2^0$ (and carry one lot of 2^1)

Repeat on the center (twos or 2^1) column

$$(\mathbf{1}+\mathbf{0})+(\mathbf{1}\text{ carried over})=\mathbf{1}+\mathbf{1}=\mathbf{10}\left(\mathbf{1}\text{ lot of }2^2+\mathbf{0}\text{ lot of }2^1\right)$$

Giving $+\,\mathbf{0} * 2^1 + \mathbf{0} * 2^0$ (carry $\mathbf{1} * 2^2$)

Repeat on the left (2^2) column

$$(\mathbf{1}+\mathbf{1})+(\mathbf{1}\text{ carried over})=\mathbf{10}+\mathbf{1}=\mathbf{11}\left(\mathbf{1}\text{ lot of }2^3\text{ plus }\mathbf{1}\text{ lot of }\mathbf{2}^2\right)$$

Giving $+\,\mathbf{1} * 2^2 + \mathbf{0} * 2^1 + \mathbf{0} * 2^0$ (carry $\mathbf{1} * 2^3$)

Add the **1** lot of 2^3 carried over

Total

$$=\mathbf{1}*2^3+\mathbf{1}*2^2+\mathbf{0}*2^1+\mathbf{0}*2^0=1100$$
$$(=8+4+0+0=12\text{ in decimals})$$

Hence, 111 + 101 = 1100.

2.3 SQUARE ROOTS

A number that is multiplied by itself is said to be the *square* of that number. For example, $3 * 3$ is the square of three (since it is the area of a square that is three units by three units) and usually written as $3^2 = 9$. Thus, 9 is the square of three, while 3 is the square root of 9, which is normally written as $\sqrt{9}$. Similarly, a number may be cubed and have a cube root, for instance, $4^3 = 64$ while $\sqrt[3]{64} = 4$.

Squares and square roots are of particular importance when dealing with coordinate systems. According to the theorem of Pythagoras (a proof of which is given in Chapter 3), in any right-angled triangle the square of the diagonal length is equal to the sum of the squares on the two shorter sides. Thus, if we have the rectangular Cartesian coordinates for two points A and B and the difference between them in the x-direction or eastings is E and the difference in the y-direction or northings is N, then (see Figure 2.2) the square of the distance from A to B is equal to $E * E + N * N$ or $(E^2 + N^2)$. Put another way, $AB = \sqrt{\left(E^2 + N^2\right)}$.

In the very special case when $E = 3$ units and $N = 4$ units (or vice versa) then AB is exactly equal to 5 since $3^2 + 4^2 = 25 = 5^2$. A similar relationship exists with $5^2 + 12^2 = 169 = 13^2$. Normally, however, the square root will be an irrational number and not an integer. Some numbers are *perfect squares* (1, 4, 9, 16, etc.) in that their square roots are exact numbers. Most numbers do not have an exact square root; hence, any system that generates square roots can only produce an answer that is sufficient for the needs in hand.

The easiest way to find the square root of a number is to use a pocket calculator! If nothing other than a pencil and paper is available, then the value can be obtained by repeated attempts at long division until

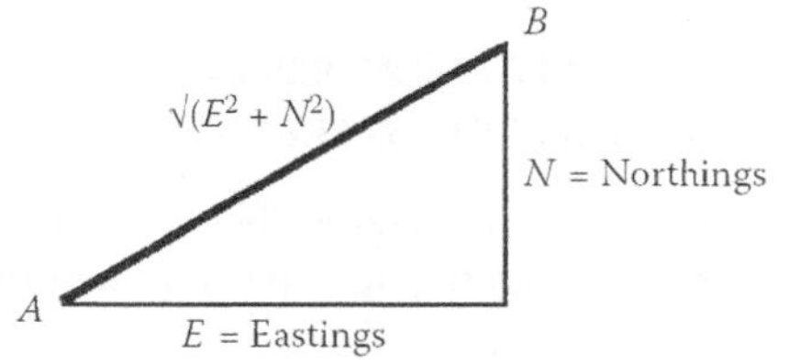

FIGURE 2.2 The distance from A to B.

an answer to the appropriate level of significant numbers is achieved. It is done by *iteration* in which one gets closer and closer to the answer until sufficient significant figures have been reached.

Example 2.4 shows the calculation of a square root using multiplication and division. It finishes up with a number that is an approximation to the true value that, like $\sqrt{2}$, is irrational and the numbers after the decimal point could go on forever. We therefore need to define a suitable level of accuracy and precision beyond which there is no merit in going. A similar problem arises in geomatics where although the calculations may be precise and accurate, even when handling uncertainty, the data that are fed into formulae may be unreliable. The fact that a distance may be calculated in meters and expressed as 123.165284 does not mean that it is accurate to a one-thousandth part of a millimeter. Such a number may be calculated to six *decimal places* (six figures after the decimal point), but it does not mean that it is reliable in terms of what is really on the ground to that degree of precision.

The extent to which such a number is reliable can be expressed in terms of its *significant figures*. This is the number of figures after which we can write zeros for all the difference it will make. If the measurement is only reliable to the nearest ten meters, we could say that rather than being 123.165284, the distance is 120 meters and is reliable to two significant figures; to the nearest five significant figures the distance would be expressed as 123.17. In this case, there has been *rounding up* in that the fifth figure (here the number 6 in "123.165284") has become 7 and the subsequent numbers have been ignored. The number 123.17 is closer to 123.165284 than is 123.16. If the original figure had been 123.164284, then it would have been *rounded down* and the best estimate of the number to five significant figures would have been 123.16.

There is, of course, a potential ambiguity when, for example, rounding 123.5 to three significant figures. Should the answer be 123 or 124? Some people would say "always round down," some "always round up," while others would go for the answer that is an even number because it can be halved. The convention used here will be that (0, 1, 2, 3, 4) round down to 0 while (5, 6, 7, 8, 9) round up to 10. Thus, 123.5 would be rounded up to 124.

Example 2.4 Finding a Square Root by Iteration

1. The first stage is to mark off the number into pairs. Thus, a number such as 27392834 becomes 27, 39, 28, 34.
2. Take the nearest square root of the first pair of significant figures (here 27). Here the best estimate is 5 (since 5 * 5 = 25). Add zeros for the remaining pairs. The first trial number is 5000. Then, working to three decimal places:
3. Calculate (1 / 2) * (Trial + number / trial)

$$= 0.5 * (5000 + 27392834/5000)$$
$$= 0.5 * (5000 + 5478.567) = 5239.284$$

4. Repeat using 5239.284 as the best estimate

$$= 0.5 * (5239.284 + 27392834/5239.284)$$
$$= 0.5 * (5239.284 + 5228.354) = 5233.819$$

5. Repeat using 5233.819 as the best estimate

$$= 0.5 * (5233.819 + 27392834/5233.819)$$
$$= 0.5 * (5233.819 + 5233.814) = 5233.816$$

To three decimal places, 5233.816 is as close to the square root of our original number as we can get. The solution has thus converged quite rapidly from 5000 through 5239.284 through 5233.819 to 5233.816.

An explanation of why this works is given in Chapter 10, Section 10.4 on linearization. It is known as *Newton's method* after the mathematician and scientist Sir Isaac Newton.

2.4 INDICES AND LOGARITHMS

Longhand calculations are tedious but there is another way than that shown in Example 2.4, whereby we can approach the problem of calculating square roots. It is based on indices. When we write the square of the number a as a^2, the number "2" is said to be the *index* of a or that a has been raised to the *power* of 2. The cube of a is a^3 with a being raised to the power of 3. If we multiply a by itself five times, then we obtain a^5—for example—2^5 or "two to the power of 5" equals 2 * 2 * 2 * 2 * 2 or 32. For any number a then if its square ($a * a$) is multiplied by its cube ($a * a * a$) we obtain $a * a * a * a * a = a^2 * a^3 = a^5$. Note that 5 = 2 + 3. As a general rule, we add the indices when multiplying the same base number raised to separate powers—see Rule 2.5 in Box 2.3.

BOX 2.3 FURTHER RULES OF ARITHMETIC

RULE 2.5: THE LAW OF INDICES

When multiplying any number raised to the power of m (a^m) by itself raised to the power of n (a^n), then we add the indices so that $a^m * a^n = a^{(m+n)}$.

RULE 2.6

Any number raised to the power of zero equals 1

$$a^0 = 1$$

Any number raised to the power of 1 is itself

$$a^1 = a$$

If we divide a to the fifth by a squared, that is, $(a * a * a * a * a) / (a * a)$ or a^5 / a^2, then the answer is $(a * a * a)$ or a^3. Here, 5 – 2 = 3. Rule 2.5 applies if n is negative as well as if it is positive. Hence, we can write a^{-n} to mean either "divide by a^{+n}" or to mean $1 / (a^n)$. A particular consequence of this is that if we divided a^n by itself then

$$\left(a^n / a^n\right) = a^n * a^{-n} = a^{(n-n)} = a^0 = 1.$$

As we stated earlier, any number raised to the power of zero equals 1. Similarly, if we divide $a^{(n+1)}$ by a^n then

$$\left(a^{(n+1)} / \left(a^n\right) = a^{(n+1)} * a^{-n} = a^{(n+1-n)} = a^1 = a.$$

This results in Rule 2.6, shown in Box 2.3.

The power of a number does not have to be an integer. Thus, for example,

$$a^{0.5} * a^{0.5} = a^{(0.5+0.5)} = a^1 = a$$

or put another way the square root of $a = a^{0.5}$. Thus, the square root of the number 10 can be expressed either as $\sqrt{10}$ or as $10^{0.5}$. Rule 2.5 above applies to all values of m and n, not just integers.

This simple rule is the basis of a system of multiplication and division known as *logarithms*. The logarithm of any number n is the *power* of a fixed number—called the *base* of the system—that gives the same quantitative value as n. Whereas any base can be used, the most common system of logarithms uses the decimal number 10 as the base. Logarithms to the base 10 are known as *common logarithms*

TABLE 2.2
Example of Seven-Figure Common Logarithms

No.	Log	No.	Log	No.	Log
1.00	**0.0000000**	4.00	**0.6020600**	7.00	**0.8450980**
1.10	**0.0413927**	4.10	**0.6127839**	7.10	**0.8512583**
1.20	**0.0791812**	4.20	**0.6232493**	7.20	**0.8573325**
1.30	**0.1139434**	4.30	**0.6334685**	7.30	**0.8633229**
1.40	**0.1461280**	4.40	**0.6434527**	7.40	**0.8692317**
1.50	**0.1760913**	4.50	**0.6532125**	7.50	**0.8750613**
1.60	**0.2041200**	4.60	**0.6627578**	7.60	**0.8808136**
1.70	**0.2304489**	4.70	**0.6720979**	7.70	**0.8864907**
1.80	**0.2552725**	4.80	**0.6812412**	7.80	**0.8920946**
1.90	**0.2787536**	4.90	**0.6901961**	7.90	**0.8976271**
2.00	**0.3010300**	5.00	**0.6989700**	8.00	**0.9030900**
2.10	**0.3222193**	5.10	**0.7075702**	8.10	**0.9084850**
2.20	**0.3424227**	5.20	**0.7160033**	8.20	**0.9138139**
2.30	**0.3617278**	5.30	**0.7242759**	8.30	**0.9190781**
2.40	**0.3802112**	5.40	**0.7323938**	8.40	**0.9242793**
2.50	**0.3979400**	5.50	**0.7403627**	8.50	**0.9294189**
2.60	**0.4149733**	5.60	**0.7481880**	8.60	**0.9344985**
2.70	**0.4313638**	5.70	**0.7558749**	8.70	**0.9395193**
2.80	**0.4471580**	5.80	**0.7634280**	8.80	**0.9444827**
2.90	**0.4623980**	5.90	**0.7708520**	8.90	**0.9493900**
3.00	**0.4771213**	6.00	**0.7781513**	9.00	**0.9542425**
3.10	**0.4913617**	6.10	**0.7853298**	9.10	**0.9590414**
3.20	**0.5051500**	6.20	**0.7923917**	9.20	**0.9637878**
3.30	**0.5185139**	6.30	**0.7993405**	9.30	**0.9684829**
3.40	**0.5314789**	6.40	**0.8061800**	9.40	**0.9731279**
3.50	**0.5440680**	6.50	**0.8129134**	9.50	**0.9777236**
3.60	**0.5563025**	6.60	**0.8195439**	9.60	**0.9822712**
3.70	**0.5682017**	6.70	**0.8260748**	9.70	**0.9867717**
3.80	**0.5797836**	6.80	**0.8325089**	9.80	**0.9912261**
3.90	**0.5910646**	6.90	**0.8388491**	9.90	**0.9956352**

and written as "log" or $\log_{10}$. In Table 2.2 it can, for example, be seen that the logarithm of the decimal number 1.60 is 0.2041200 and of 8.00 is 0.9030900. We write this as $\log_{10}(1.60) = 0.2041200$, which means that $10^{0.2041200} = 1.60$. Likewise, $8.00 = 10^{0.9030900}$. Note also that $\log_{10}(5) = 0.6989700$ which when added to $\log_{10}(1.60)$ gives 0.2041200 + 0.6989700 = 0.9030900 which is log10(8), just as 5 * 1.6 = 8.

For many scientific purposes, a different base is used, namely, the number "e" (also called the *Euler number* after the 18th-century Swiss-born mathematician Leonhard Euler) that has a decimal value that is approximately equal to 2.7182818285.

Logarithms to the base "e" are known as *Natural* or *Naperian* logarithms after the 16th-century scientist and mathematician John Napier. They are written as "ln" or as $\log_e$. The quantity "e" occurs in a variety of mathematical formulae relating to natural phenomena (hence, the name *natural logarithms*). Its value can be calculated from the formula

$$"e" = 1 + 1/1! + 1/2! + 1/3! + 1/4! + \cdots\cdots, \text{ and so on, where}$$

$$2! \text{ (spoken as two factorial)} = 1 * 2 = 2,$$

$$3! = 1 * 2 * 3 = 6,$$

$$4! = 1 * 2 * 3 * 4 = 24, \text{ and so forth.}$$

In general, *n* factorial (where *n* is an integer) =

$$n! = 1*2*3*\cdots\cdots*(n-2)*(n-1)*n$$

We will consider aspects of natural logarithms in Chapter 10. For now, we will focus on common logarithms or "logs." These use the base number 10.

Numbers between 1 and 10 have a logarithmic value between 0 and 1; between 10 and 100 the logarithmic value will be between 1 and 2, and so forth. The value for other numbers can either be calculated, for instance, using a pocket calculator, or obtained from special tables. These will show, for example, that the value for the number 5 is given as 0.6989700. Thus, the logarithm of $5 = \log_{10}(5) = 0.6989700$. Put another way, $5 = 10^{0.6989700}$.

Since $50 = 10 * 5 = 10^1 * 10^{0.6989700}$, then $50 = 10^{1.6989700}$. Thus, the log of

$$50 = \log_{10}(50) = 1.6989700$$

Similarly, the log of

$$500 = \log_{10}(500) = 2.6989700, \text{ and so on.}$$

As an example of the use of logarithms, consider $50^2 = 2500$. Since $50 = 10^{1.6989700}$ and the logarithm of 50 = 1.6989700, then

$$50 * 50 = 10^{1.6989700} * 10^{1.6989700} = 10^{1.6989700+1.6989700}$$
$$= 10^{3.3979400} \text{ by the law of indices.}$$

Log 50 + log 50 = 1.6989700 + 1.6989700 = 3.3979400. This is the logarithm of the number 2500 (see the logarithm of 2.5 in Table 2.2). Thus, adding logarithms gives the same answer as multiplying the original numbers. This gives rise to Rule 2.7 in Box 2.4.

As another example, the logarithm of the square root of 50 = half of log 50 = 0.849485. This represents the number $10^{0.849485}$, which is the logarithm of the decimal number 7.071068. Thus, the square root of 50 = 7.071 to three decimal places.

BOX 2.4 RULES FOR LOGARITHMS

RULE 2.7

Adding two logarithms means that we are multiplying together two numbers. Subtracting two logarithms means that we are dividing.

Note: To use logarithms, all numbers must be treated as positive. There is no such thing as the logarithm of a negative number.

RULE 2.8

When using logarithms (common or natural),

a. The log of $a * b = \log(a * b) = \log a + \log b$
b. The log of $a / b = \log(a / b) = \log a - \log b$
c. The log of $a^n = \log(a^n) = n * \log(a)$

Note: $\log(a \pm b)$ is not the same as $\log(a) \pm \log(b)$. Adding (or subtracting) two logarithms means multiplying (or dividing) the two numbers.

As a further example, log 100 = 2 since 100 = 10^2. Also, log 50 = 1.6989700. To divide two numbers, we subtract their logarithms so that:

$$\log(100/50) = \log 100 - \log 50 = (2 - 1.698700) = 0.3010300$$

This represents $10^{0.3010300}$ and is the logarithm of the number "2."

For numbers less than 1 and greater than 0 the power will be less than zero (see Box 2.5). For instance, 0.5 = 5 * 10^{-1} while 0.05 = 5 * 10^{-2}. Hence, log (0.5) = −1 + 0.6989700. This can be expressed in one of three ways: either

$$\log(0.5) = -0.3010300$$

or, we can separate the substantive number from the decimal point and write

$$\log(0.5) = \bar{1}.6989700 (\text{read as " bar one plus } 0.6989700\text{"}$$
$$\text{where " bar one " or "}\bar{1}\text{" means " minus 1"}).$$

Alternatively, we can add ten and call it 9.6989700 knowing that for most calculations in geomatics, we are not dealing with numbers as large as 10^{10}. So

$$\log(0.5) = 9.6989700$$

Thus, log (0.05) can appear either as −1.3010300 or else bar 2 plus 0.6989700 (= $\bar{2}.698970$) or as 8.698970.

The advantage of the second and third systems is that all logarithms are positive numbers to the right of the decimal point and all multiplications can be done by addition. On the other hand, most pocket calculators use the negative number approach.

BOX 2.5 LOGARITHMS TO THE BASE 10

$$\begin{array}{ll} 10^0 = 1 & \log(1) = 0 \\ 10^1 = 10 & \log(10) = 1 \\ 10^2 = 100 & \log(100) = 2 \\ 10^3 = 1000 & \log(1000) = 3 \\ 10^4 = 10000 & \log(10{,}000) = 4 \\ 10^5 = 100000 & \log(100{,}000) = 5, \text{ and so on.} \end{array}$$

Likewise,

$$\begin{array}{ll} 10^{-1} = 0.1 & \log(0.1) = -1 \\ 10^{-2} = 0.01 & \log(0.01) = -2 \\ 10^{-3} = 0.001 & \log(0.001) = -3 \\ 10^{-4} = 0.0001 & \log(0.0001) = -4 \\ 10^{-5} = 0.00001 & \log(0.00001) = -5, \text{ and so on.} \end{array}$$

There is no real value for log (0)—it would be minus infinity.

Note that a negative logarithm represents a positive number between 0 and 1. There can be no such thing as the logarithm of a negative number since plus and minus relate to multiplication and division and the location of the resulting decimal point. The basic rules for operating logarithms are given in Box 2.4.

Some calculations become cumbersome when using logarithms. In Example 2.5, we apply logarithms to calculating a distance between two points. The process is tedious, especially when pocket calculators are at hand, because it means we have to keep switching between logarithms and their inverse (known as *antilogarithms*). Their use is, of course, also dependent on access to tables of logarithms, known as *log tables*.

Example 2.5 Calculating Distances Using Logarithms

Consider $\sqrt{(1243.18^2 + 656.91^2)}$. To square 1243.18, we must multiply it by itself.

$$\log(1243.18) = 3.0945340 \text{ (using seven-figure logarithms)}$$

Hence, the log of $1243.18^2 = 2 * \log(1243.18) = 6.1890680$. This represents the logarithm of the number 1545496. Similarly, $\log(656.91^2) = 2 * 2.8175059 = 5.6350118$. This represents the logarithm of the number 431530.8. Using seven-figure logarithms, we have

$$(1243.18^2 + 656.91^2) = (1545496 + 431531) = 1977027$$

Now,

$$\log(1977027) = 6.2960126$$

To take the square root, we halve the logarithm

$$\text{Half } \log(1977027) = 3.1480063$$

To two decimal places, this is the logarithm of 1406.07. Hence,

$$\sqrt{(1243.18^2 + 656.91^2)} = 1406.07.$$

One consequence of Rule 2.8(c) in Box 2.4 is that

$$\log_{10}(e^y) = y * \log_{10}e.$$

Also,

$$\log_e(e^y) = y \log_e(e) \text{ or we can write this as } \ln(e^y) = y \ln(e)$$

But,

$$\log_e(e) = \ln(e) = 1 \text{ just as } \log_{10}(10) = 1$$

Hence,

$$\ln(e^y) = \log_e(e^y) = y \log_e(e) = y$$

As a result,

$$\log_{10}(e^y) = \log_e(e^y) * \log_{10}(e)$$

Replacing e^y by a,

$$\log_{10}(a) = \log_e(a) * \log_{10}(e)$$

Example 2.6 Calculating the Value of Logarithms

In Chapter 10, Section 10.3, we derive a formula known as *Taylor's expansion*, which allows us to calculate the value of natural logarithms. It takes the form

$$\ln(1+z)=\log_e(1+z)=z-z^2/2+z^3/3-z^4/4+z^5/5-$$

Using the modulus for common logarithms, we obtain

$$\log_{10}(1+z)=0.434294482*(z-z^2/2+z^3/3-z^4/4+z^5/5-\cdots)$$

If z = 0.2, then considering the first eight terms:

$$\log_{10}(1.2)=0.434294482*(0.2-0.02+0.0026667-0.0004+0.000064$$
$$-0.0000107+0.0000018-0.0000003\cdots.)=0.0791812$$

This confirms the value that we quoted in Table 2.2. Other logarithms can be calculated in a similar manner.

"e" has the value 2.718... and, hence, we can calculate the value of $\log_{10}(e)$. This number is known as the *modulus* of common logarithms.

$$\log_{10}(e)=0.4342944819$$

Thus,

$$\log_{10}(a)=0.434294482 * \log_e(a)=0.434294482\ \ln(a)$$

This applies for any positive value of *a*. Thus, we can convert common logarithms into natural logarithms and vice versa. We show an application for this in Example 2.6.

Logarithms are most effective when there are several multiplications or divisions or when dealing with certain statistical distributions. They also appear in integral calculus, as we will see in Chapter 10.

2.5 ORDERS OF MAGNITUDE

Order of magnitude is usually understood to denote the scale of numbers when comparing them to each other. By convention, one order of magnitude refers to numbers within 10 times from each other, if a number is more than 10 times larger or smaller, it is referred to as being in a different order of magnitude. In some applications the numbers encountered are all within the same order of magnitude. For example, height of adult humans typically are no more than 2 or 3 times larger or smaller than each other. In other situations, the numbers encountered can spread into multiple orders of magnitude, that is, multiple powers of 10.

A good example is the size of planets and stars (Figure 2.3). In the solar system, the largest planet Jupiter is roughly 30 times the diameter of Mercury. The Sun is roughly 10 times the diameter of Jupiter. If we measure by volume or weight the differences between these celestial bodies are even greater: the Sun would be over 800 times the volume of Jupiter, and Jupiter would be over 25,000 times the volume of Mercury. Using the terminology of Order of Magnitude, one would say that Jupiter is 4 orders of magnitude larger than Mercury by volume.

In real-life there is often a lack of understanding of Order of Magnitude, because the large differences in the values of data crossing into multiple orders of magnitude cannot be easily visualized.

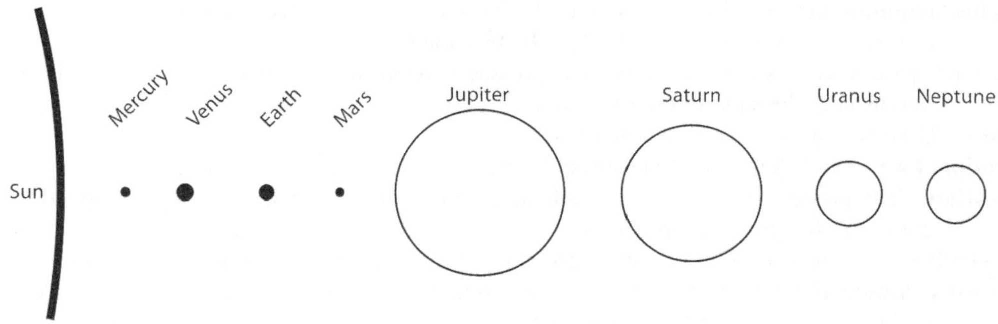

FIGURE 2.3 Size differences between the planets.

When displayed in a line graph or bar charts, either the larger values would extend "off the charts," or the smaller values are visualized as zero. For such data the trick is to apply the logarithmic scale. This means visualizing the logarithm of each value in place of the original values. This takes advantage of the characteristic of logarithm that turns multiplicative calculations into additive calculations. After taking the logarithm, values originally in different orders of magnitude typically end up being in the same order of magnitude. There are many geography-related applications that deal with data spanning multiple orders of magnitude, such as population density, area or population size of countries, amount of annual rainfall between different regions, to name a few. Such quantities are best visualized and analyzed in logarithmic scales, and it essentially converts non-linear, exponential distributions to more-or-less linear distributions in statistics (see Chapter 14).

GLOSSARY

Antilogarithm The inverse of a *logarithm*. If, using four-figure logarithms, log (2) = 0.3010, then the "antilog" of 0.3010 is the number 2.

Binary system A system of counting using only two numbers, 0 and 1.

Bit Binary digits 0 and 1.

Byte A number made up of 8 *bits*, hence in decimal terms a number from 0 to 255.

Common logarithm The *power* to which the number 10 must be raised to equal a given number. For instance, the common logarithm for the number 100 is 2 since $100 = 10^2$. This is written as $\log_{10}(100) = 2$.

Complex number A number containing a real and an imaginary part in the form $(x + iy)$ where $i = \sqrt{(-1)}$. The part x is *real*, y is real, but iy is an *imaginary number.*

Decimal places The number of digits following a decimal point.

Decimal system A method of counting in quantities of ten as distinct from, for example, *binary.*

Denominator The lower part of a *fraction.* It is the denomination or category of that fraction, for instance, the number "33" in the fraction 6 / 33.

Distribution Describes how values are spread out across a range of possibilities.

Exponent A number or expression, usually written as a superscript that indicates the *power* to which a number or quantity has been raised, for instance, the number 1000 is ten cubed or 10 * 10 * 10 or 10 to the power of 3 or 10^3, where the 3 is the exponent or *index.*

Factor Sometimes known as a divisor or submultiple, any quantity that divides exactly into a larger quantity—for instance—1, 2, 3, 4, and 6 are all factors of the number 12.

Factorial A function that is equal to the first *n* numbers multiplied together. For example, factorial 5 (written as 5!) = 5 * 4 * 3 * 2 * 1 = 120.

Fraction The ratio between two integers, for example, 6 / 33.

Highest common factor The largest quantity that is a factor of both the *denominator* and *numerator* of a *fraction*, for example, 3 in the fraction 6 / 33.

Imaginary number A *complex number* expressed in terms of the square root of a negative number, usually the square root of $-1 = \sqrt{(-1)} = i$.

Index Used here as the *power* of a number.

Irrational number A number that cannot be expressed as a ratio of two integers.

Iteration The process of repeating a mathematical operation, where each step is applied to the answer of the previous operation.

Logarithm The *power* to which some base number must be raised to equal a given number.

Lowest common denominator The smallest number into which all the *denominators* can be divided—for instance 6 / 33 and 4 / 55 have a lowest common denominator of 165 (which is divisible by both 33 and by 55). Also known as the least common denominator.

Natural logarithms The *power* to which the number "e" must be raised to equal a given number. "e" is the number $1 + 1 / 1! + 1 / 2! + 1 / 3! + 1 / 4! + \cdots$ where 4! means four *factorial* or $4 * 3 * 2 * 1$. "e" is approximately equal to 2.7182818285.

Numerator The top number in a *fraction* indicating how many of a certain *denominator*, for instance, the number "6" in the fraction 6 / 33.

Order of magnitude Describes the scale of numbers in relation of each other. Two numbers are said to be "within one order of magnitude" if their ratio is between 1 / 10 and 10.

Perfect square An integer that is the square of another integer.

Power of a number The number of times by which a number is multiplied by itself.

Prime number An integer that has no factors other than itself and 1.

Pythagoras The theorem that states that in any right-angled triangle the square on the hypotenuse (the length of the diagonal line) is equal to the sum of the squares on the other two sides.

Rational number A number that can be expressed as the ratio of two integers.

Real number A number that can take any value, not just integer, and is not an *imaginary number*.

Rounding up and down Deciding whether 0.5 should be 1 or 0, for instance, the decimal number 8.1465 can be rounded up to 8.147 or down to 8.146.

Significant figures The number of figures, starting from the left, after which we can write zeros for all the difference it will make to the accuracy of the answer.

Square root A number which, when multiplied by itself, gives a given quantity. Usually written with the symbol $\sqrt{}$. Thus, 4 is the square root of 16 or $4 = \sqrt{(16)} = 16^{0.5}$ since $4 * 4 = 16$.

3 Data Inside Computers

3.1 DIGITAL DATA

In order to better manage data and program algorithms in computers, it is necessary to understand how computers store and manipulate data. This section is intended to examine the basic structure of the modern computer system as it relates to geographic information.

Computers appear to be able to store all kinds of data, from numbers, text, and categories. As briefly mentioned in Chapter 2, computers fundamentally store all data in a binary system; this is decided by the fact that each elementary electronic device that store data inside computers can support two states: on or off, which are designated to represent 1 or 0 in the binary system. Each such device is said to store a *bit* of information. For example, to represent a decimal number 5 we use three binary digits 101. To store three digits 101 in a computer, three "bits" are necessary. By convention eight bits are called a *byte*. The number of possible data values represented by a byte of information is 2^8, or 256. The possible values translated into decimal numbers for a byte of data is from 0 to 255. The capacity of data storage devices and computer *memory* is measured in *kilobytes*, *megabytes*, *gigabytes*, *terabytes*, and so on.

We say that the *base* of the binary numeral system is 2, and the base of the decimal numeral system is 10. To understand how numbers are represented in the binary format, let us consider a decimal 5, which we already know is 101 in the binary form. The right-most digit, 1, corresponds to one, just like the right-most digit in the decimal system. The next digit in the decimal system corresponds to tens. In the binary system, the next digit corresponds to "twos," while the third digit from the right corresponds to "fours." In the binary number 101 we have one, plus zero twos and one fours, totaling to $4 + 0 + 1 = 5$.

As we go from right to left, each subsequent digit doubles its value: 1, 2, 4, 8, 16, …. As we go to the right the value is halved: 8, 4, 2, 1. If we continue halving the value, the next one to the right of 1 will be 1 / 2, then 1 / 4, then 1 / 8, and so on. This is how we can represent numbers with a decimal point that are not whole numbers. For example, the number 5.75 in binary is 101.11; it has one four's, zero two's, one unit, one half, and one quarter: $4 + 0 + 1 + 1 / 2 + 1 / 4 = 5.75$. To store it we need five bits plus one additional bit for the plus or minus sign, and a few more bits to specify the position of the point that is between the "101" and "11." The position of the point can float according to the value stored in additional bits. Numbers stored this way are called *floating-point numbers.*

There are several standard formats (e.g., IEEE 754) to store floating-point numbers, the most common are half, single, and *double precision* formats. When one creates databases to store data, each data field storing floating-point numbers need to have a specified precision level. Databases are covered in Section 3.2.

Half precision floating-point number occupy 16 bits, which is two bytes. It has 11 bits for the digits of the number, and 5 extra bits for the location of the floating point. Let us focus on the 5 extra bits for the floating point first. The minimum value that can be stored in five bits is 00000, the maximum is 11111, which is a decimal 31. Therefore, the floating point can be shifted as far as 15 binary positions to the left (binary 0.000000000000001) or to the right (binary 1000000000000000.0). If we fill all available bits with ones, and shift the floating point all the way to the right, we will obtain the binary number 1111111111100000, which is the decimal number 65,504. Although the range from −65,504 to 65,504 is limited, depending on the application computers can handle half precision numbers more than 10 times faster.

DOI: 10.1201/9781003464402-4

To simplify writing numbers with many leading or trailing zeros, such as 0.000000000000001 in the decimal system, a convenient notation called the Scientific Notation can be adopted. For example, the following expression simplifies writing of 0.000005:

$$0.000005 = 5 \times 10^{-6}$$

To make sense of this notation, mentally replace the number 5 with 5.0 so that it is clear where the decimal point is. Then move the decimal point to the right or to the left as indicated by the exponent, adding zeros where necessary. By convention the letter E is used to indicate that the following number is the exponent. A few examples of using this convention: 0.000005 = 5E-6, 500 = 5E2, 425 = 4.25E2. This notation is widely used in computer programs and calculators.

Precision: The number of available bits to store a number determines approximately how many decimal digits can be accommodated. This means when one sees a number on a computer screen, not all the decimal digits may be accurate. The higher the precision of the digital storage, the more accurate decimal digits can be accommodated.

For example: the decimal number 999 takes 10 bits of storage, its binary digits are: 1111100111. If there are 12 bits available, we can store only about three decimal digits disregarding leading and trailing zeros. If we are lucky, we can store up to five, as with 65,504. For example, we can store the number 54,800, which is 1101 0110 0001 0000 in the binary form. Here the digits 1101 0110 0001 are stored in bits, while the last four zeros are produced by shifting the floating point. A slightly bigger number, 54,801, cannot be stored exactly. The binary form of 54,801 is 1101 0110 0001 0001. By shifting the floating point we can only produce leading or trailing zeros, but now we need the last four bits to be 0001. As a result, the number 54,801 will be stored as 54,800. The next number after 54,800 that can be stored exactly is 54,832, which in binary is 1101 0110 0011 0000.

A *single precision* (twice the half precision) floating-point number takes 32 bits of storage, which is 4 bytes. It includes 24 bits for binary digits and 8 bits for the location of the floating point. The range of allowed numbers is from about $\pm 1.18 \times 10^{-38}$ to $\pm 3.4 \times 10^{38}$. This format can store numbers with up to seven decimal digits exactly (except for certain fractions). Numbers with more than seven decimal digits may or may not be stored exactly. A number 33,554,433 has eight consecutive decimal digits, and it will be stored only approximately. Its binary form is 10 0000 0000 0000 0000 0000 0001 and it does not fit in 24 bits of the single precision format.

There are ways to represent numbers in the computer that preserve all the decimal digits without loss of precision. For instance, one byte of memory can hold any of 256 distinct binary numbers, namely binary numbers from 0 to 11111111, or from 0 to 255 in the decimal form. Therefore, the capacity of one byte is large enough not only to encode ten decimal digits as we see them on our keyboard, but also encode 26 uppercase and 26 lowercase letters of the Latin alphabet plus a number of special characters and punctuation marks. There are different standard *mappings* (correspondences) between numbers and symbols of various alphabets. One of such standards is called *ASCII*, which stands for American Standard Code for Information Interchange. According to this standard, a capital letter "A" corresponds to the decimal number 65, the symbol "9" (nine) maps to the decimal number 57. A *white space* is encoded by the number 32. If, for instance, we want to store the word "GIS" in a computer, we can use ASCII and write three numbers, 71, 73, 83, in three consecutive bytes of memory. We will need three bytes of memory to store a three-letter word. In general, we can store text of arbitrary length in a sequence of bytes. These sequences of bytes that store text in computers are often referred to as *strings*.

In a similar fashion, we can encode the number 0.1 with ASCII. To do this we think of this number as of a word. The word "0.1" has three characters, so we need three bytes of memory where we write three numbers: 48, 46, 49. The numbers 48 and 49 correspond to "0" and "1," while 46 corresponds to the period. Since strings can have arbitrary lengths, we can store any number as long as it can be written as text without any loss of precision; such numbers have a finite number of decimal digits.

Binary number storage formats may differ between computer systems developed by different vendors, but all kinds of computer systems understand textual data due it being the most popular means to convey information. Therefore, using textual representation is convenient for information exchange between different computer systems. One might wonder why we even need floating-point formats when ASCII strings eliminate all restrictions altogether for storage of numbers. This is because given the limited number of physical storage units inside each computer, binary representations of numbers is much more efficient than storing individual decimal digits as text. Storing numbers as text strings is inefficient by a factor of 2.5: storing an integer number from 0 up to 999,999,999 will take 9 bytes as text string, but 4 bytes in the binary integer format.

Locations on Earth are most commonly specified by their longitude and latitude coordinates, as illustrated in Chapter 1. These are two angles measured in degrees from a reference point on the equator where the latitude and longitude are both zero. The longitude can be a positive or a negative number that extends from −180 to +180 degrees. This range covers the entire equator, which has the length of about 40,075 km. To find the distance along the equator that corresponds to one degree of longitude we match the range [−180, 180] with the length 40,075 km, and find that the length is 40,075 km per 360 degrees. Dividing both numbers by 360 we find the length, **L**, per one degree:

$$\mathbf{L} = \frac{40{,}075}{360} = 111.31944(4)\ (\text{km})$$

If we use whole numbers to represent longitude our accuracy will be only up to about 56 km. The number 56 km is roughly a half of 111.31944(4) km, and this means that we will be no more than 56 km off by specifying longitude as a whole number. What if we need a precision of one meter? To find the angle of longitude corresponding to one meter along the equator we divide 360 degrees by 40,075,000 meters:

$$\text{One meter angle} = \frac{360}{40{,}075{,}000} = 0.000008983156581409\overline{86}$$

This tells us that a change of about nine millionths of a degree moves a location on the equator by one meter. Suppose this precision is all we need and let us try to represent a point on the equator far away from zero degrees. Our location will be 150.111763 degrees of longitude. In the binary form this number has the digits:

100 1011 0000 1110 0100 1110 0011 1111 1111 1110 1111 0011 1001

and it takes 51 bits of storage. In order to achieve at least one-meter precision, geographic coordinates are typically stored in the double precision floating-point format. A number in the double precision format occupies eight bytes of memory, it allocates 53 bits for storing binary digits, and 11 bits for storing the position of the floating point. For example, the two values of longitude: 123.456789012344 and 123.456789012345 have the following representation in the double precision format with 53 binary digits:

1 1110 1101 1101 0011 1100 0000 0111 1111 1011 0100 1100 0010 0011

1 1110 1101 1101 0011 1100 0000 0111 1111 1011 0100 1100 0110 1001

TABLE 3.1
Maximum and Minimum Possible Values that Can Be Stored by Floating-point Numbers

Format	Bytes	Digits	Exponent	Decimal Digits	Minimum Val.	Maximum Val.
Half Precision	2	11	5	3–5	$\pm 5.96 \times 10^{-8}$	±65,504
Single Precision	4	24	8	6–9	$\pm 1.4 \times 10^{-45}$	$\pm 3.4 \times 10^{38}$
Double Precision	8	53	11	15–17	$\pm 4.5 \times 10^{-324}$	$\pm 1.8 \times 10^{308}$

respectively. The difference of 0.00000000001 degrees of longitude between these two values corresponds to about 0.00111 millimeters along the equator.

The summary of different floating-point formats and their capabilities are given in Table 3.1.

Bytes—the number of bytes of memory required.
Digits—the number of bits available to store binary digits.
Exponent—the number of bits available to store the exponent (shifting the floating point).
Decimal Digits—the number of consecutive decimal digits in a number that can be represented by the floating-point format.
Minimum Val.—the smallest absolute value (positive or negative) that can be stored in the floating-point format.
Maximum Val.—the largest absolute value (positive or negative) that can be stored in the floating-point format.

Data Types: Besides numeric values and text, there are many other data types, such as geometric shapes, dates, raster data, colors, categorical data, and so on. In computers all data ultimately must be converted to sequences of bytes, which has ramifications on the precision of the data stored. We describe a few data types commonly used in geographic information system (GIS) applications below.

Date Data Type: Dates are commonly stored in the binary format as an 8-byte integer (a whole number) that represents the number of time units since a certain reference point in time depending on the particular *operating system*. For example, for some operating systems this reference point is the midnight of January 1, 1970, and the time unit is millisecond. Date values before the reference point would be stored as negative numbers. Date values are displayed in many possible formats depending on user preference or the conventions of each country, typically consisting of multiple integers representing the year, month, day, hour, minute, second, as well as time zone. It is the job of computer operating systems to convert date and time data between different display formats, and to calculate the commonly used Gregorian calendar to obtain day of week, day of year, leap years, etc. (Table 3.2).

It is possible to encode only time and not include the date. When an accurate time is necessary it can be recorded as several integers representing seconds and a fraction of a second. Dates and time can be also stored as text strings such as "Wed May 12 14:25:39 2015," but typically if one wants to calculate quantities such as elapsed days between two dates, one needs to first convert such text strings to the binary date format.

Integer Data Type: Integers are stored in binary format without the potential loss of precision like in floating-point number formats. Not having the need to store the position of the floating point enables integer data formats to maximize the use of available bits, allowing integer formats to store slightly more *significant digits* than corresponding floating-point number formats. Single precision *unsigned integer* format can store numbers from 0 to 65,535, while the signed integer format can

TABLE 3.2
Representation of Date and Time

Number	Meaning	Range of Values
1st integer	Seconds	0–59
2nd integer	Minutes	0–59
3rd integer	Hours	0–23
4th integer	Day of month	1–31
5th integer	Month	0–11
6th integer	Year	Subtract 1900
7th integer	Day of week	0–6
8th integer	Day of year	0–365
9th integer	Daylight saving time	0 or 1
Text string	Time zone	Abbreviation, e.g., "PST"
Long integer	Offset from universal time (sec.)	0–86,400

store from –32,767 to 32,767. The double precision signed integer format would be able to store numbers as large as 2,147,483,647. This size limitation of the integer data type limits the type of data that can be stored as integers, and as a result floating-point numbers or text data representing integers are often used instead.

Text Data Type: Text in the computer memory is stored as sequences of bytes, also called strings. There are several standard ways to represent text characters with numbers. ASCII is one of the most popular standards suitable to represent the Latin alphabet, decimal numerals, punctuations and special characters, such as space, end of line, the tab character, and so on. Based on ASCII one element of text—a character—is represented in memory as one byte (only 7 bits are used). This limits the number of possible characters that can be stored using the ASCII standard and makes it unsuitable for representing text written in other writing systems. To overcome this limitation, the *Unicode* standard is introduced which uses up to four bytes of memory to represent one character of text. The Unicode standard allows characters of any world language to be represented.

Color Data Type: colors on digital images are stored as numbers. A color is first broken down into components much like the human eye decomposes any color into red, green and blue constituents, and the intensity of each component is represented by a number. A common way of encoding color is to record three intensities for red, green, and blue components as three one-byte unsigned integers in three consecutive bytes of memory. For example, the color red is encoded as:

11111111 0000 0000 0000 0000

The color cyan is encoded as:

0000 0000 1111111111111111

When color is used to shade semitransparent objects on screen, one extra value is added to represent transparency. This value is often called *alpha*, and it is stored together with intensity values. Half-transparent green can be encoded as:

0000 0000 11111111 0000 0000 01111111

TABLE 3.3
Category Data Example: "Fire Danger"

Category Name	Numeric Value
n / a	0
Low	1
Medium	2
High	3

Here the first three bytes represent the intensities of RGB, and the fourth byte is set to a decimal 127 which is about 50% of the maximum possible value 255. This format is often called RGBA, where A stands for alpha.

Categorical Data Type: categorical data are, not surprisingly, represented as numbers (often integers) in computers. The author of such data designates a unique number to represent each category. For example, the possible categories of "Fire Danger" and their numerical representations in the computer might be as shown in Table 3.3.

Geometric Object Data Type: The fundamental building block for geometric objects in GIS are points. A point in two-dimensional space has two coordinates represented by two floating-point numbers. The units of these coordinates as well as the coordinate system can vary from application to application. As we saw earlier, a point on the Earth's surface can be represented by the longitude and latitude pair of angular coordinates, where both numbers are in the units of degrees. Alternatively, the coordinate system can be Cartesian based on a specific *map projection* which represents the Earth's surface in a flat two-dimensional space, and the coordinate values would be based on distance units such as meters, kilometers, etc. For reference, Chapter 12 covers the topic of map projections.

In three-dimensional space each point has three coordinates, and each coordinate is typically encoded as a floating-point number. A GIS software might allow using different units for different coordinates. For example, two coordinates might encode latitude and longitude in degrees, whereas the third coordinate might be encoded in meter units of elevation.

More complex geometric objects such as lines and polygons can be built from points. For example, a polygon can be represented as a set of lines connected in a defined sequence. Each line, in turn, is represented by two endpoints in a specific order. Let us consider an example where we represent a rectangle inside the computer memory. This rectangle will reside in a two-dimensional space, and we represent numbers in the double precision floating-point format. Each corner of the rectangle is a point made of two numbers. The first number is the horizontal coordinate, the second number is the vertical coordinate. Therefore, our sequence of numbers will have eight elements:

$$\{x_1, y_1, x_2, y_2, x_3, y_3, x_4, y_4\}$$

The "orientation" of our rectangle also needs to be specified; this is done implicitly via the sequence of the four points. Consider that the rectangle represents an area on the map which will be filled with a specified color. As we advance from one point to the next in the order they are stored, we are moving along an edge of the rectangle. As we move along the edge downward, the area of the rectangle to be filled with color is on our right (this is the *right-hand rule* that denotes the orientation of a two-dimensional area), and if we move upward, the inside of the rectangle is on our left. If we store the points in the opposite order, the rectangle would represent a hole in the plane so that color would be applied everywhere on the outside of the rectangle rather than inside.

Memory: We have already discussed how digital data are represented inside the computer memory. Now let us look at a bigger picture. There are several types of memory. A *volatile memory* is the one that requires electric power to hold data. The volatile memory is usually called just memory or *RAM* (Random Access Memory). When we turn a computer off, all data stored in a volatile memory are gone. When we turn the computer back on, the data that was there before are all forgotten. A *non-volatile memory* on the other hand, is able to keep data even when electric power is off. A familiar example is a hard disk or a USB memory stick. Each computer has both volatile memory and non-volatile memory. The RAM in a computer is located within close proximity to the *CPU* (Central Processing Unit) and can be utilized by the CPU at much faster speeds than the non-volatile memory. However, the RAM in a computer has more limited volume compared to non-volatile memory. Data, including intermediate calculation results, are only stored in volatile memory temporarily while *software programs* are running. After a program finishes running, it will store the final results of the algorithm into a non-volatile memory for longer term storage, while freeing up the RAM memory for other programs to run.

When a computer software needs to access a number recorded in memory, it needs to know exactly where that number is located. Just like people have their residence addresses, each byte of memory also has its address. There are several memory devices in a computer, and memory in each device is separated into individual segments. Within one such segment each byte is assigned a sequential number from the beginning of the segment. Depending on whether the type of operating system is 16-bit or 32-bit, *memory addresses* are referenced using either 16-bit (2 bytes) or 32-bit (4 bytes) integers. Using a 16-bit operating system limits software programs to a much smaller amount of RAM memory than using a 32-bit operating system. Ability to use more RAM can improve efficiency and running speed of software programs that are computationally intensive.

3.2 DATABASES

The *spreadsheet* is the most common form of repository for day-to-day use to keep track of data. It organizes data by rows and columns, where each row contains all data related to an individual *entity*, the entity being a real-life person, object, or event. The entity described in each row is also called a *data record*. To distinguish a spreadsheet from a simple list, there are columns in the spreadsheet that organizes each kind of characteristics of the entities, or a *data field*.

For example, we have a spreadsheet of cities in our state or province (Table 3.4). The first column or field would contain the names of the cities, the second column would contain the population number in each city. The third column could be the future date one plans to visit the city. A spreadsheet allows the records to be sorted by one column's values, or a combination of several columns. In the above example, we could sort the spreadsheet by the population number from high to low,

TABLE 3.4
Example of a Spreadsheet

City Name	Population	Planned Visit Date
Addis Ababa	1,600,000	March 2026
Newark	350,000	July 2025
Turin	1,200,000	December 2025

this will shift the biggest city in the list to the top row. Or, we could sort the spreadsheet by the next date of visit, this will shift the city to be visited earliest to the top row. Or, we could simply sort the names of the city by alphabetical order, therefore a city whose name starts with "A" would be placed higher than one whose name starts with "T".

Databases are computer software systems and data storage files that manage data on computer systems. A spreadsheet is the simplest type of database, it is also sometimes called a *flat-file database*. Besides the flat file database, the most commonly used type of database is the Relational Database. This type of database is used by virtually all business enterprises, institutions, and everybody else who need an efficient solution to organize large amounts of data. GIS databases are typically a *relational database* with inclusion of binary data to encode the coordinates of spatial objects. These are often called *Object-Relational databases*.

The word "relational" signifies that the database is organized along the lines of relationships between different kinds of entities. A relational database typically consists of many tables that resemble spreadsheets, but each table is streamlined to store information about one kind of entity. Multiple kinds of entities would be stored in separate tables, and the relationships between different types of entities are encoded via *Primary Keys* and *Foreign Keys*. A primary key is a column in a table that uniquely identifies an entity. A foreign key is a column in a table that stores the unique identifier of a different entity. Usually, the designers of a relatively complex database would publish an *Entity-relationship diagram*, which graphically describes each entity table in the database and how their foreign key columns relate to the primary key columns of other entity tables (Figure 3.1). The design of a database is called a *Database Schema*, which consists of all the entity tables, their primary and foreign key columns, and the data types of all columns. Once created, a relational database's schema is usually stable without frequent changes. The database can then allow people who have ownership of individual entities' data to make *database updates* on the entity tables, and enable analysis of the data by decision-makers via *database reports*.

In our example in Table 3.4 pertaining to planned visits to cities in future dates, suppose that the following updates are needed as time goes on:

1. City's name has been changed by the authorities.
2. Planned visit dates are changed, or new planned visit dates are added.
3. Population of a city has been updated through a new census.

Once the above data updates have been made, we can run a report of the full data in the database which will give us a new list of planned visit dates to various cities (Table 3.5).

From the report (Table 3.5) we notice that there are two new cities added to the list, but upon closer inspection we discover that someone entered a variant spelling for the city Turin. This confusion can be avoided if the database is designed as a relational database rather than a flat file. To do so, we need to separate the two different real-life entities in the data into two separate tables: one for cities and another for planned visits. The relationship between these two entities is a *Many-to-Many relationship*, since one can visit the same city multiple times, and one can also visit multiple cities

TABLE 3.5
Normalization Example: Inconsistent Data

City Name	Population	Planned Visit Date
Addis Ababa	1,600,000	March 2026
Newark	450,000	July 2026
Turin	1,200,000	December 2025
Torino	1,100,000	July 2027
San Juan	420,000	August 2025

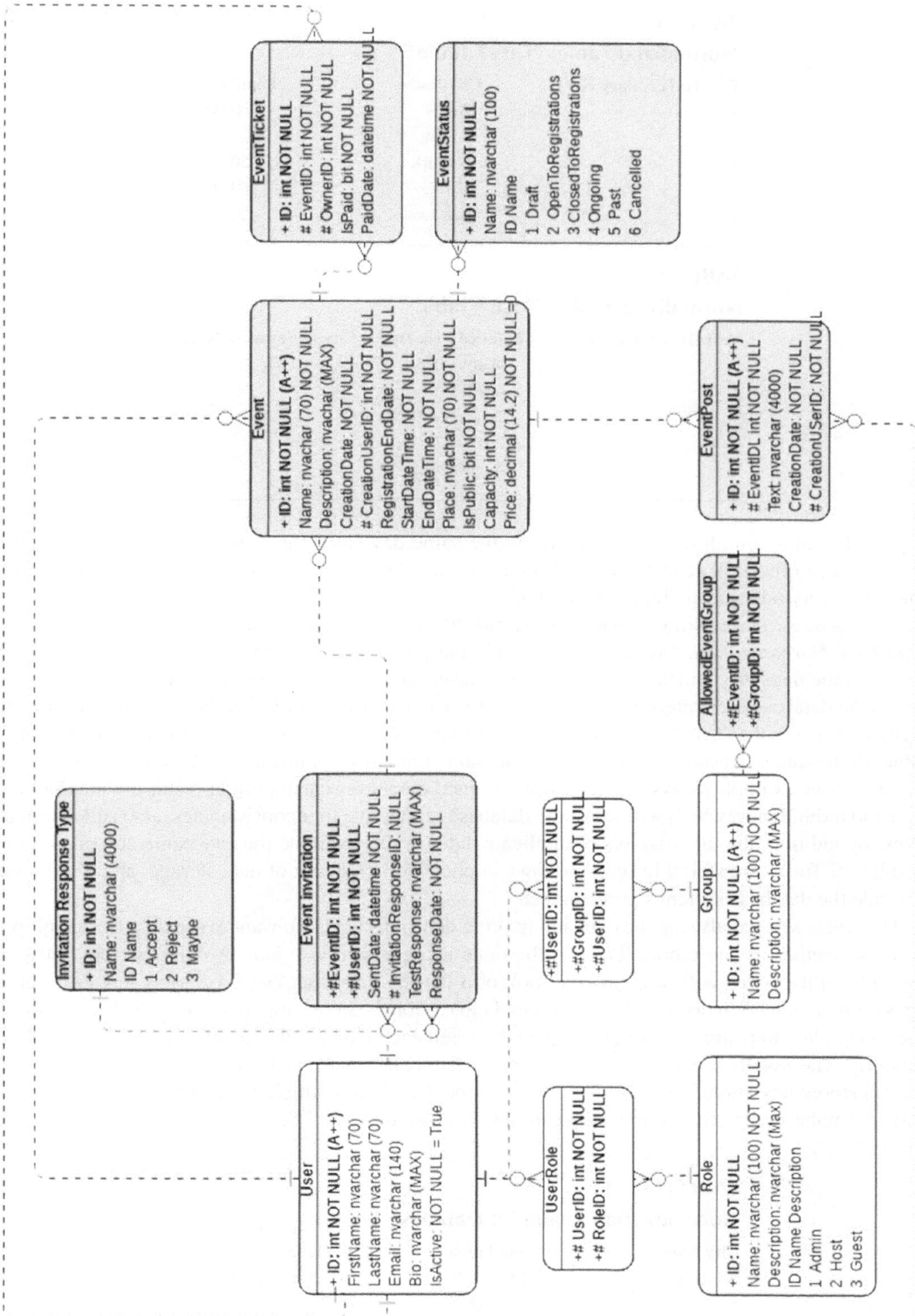

FIGURE 3.1 Example of an entity-relationship diagram.

TABLE 3.6
Normalized Table: "City" Table

City ID (Primary Key)	City Name	Population
1	Addis Ababa	1,600,000
2	Newark	450,000
3	Torino	1,100,000
4	San Juan	420,000

TABLE 3.7
Normalized Table: "Visit" Table

Visit ID (Primary Key)	Planned Visit Date	City ID (Foreign Key)
1	March 2026	1
2	July 2026	2
3	December 2025	3
4	July 2027	3
5	August 2025	4

within the same month (though unlikely on the same day since the cities are not geographically close to each other). To redesign the flat file as a true relational database, its schema would be split into two entity tables as in Tables 3.6 and 3.7.

This process of breaking down a larger flat file database into multiple entity tables is called *Database Normalization*. Normalization avoids the problem of redundant data being entered and stored in the database, which can cause discrepancies such as the city name variance we saw above. When the database user enters another entry for the city of Turin, they will notice that there is already a city of Turin in the "City" entity table. They will instead only enter a new entry to the "Visit" entity table. Both planned visits to Turin will have the same foreign key value in the "City ID" column.

The above example shows that a poorly designed database schema might result in data duplication and redundancy, which will make the database vulnerable to inconsistencies, also called anomalies. In addition, the fact that lots of duplicate data (in this example the city name and population number of Turin) are stored in separate rows implies inefficient use of data storage, and it will also degrade the database system's performance.

However, when analyzing the data and making decisions we often want to see data from multiple entities together in one report. This can be done using a *database join*. A database join between multiple entity tables will re-create the look of a flat file database. Database joins rely on foreign keys being matched to primary keys in related tables. Joins can be done with many tables as long as the foreign key to primary key relationships between each pair of related entity tables are followed properly. The results of such database joins are also called *Database views*, which, by definition, are not stored anywhere but are only "views" of the data from multiple tables arranged in a specific way. A database view from our earlier example is given in Table 3.8.

TABLE 3.8
Normalization Example: Database Report

City Name	Population	Planned Visit Date
Addis Ababa	1,600,000	March 2026
Newark	450,000	July 2026
Torino	1,100,000	December 2025
Torino	1,100,000	July 2027
San Juan	420,000	August 2025

3.3 DATABASE QUERY

Now we briefly describe the power behind database systems to manage mass quantities of data, and to retrieve a specific piece of data from millions of database records. This tool is the *Database Index*, it is the mechanism databases use to quickly find records that meet certain criteria. The database index works like a dictionary or the index in the back of a book, it relies on pre-sorted lists of references to individual row records in a database table. For any data column that is eligible to build an index for, it must have a data type that can be sorted and compared. Most common data types fit this criterion, such as numbers, text strings (to be sorted in the alphabetical order), or dates. Only very large text strings (called "memo" data type in some database systems) or Binary Large OBject data types (called *BLOB* data type) are not sortable and therefore cannot be used to build an index or a *database query*. This means that the database user cannot search for, or query specific records using values in the memo or BLOB data types.

With the aid of an index, the ability to search for specific data values in a database table relies on the highly efficient *binary search* algorithm (example in the following paragraph). In order to build the required index, it takes additional space in the computer memory, volatile or non-volatile, just like building a book index. Indexes must be updated whenever the values cataloged by the index have been updated, in order to ensure efficiency of searching. In modern database software systems, temporary indexes in volatile memory are often built on-the-fly when responding to database queries, but this is not feasible if the number of records in the data table is larger than a certain threshold. If the needed indexes are not present on a data table with millions of records, the database software system, also known as *DBMS* (Database Management System) will have poor performance when responding to queries.

Here we demonstrate the binary search algorithm—the most efficient way to search the index in the back of a book via a series of *iterations*. The words are sorted in the alphabetical order, this minimizes the number of times the user needs to check an entry in the index when looking for a specific word. For example, if the index has 100 pages and we're looking for the word "New York":

- step 1: flip to the middle of the index and see that the top of page 55 has "P**"
- step 2: flip to the dictionary before page 55 to page 30, and see that the top of page 30 has "K**"
- step 3: flip to the dictionary before page 55 and after page 30, to page 42, and see that the top of page 42 has "No**"
- step 4: flip to the dictionary before page 42 and after page 30, to page 38, and see that the top of page 38 has "Mu**"
- step 5: flip to the dictionary before page 42 and after page 38, to page 40, and see that the top of page 40 has "Na**"
- step 6: flip to the dictionary before page 42 and after page 40, to page 41, and see that "New York" is on this page. Done.

Therefore a 100-page dictionary takes an average of six to eight steps to compare the word found at each step with the word one is searching for. On average the number of search steps is proportional to the logarithm of the number of pages in the dictionary. In the database system this means the number of value comparisons the DBMS makes before finding the correct row record is proportional to the logarithm of the total number of records in the table. For normal sized databases this query process can go very quickly with the speed of computers these days.

As mentioned earlier, databases commonly used to store GIS data are Object-Relational databases, which are relational databases that have special *object data types* inserted in order to store spatial data such as points, lines, polygons, rasters, etc. The columns storing geometric data are usually called the "Shape" column, and typically utilizes the BLOB (Binary large object) data type. In general, object data types are designed to store data with complex internal structures, such as a geometry of polygon that contains multiple *rings* (in case of multiple disjointed areas, or holes within a polygon's outer boundaries), each ring containing a sequential list of points, and each point containing a *tuple* of coordinates. Such a tuple of coordinates can be (X, Y, Z, M) when we also include linear reference measures M (see Chapter 1). Object data types can be either binary or string-based; the binary format occupies much less storage space in the computer memory and is much faster to process by the software platforms designed to work with the specific binary data format. However, binary object data formats typically are not compatible with different kinds of software platforms. It would look like gibberish when one attempts to read it in a text editor. Text string-based object data on the other hand, are compatible with many software platforms from different vendors, and are standardized by software consortiums to facilitate collaboration across the internet. *XML* and *JSON* are two popular types of object data formats that stores each record as text strings.

3.4 DATABASE QUERY LANGUAGE

Database searches, also called queries, are the most basic functions of database management systems. The most common way to write queries in order to retrieve desired data from databases is via the Structured Query Language (*SQL*), an industry standard created in the 1970s. This standard enables users to request data from databases in the following ways:

- Use logical expressions as criteria to retrieve data.
- Use database joins to retrieve data from multiple entity tables into database views.
- Use calculation formulas to process data into new values.
- Aggregate data values to produce statistics such as sum, average, standard deviation, etc.
- Establish sort order and formatting of output data.

Queries compliant with the SQL standard are written as plain text and contain commands (select, insert, update, delete, etc.), the specification for the output, and the set of logical expressions to restrict the scope of data. They might also include custom functions, joins, sort order, and other instructions to the DBMS. For example, the following query is written in SQL and it requests the name and ID of cities built after 1905:

```
select ID, Name from cities where built_year > 1905;
```

Below is a more detailed list of capabilities of the SQL language:

1. Filters: also called the "WHERE clause," they are logical conditions that the data retrieved by the DBMS must satisfy. Filters typically ask for specific values or ranges of values in a database column, but multiple such logical expressions written against multiple related data tables can be combined using AND, OR, or NOT operators.
2. Database views: these are virtual tables created in response to a query to represent data retrieved, formatted, or derived from database tables.
3. Table joins: as discussed earlier, the process of combining data from multiple related tables into a single database view is called table join.
4. Calculated fields: these are virtual fields in a database view that hold calculation results from data in other data fields.
5. Formatted data: these are virtual fields in a database view that hold data transformed to different data types or formats to meet the needs of users.
6. Renamed fields: these are virtual fields in a database view with new field names that hold retrieved data.

7. Sort orders: database tables contain rows of data without a specific order. SQL language contains commands to sort retrieved data using one or more database fields. If there are multiple database fields in the sort order instruction, there is a priority ranking specified, and each field can be sorted in ascending or descending order.
8. Aggregation: SQL language can calculate summary fields to distill overall trends in the data. These summary calculations can be based on grouping instructions that divide a whole data table into sub-groups. These summary fields are typically based on statistical measures: mean (average), median, standard deviation, sum, variance, etc.

In parallel with relational database tables, GIS datasets can be queried using *spatial queries*, and can also be joined via *spatial joins*. GIS software also rely on indexes known as *spatial indexes* in order to facilitate fast searching of records by spatial location. These are similar functionality as their counterparts in relational databases, albeit more complex behind the scenes due to utilization of geometric algorithms instead of simple numerical comparisons for typical relational databases.

For example, most GIS software enable users to search for records based on how far they are in Cartesian distance from specific spatial objects, or whether they overlap with or is located inside of specific spatial objects. Such queries are called spatial queries.

GLOSSARY

Alpha Number used to store the level of transparency of a color value.

ASCII Stands for American Standard Code for Information Interchange. It's a 7-bit character code where each individual bit represents a unique character or symbol.

Base The inverse of a *logarithm*. If, using four-figure logarithms, log (2) = 0.3010, then the "anti-log" of 0.3010 is the number 2.

Bit A unit of computer data that can store one of two possible values: 0 and 1.

Binary search Searching algorithm used in a sorted list of values by repeatedly dividing the target interval in half.

BLOB Stands for Binary Large Object, a type of data formatting used in database management systems to manage a large amount of binary data as a single value.

Byte A unit of computer data that contains 8 bits.

CPU Stands for Central Processing Unit, is the main part of a computer that processes instructions and performs calculations. It is often called the "brain" of a computer.

Data field A container that stores data of a specific kind in a database record. A data field is usually represented as a column in a table.

Data record A collection of related information in a database, usually represented as a row in a table.

Database An organized collection of data based on the use of a database management system (DBMS).

Database index Data structure that speeds up the process of finding data in a database.

Database join Database management system capability to combine rows of data from multiple tables based on their mutual relationships.

Database normalization Process of organizing data in a database to reduce redundancy and improve data consistency.

Database query A tool to request for data from a database.

Database report A document that presents data from a database in an organized way to facilitate utilization of data.

Database schema A design or blueprint of how data are structured in a database.

Database update Modifying or changing the existing data or inserting new data in a database.

Database view A virtual table that provides a subset of data from a database in a user-friendly format.

DBMS Stands for Database Management System, a computerized system that stores, retrieves, updates, and analyzes data.

Double precision A floating-point number format that usually occupy 64 bits or 8 bytes of computer memory.

Entity A person, place, thing, or object that can be uniquely identified and have data stored about it.

Entity relationship diagram Visual model that shows how entities relate to each other in a database design, or schema.

Flat-file database A simple database in the form of one data table with rows and columns.

Floating-point number A format to store numbers in computer memory that allows a decimal point in any location, as opposed to integer numbers.

Foreign key A column in a relational database table that references the primary key of another table.

Gigabyte Equals to a billion bytes of binary data.

Half precision A floating-point number that occupies 16 bits of computer memory.

Iteration Repetition of a process where the output of each step is used as the input for the next step.

JSON Stands for JavaScript Object Notation, is a text-based format for storing and exchanging data among different computing platforms.

Kilobyte Equals to a thousand bytes of binary data.

Map projection A method for transforming the location of features on the curved surface of the Earth onto a flat or plane surface.

Mapping A system of specification to associate members of one group of values to another group of values. The ASCII code system is a kind of mapping.

Megabyte Equals to a million bytes of binary data.

Memory A physical storage mechanism of data inside a computer.

Memory address A unique identifier for a location in a computer's memory.

Non-volatile memory Computer memory that stores data even when the power is off, used for long-term data storage, such as disk drives and USB drives.

Object data type A computer data type that has complex internal structures, usually in the form of encapsulating many data elements within one unit of data. Typically, object data types are stored in the binary format unless it is necessary to transmit data between different computing systems.

Object-relational database A database that combines relational and object-oriented database features.

Operating system A foundational computer program that controls a computer system's hardware and software and manages how software programs are run.

Precision Level of exactness in a number's digits.

Primary key Column or combination of columns in a database table that uniquely identifies each row.

RAM Stands for Random Access Memory, is a computer's short-term memory.

Relational database A database that organizes data in multiple data tables with predefined mutual relationships.

Right-hand rule A rule followed when constructing polygon shapes from rings: exterior rings of a polygon are counterclockwise, and rings forming holes inside a polygon are clockwise.

Ring As a sub data element of a polygon, it is a collection of contiguous line segments whose start and end points are the same, forming a ring shape.

Spreadsheet A worksheet consisting of cells arranged in a table form where data can be entered, calculated, analyzed, or presented.

Significant digit Also known as significant figures, are specific digits within a number that are necessary to decide the precision of the measurement.

Single precision A floating-point number that occupies 32 bits of computer memory.

Software program A set of instructions that tells a computer how to perform tasks.

Spatial index A data structure like a database index that references spatial data records by location and shape to enable fast retrieval of data with specific spatial properties by a GIS database system.

Spatial join A functionality of GIS software that combines the data records from multiple spatial layers via spatial relationships.

Spatial Query A request for information from a spatial database based on geographic criteria, like a database query.

SQL Structured Query Language—programming language used to manage data, especially in a relational database management system.

String A sequential group of characters or symbols that constitute a data element in a computer system.

Terabyte Equals to a trillion bytes of binary data.

Tuple A finite sequence of values or elements that are ordered.

Unicode A text encoding standard maintained by the Unicode Consortium designed to support the use of text in all of the world's writing systems.

Unsigned integer A whole number that is always non-negative.

Volatile memory Computer memory that stores data temporarily and loses the data when the power is off.

White Space A number of character symbols that are rendered as a white space when displayed by a computer; examples are the space character, tab character, etc.

XML Stands for Extensible Markup Language, is a text-based format for storing and exchanging data among different computing platforms.

Unit II

Measuring Space

Space and time are two fundamental aspects of the physical world around us. Being able to fully understand space has been one of the earliest developments of mathematics a few thousand years ago. This endeavor to develop methods to measure and make sense of space is known as Geometry. Unit II in the book is about understanding the mathematical properties of three-dimensional space and mathematical calculations involving different kinds of commonly found spatial objects such as points, straight or curved lines, areas on a flat plane, areas in a three-dimensional structure like the surface of the earth, etc. Real-life geographic information system (GIS) applications will be discussed in the context of these types of spatial objects. Most real-life GIS-related problems involving manipulation of space can be solved with mastery of the mathematical concepts and methods covered in Unit II.

Unit II starts with planar geometry, that is, the mathematical tools to describe and calculate two-dimensional space using Cartesian coordinates. Simpler types of three-dimensional shapes such as spheres and cones are also discussed. In Chapter 4, we introduce the fundamental concepts of planar geometry through the mathematical tool of algebra. In Chapter 5, we introduce geometrical methods that describe and manipulate common shapes—types of spatial objects that are encountered most often by GIS practitioners. In Chapter 6, we introduce Trigonometry—the mathematical tools to make sense of angles. In Chapter 7, we discuss real-life GIS applications that combine planar geometry with computing methods and algorithms.

DOI: 10.1201/9781003464402-5

4 Algebra
Treating Numbers as Symbols

4.1 THE THEOREM OF PYTHAGORAS

We can apply the ideas of arithmetic to unknown numbers. Algebra is the body of mathematical knowledge that deals with symbols. It deals with constants (such as the number e = 2.718…. mentioned in Chapter 2) and variables, such as (x, y) that can be the coordinates of a point with x and y taking any value. Each value of x and y is a number that can be added, subtracted, multiplied, or divided, and, hence, algebra may be regarded as generalized arithmetic. Algebra is about treating and manipulating numbers as symbols.

The proof of Pythagoras' theorem given in Box 4.1 uses a number of operations that were discussed in Chapter 2, namely,

1. $a^m * a^n = a^{(m+n)}$
2. $a^m / a^n = a^m * a^{-n} = a^{(m-n)}$
3. $\left(a^m\right)^n = a^{(m^* n)}$

The proof also introduces the use of brackets so that

$$(a+b) * (c+d) = a * (c+d) + b * (c+d)$$

or

$$(a+b) * (c+d) = (a+b) * c + (a+b) * d$$

This in both cases is equal to

$$(a * c) + (a * d) + (b * c) + (b * d)$$

It also uses indices so that

$$\begin{aligned}(x+y)^1 * (x+y)^1 &= (x+y)^{(1+1)} \\ &= (x+y)^2 = x^2 + y^2 + 2xy\end{aligned}$$

It also makes the point that we can often dispense with the "multiply" and "divide" symbols by writing the above as

$$(a+b)(c+d) = ac + ad + bc + bd$$

DOI: 10.1201/9781003464402-6

BOX 4.1 ONE PROOF OF PYTHAGORAS' THEOREM

Consider a simple right-angled triangle with sides of length x, y, and z (the hypotenuse) as in Figure 4.1a. Copy the triangle three times and rotate it each time to form Figure 4.1b. This builds a square on the hypotenuse or long side.

The triangle in Figure 4.1a is half a rectangle with sides x and y. Hence, its area is half of x times y or ½ xy. The area of the outer square in Figure 4.1b is $(x + y) * (x + y) = (x + y)^2$. We can write this as $x * (x + y) + y * (x + y)$ or as $x * x + x * y + y * x + y * y$ or as $x^2 + y^2 + 2xy$ using the axioms and rules laid down in Chapter 2. The area of the outer square ABCD therefore $= x^2 + y^2 + 2xy$. It also equals four triangles plus the inner square whose sides are of length z. Four times "1 / 2 xy" (= $2xy$) plus the inner square area of $z * z$ gives a total area $= 2xy + z^2$.

Thus, $2xy + z^2 = x^2 + y^2 + 2xy$ or, subtracting $2xy$ from both sides of the equation,

$$z^2 = x^2 + y^2$$

The square on the hypotenuse is equal to the sum of the squares on the other two sides.

Using indices, we can also write

$$\frac{(a+b)}{(c+d)} = (a+b) * (c+d)^{-1}$$

using the –1 index to mean "divide by" or "one over."

Pythagoras deals with the sum of two squares. An equation, in which the highest power of any term is 2, is known as a *quadratic*. A particular quadratic form that is often important is known as the *difference of two squares*. The expression $(x^2 - y^2)$ can be *factorized* (broken into factors) into the form $(x - y)(x + y)$. Multiplying these two factors together yields

$$(x-y)(x+y) = x^2 + xy - xy - y^2 = (x^2 - y^2)$$

We will have cause to use this on a number of occasions.

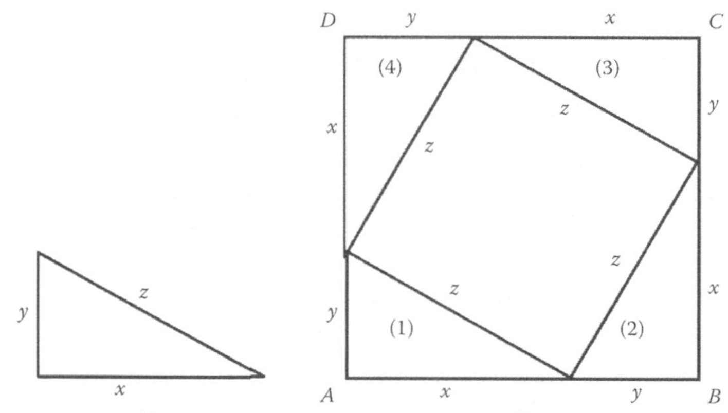

FIGURE 4.1 **(a, b)** Rotating a triangle to prove Pythagoras' theorem.

4.2 THE EQUATIONS FOR INTERSECTING LINES

The rules listed in Chapter 2 give powerful opportunities to solve problems. Consider a rectangular coordinate system with two lines A to B and C to D that intersect at P (as in Figure 4.2).

The *origin* of the coordinate system is at point O and the coordinates of

$$A = (x_A, y_A); B = (x_B, y_B); C = (x_C, y_C); \text{ and } D = (x_D, y_D)$$

Let the line AB cut the x-axis at point Q $(d, 0)$. d is said to be the *intercept* on the x-axis. (Note that superscripts are used as indices, subscripts such as the $_A$ in x_A are used as identifiers.)

Consider first a straight line through the origin (Figure 4.3a). If the coordinates of any point on this line are (x, y), then the ratio between y and x (or y / x) will always be the same. This is the slope of the line—let us call it m. We then have $y = mx$ for all points on the line, with m being a constant number and x and y being variables.

If the line slopes backward (so that as y increases, so x decreases, and vice versa) then m will be a negative amount (Figure 4.3b).

Lines do not necessarily have to pass through the origin but they can always be made to do so by changing the position of the origin. Thus, in Figure 4.2, if the origin were at point Q where the line AB crosses the x-axis (which is the line where $y = 0$) then AB will pass through this *false origin*. To achieve this, all x values must be reduced by the amount $OQ = d$.

Thus, we can express the equation of the line AB using the original coordinates in the form $y = m(x - d)$ or more generally $y = mx + c$ where $c = -m * d$. m and c are constants for the line. When we know m and c, we can calculate the value of y given x or vice versa. The line $y = mx + c$ crosses the vertical y-axis where $x = 0$ and $y = c$. It crosses the x-axis where $y = 0$. At this point $x = -c / m$.

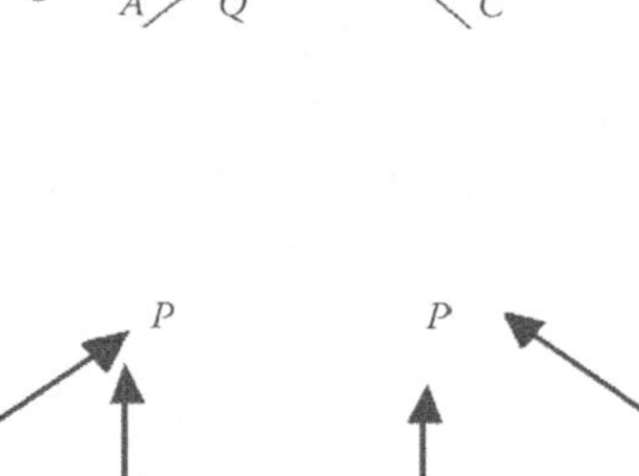

FIGURE 4.2 Intersecting lines.

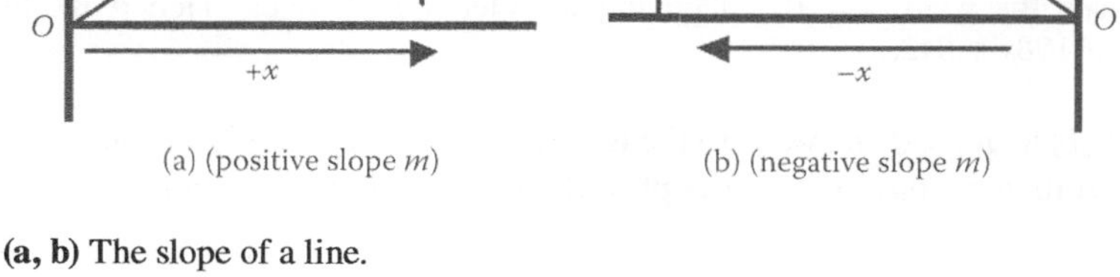

FIGURE 4.3 **(a, b)** The slope of a line.

We can use the coordinate values to calculate m and c. Since the line AB passes through (x_A, y_A) and (x_B, y_B), these values must satisfy the conditions that

$$y_A = m_1 x_A + c_1$$
$$y_B = m_1 x_B + c_1$$

We can subtract the lower numbers from the upper so that

$$\begin{aligned} y_A - y_B &= (m_1 x_A + c_1) - (m_1 x_B + c_1) \\ &= m_1 x_A + c_1 - m_1 x_B - c_1 = m_1 x_A - m_1 x_B \\ &= m_1 (x_A - x_B) \end{aligned}$$

Hence,

$$m_1 = (y_A - y_B) / (x_A - x_B) \quad \text{and} \quad c_1 = y_A - m_1 x_A = y_B - m_1 x_B$$

Thus, given the coordinates of any two points, we can calculate the parameters of the line that joins them (Example 4.1).

The relationship $y = mx + c$ is a general expression for a straight line in rectangular Cartesian coordinates. y is said to be the *dependent* variable and x the *independent*, meaning that for any chosen value of x, we can obtain a unique value of y. In fact, we could have written the equation in the form $x = ny + d$ (where $n = 1 / m$ and $d = -c / m$). x would then be dependent on y.

In Chapter 13, we discuss parametric forms, which in the case of a straight line would be $x = at + c$, $y = bt + d$ where t is the independent variable or parameter and both x and y are dependent variables. a, b, c, and d are constants.

Example 4.1 The Line Joining Two Points

Consider two points AB with coordinates

$$A(1234.56, 2345.67) \quad \text{and} \quad B(1296.32, 2417.38)$$

Then,

$$\begin{aligned} m_1 &= (y_A - y_B) / (x_A - x_B) \\ &= (2345.67 - 2417.38) / (1234.56 - 1296.32) \\ &= (-71.71) / (-61.76) = +1.161108 \\ c_1 &= y_A - m_1 x_A = 2345.67 - 1.161108 * 1234.56 \end{aligned}$$

or

$$c_1 = y_B - m_1 x_B = 2417.38 - 1.161108 * 1296.32$$

In both cases, this gives $c_1 = 912.21$ (giving an independent check). Hence, the line joining A to B is $y = 1.161108x + 912.21$.

When $t = (-d / b)$ then $y = 0$ and we have the intercept of the line on the x-axis given by $x = c - ad / b$. When $t = (-c / a)$ then we have the intercept of the line on the y-axis, namely, $y = d - bc / d$. The slope of the line will be a / b.

We can now calculate the point of intersection of two straight lines, as shown in Example 4.2. Returning to Figure 4.2, we can describe the two lines,

$$\text{the first } (AB) \text{ as } y = m_1 x + c_1$$

$$\text{the second } (CD) \text{ as } y = m_2 x + c_2$$

The lines intersect where $(m_1x + c_1) = (m_2x + c_2)$ or

$$x = (c_2 - c_1)/(m_1 - m_2)$$

Here,

$$\begin{aligned} y &= m_1(c_2 - c_1)/(m_1 - m_2) + c_1 = m_2(c_2 - c_1)/(m_1 - m_2) + c_2 \\ &= (m_1c_2 - m_2c_1)/(m_1 - m_2) \end{aligned}$$

In Figure 4.4, let the line AB be $y = mx + c$. Consider the line $y = mx + c + 1$. For this line, for any value of x, the value of y is precisely 1 unit above the value in the line AB. This means that this new line is parallel to AB but 1 unit above it. More generally, every line $y = mx + d$, where d has any fixed value, will be parallel to AB. m is a measure of the slope or angle and has the same value for all parallel lines.

Example 4.2 The Point of Intersection of Two Lines

Following on from the example in Example 4.1, for two points CD with

$$C(1300.24, 2351.77) \quad \text{and} \quad D(1212.45, 2431.78)$$

we have

$$m_2 = (-80.01)/(87.79) = -0.9113794 \quad \text{and} \quad c_2 = 3536.78$$

Or for the line CD

$$y = -0.9113794\ x + 3536.78$$

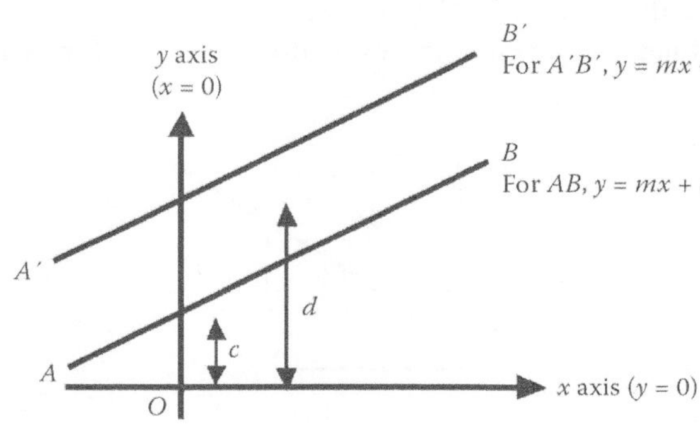

FIGURE 4.4 Parallel lines.

The point of intersection (P in Figure 4.2) between the lines AB and CD must satisfy both these conditions. Then,

$$\text{For } AB: y = 1.161108x + 912.21$$

$$\text{For } CD: y = -0.9113794x + 3536.78$$

Hence, at point P:

$$1.161108 * x + 912.21 = -0.9113794 * x + 3536.78$$

Adding 0.9113794 * x to both sides of the equation and taking 912.21 away from both sides gives:

$$1.161108 * x + 0.9113794 * x + 912.21 - 912.21$$
$$= -0.9113794 * x + 0.9113794 * x + 3536.78 - 912.21$$

which results in

$$2.072487 * x = 2624.57$$

Dividing both sides by 2.072487

$$x = 1266.39$$

Hence,

$$y = 1.161108 * 1266.39 + 912.21$$

or

$$y = 2382.62$$

The coordinates of P are therefore (1266.39, 2382.62). Thus, using the principles of arithmetic given at the start of Chapter 2, we have found the coordinates of the point of intersection of two lines.

Furthermore, in Figure 4.5 let the line through AB have the equation

$$ax + by + c = 0$$

Consider any point C that does not lie on that line and let C be (x_C, y_C). At point A, the y coordinate is y_C, hence, the x coordinate is $x_A = -(by_C + c) / a$.

At point B, the x coordinate is x_C, hence, the y coordinate is $y_B = -(ax_C + c) / b$.

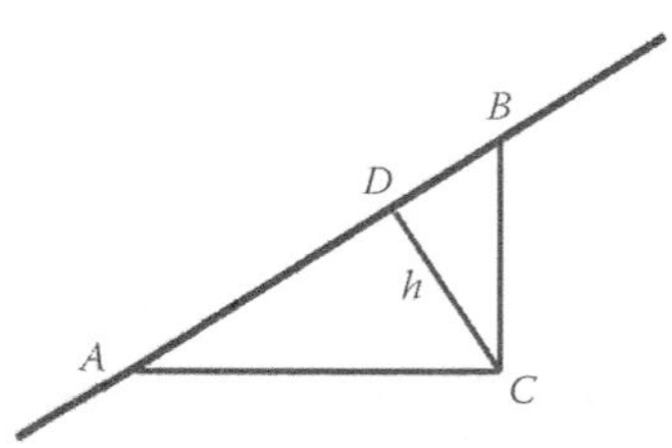

FIGURE 4.5 Distance offset from a line.

By Pythagoras,

$$\begin{aligned} AB^2 &= BC^2 + AC^2 \\ &= \left\{-\left(ax_c + c\right)/b - y_c\right\}^2 + \left\{x_c + \left(by_c + c\right)/a\right\}^2 \\ &= \left\{ax_c + by_c + c\right\}^2 * \left\{1/a^2 + 1/b^2\right\} \end{aligned}$$

Now $\{1 / a^2 + 1 / b^2\} = \{a^2 + b^2\} / \{a^2b^2\}$. Hence,

$$AB = \left\{ax_c + by_c + c\right\} * \left\{\sqrt{a^2 + b^2} / ab\right\}$$

$$\text{The area of triangle } ABC = 1/2\, AC * BC = 1/2\left\{x_c + \left(by_c + c\right)/a\right\} * \left\{\left(ax_c + c\right)/b + y_B\right\}$$
$$= 1/2\left\{ax_c + by_c + c\right\}^2 / ab$$

It also equals ½ $AB * h$ where h is the perpendicular line from C to the line AB in Figure 4.5. Hence,

$$h * \left\{ax_c + by_c + c\right\} * \sqrt{a^2 + b^2} / ab = \left\{ax_c + by_c + c\right\} * \left\{ax_c + by_B + c\right\} / ab$$

or

$$h = \left(ax_c + by_c + c\right) / \sqrt{a^2 + b^2}$$

This gives a simple formula for calculating how far a known point is away from a known line. The value is, of course, zero if point C is actually on the line AB.

A further example of an application for the equation $y = mx + c$ arises when drawing lines on a map that cross the map sheet edge (at P in Figure 4.6) or any square window drawn around an area — a process known as *clipping*. The lines that cross the boundary of the area being displayed must be clipped so that all sections of the original data that lie outside the area under consideration can be discarded.

One way to compute the sections of a line to be included is given in Example 4.3. There is, however, an alternative approach to the truncation of the line AB that does not require the calculation of the equation $y = mx + c$. It is based entirely on the idea of *scale*. Consider the two triangles APQ and ABR in Figure 4.7.

The two triangles APQ and ABR in Figure 4.7 are exactly the same shape. They only differ in size and are said to be similar. *Similar triangles* do not necessarily have one angle a right angle; the key point is that they have corresponding angles of the same size. In fact, they just differ in terms

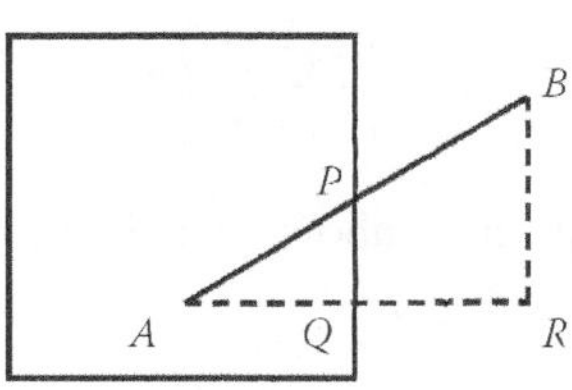

FIGURE 4.6 Clipping to a window.

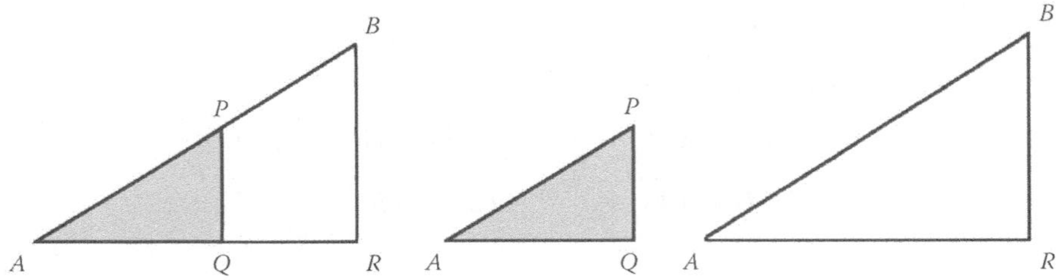

FIGURE 4.7 Similar triangles.

of scale. This means that all linear distances are in the same proportion or ratio. If the *scale factor* between them is some number s then:

$$AP = s * AB; \quad AQ = s * AR \quad \text{and} \quad PQ = s * BR.$$

Alternatively,

$$\frac{AP}{AB} = \frac{AQ}{AR} = \frac{PQ}{BR} = s$$

Example 4.3 Intersection at a Map Sheet Edge (1)

Referring to Figure 4.6 and using the coordinates of A and B as before with A (1234.56, 2345.67) and B (1296.32, 2417.38). In Example 4.2 we showed that

$$y = 1.161108 * x + 912.21$$

Now, if the edge of the map sheet is on a grid line QP whose x value is $x_P = 1250$, then at that point P

$$\begin{aligned} y_p &= 1.161108 * 1250 + 912.21 \\ &= 2363.60 \end{aligned}$$

A computer-driven graphical plotter could be made to move from A (1234.56, 2345.67) to P (1250.00, 2363.60). It would then stop plotting at the map sheet edge, clipping the line AB at P.

This will hold for similar triangles even when the triangle does not contain a right angle. In the case of Figure 4.6, we merely have to scale down BR to PQ to get the y value of P, as shown in Example 4.4.

4.3 POINTS IN POLYGONS

If we express the equation of the line through the two points A (x_A, y_A) and B (x_B, y_B) in the form

$$(y - y_A) = \frac{(y_B - y_A)}{(x_B - x_A)} * (x - x_A)$$

then it is easy to test whether any point is above or below the line AB. To do so, take any point C (x_C, y_C) and calculate

$$y = y_A + \frac{(y_B - y_A)}{(x_B - x_A)} * (x_C - x_A)$$

Example 4.4 Intersection at a Map Sheet Edge (2)

Using the same numbers as in Example 4.3, then if R has the same x value as B and y value as A, then the distance

$$AR = x_B - x_A = 1296.32 - 1234.56 = 61.76$$
$$BR = y_B - y_A = 2417.38 - 2345.67 = 71.71$$
$$AQ = x_p - x_A = 1250.00 - 1234.56 = 15.44$$

Hence,

$$AQ / AR = 15.44 / 61.76 = 0.25 = s$$
$$QP = s * BR = 0.25 * 71.71 = 17.93$$

or

$$PQ = y_P - y_A = y_P - 2345.67 = 17.93$$

Hence,

$$y_P = 2345.67 + 17.93 = 2363.60$$

Once again the coordinates of P are (1250.00, 2363.60).

If the answer is less than y_C then C is above AB and if it is greater, then C lies below the line from A to B. Simple tests like this are useful when determining if two points are on the same or opposite sides of a line. Such tests arise in hidden line and surface removal, as discussed in Chapter 11.

This also provides one way to test whether a point $P(x_P, y_P)$ lies within or outside a polygon. Consider the polygon $ABCDEF$ in Figure 4.8 where the coordinates of A are (x_A, y_A), of B are (x_B, y_B), and so forth. First, one should check whether x_P lies between the maximum and minimum values of x for the whole polygon, which in this case means that $x_A \leq x_P \leq x_E$ and similarly $y_F \leq y_P \leq y_D$ in order to reduce the amount of computation. (The notation $a \leq b$ means that a is less than or equal to b.) Testing whether something is greater or less than a specified value is a relatively trivial computation.

Next consider the section of the horizontal line through P that has x values greater than x_P. Let this be PQ where Q is such that $x_Q > x_E$ and $y_Q = y_P$. The number of occasions that this line crosses a side of the polygon must be counted.

In Figure 4.8, a series of checks must be carried out on each of the lines AB, BC, CD, and so forth, that form the polygon. In the case of AB, y_P (the y value for the horizontal line from P to Q) is greater than y_A and less than y_B so the line PQ will, if extended, intersect AB. From the coordinates of A and B we can work out the point on the line AB where the y value $= y_P$; the x value will be seen to be less than x_P and it can be ignored.

For the lines BC, CD, and DE their intersection with PQ has an x value greater than x_P and therefore they will count. Each of the lines EF and FA are wholly below P and can be ignored.

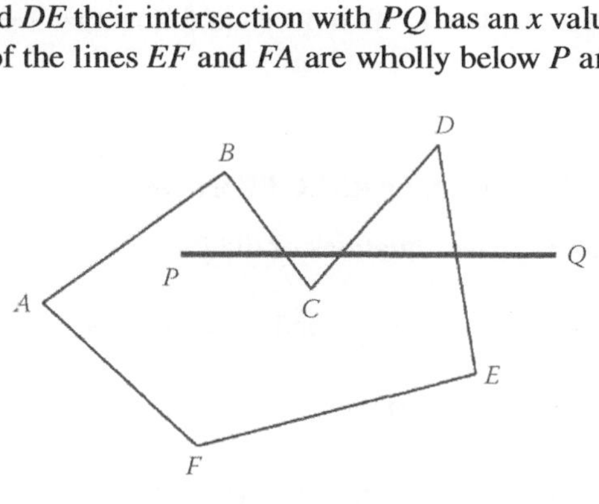

FIGURE 4.8 Point-in-polygon.

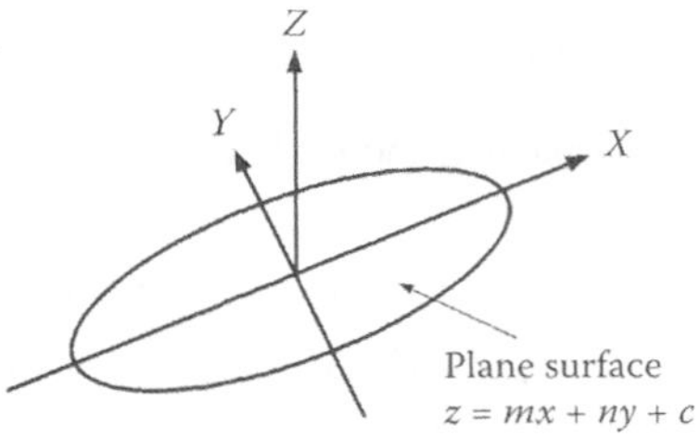

FIGURE 4.9 A plane surface.

At the end, if the number of lines that are crossed to the right of P is an even number, then P will be outside the polygon; if the number is odd (as in this case), point P will be inside the polygon. Care must be taken if the line PQ goes exactly through one of the corner points as this can upset the count; this can, however, be avoided by a minor adjustment in the coordinates of P; for example, if the vertices are given to two decimal points (0.01) then by treating P to three decimal points (rounding it to 0.005), the problem will not arise.

4.4 THE EQUATION FOR A PLANE

We can extend the equation of a line to describe a plane. Consider the equation

$$z = mx + ny + c$$

For every fixed value of y (for example, $y = d$),

$$z = mx + (nd + c) = mx + c'$$

where $c' = nd + c$ and is a constant. This is the equation for a straight line in the plane $y = d$. Similarly, for every fixed value of x we have a set of straight lines. If z is regarded as the axis in the third dimension then $z = mx + ny + c$ must represent a *plane* surface (Figure 4.9).

4.5 FURTHER ALGEBRAIC EQUATIONS

Algebraic operations often entail rearranging equations, as illustrated in Box 4.2. Provided that we carry out the same operation on both sides of the equation by adding, subtracting, multiplying, or dividing by the same amount, we will not alter the basic relationship. This is a very important principle that helps solve many mathematical equations. The only thing that we cannot do is divide both sides by zero for although 2 * 0 = 1 * 0, it does not follow that 2 = 1.

BOX 4.2 THE INTERSECTION OF TWO PLANES

When two planes intersect we have equations of the form

$$z = m_1x + n_1y + c_1 \quad \text{and} \quad z = m_2x + n_2y + c_2$$

Subtracting one equation from the other:

$$z - z = m_1x + n_1y + c_1 - (m_2x + n_2y + c_2)$$

or

$$0 = m_1x + n_1y + c_1 - m_2x - n_2y - c_2$$

Adding n_2y to both sides of the equation:

$$n_2y = m_1x + n_1y + c_1 - m_2x - n_2y - c_2 + n_2y$$

Subtracting n_1y from each side of the equation:

$$\begin{aligned} n_2y - n_1y &= m_1x + n_1y + c_1 - m_2x - n_2y - c_2 + n_2y - n_1y \\ &= m_1x + c_1 - m_2x - c_2 \end{aligned}$$

or

$$(n_2 - n_1)y = (m_1 - m_2)x + (c_1 - c_2)$$

Dividing both sides by $(n_2 - n_1)$

$$y = \left\{\frac{m_1 - m_2}{n_2 - n_1}\right\}x + \left\{\frac{c_1 - c_2}{n_2 - n_1}\right\}$$

Replacing $((m_1 - m_2) / (n_2 - n_1))$ by m and $+ ((c_1 - c_2) / (n_2 - n_1))$ by c.
The result is

$$y = mx + c$$

This has already been shown to be the equation of a straight line. Thus, two planes intersect in a straight line.

All of the above relationships are examples of equations. An *equation* is a formula that asserts that two expressions have the same value. Thus, $y = mx + c$ is an *identical equation* or *identity* that is true for all values of the variables that lie along the line that is determined by the constants m and c. A *conditional equation* is one that is only true for certain values of the variables. For instance, $x^2 - x = 2$ is a conditional equation that is only true when $x = 2$ or $x = -1$.

Two particular forms of equations commonly occur in geomatics—simultaneous equations and quadratic equations. A *simultaneous equation* is one of at least a pair of equations that must be satisfied simultaneously. An example was given in Example 4.2 where there were two straight lines that intersected. At the point of intersection P, the values of x and y had to satisfy both the equation for the line AB and the equation for the line CD simultaneously. If there are two unknowns (here x and y), then there must be at least two independent equations in order that their points of intersection can be found and the equations solved. For three unknowns (x, y, z), then there must be at least three independent equations each of which must be satisfied. (In Chapter 15, we consider what happens when there are more than the minimum number of equations.)

A *quadratic* is a function in which the variables may be "raised to the power of two" or be of "second degree"—for example—x^2, y^2, or xy (but not xy^2 which is of the third degree). $x^2 + y^2 = r^2$ is an example of a quadratic expression, one interpretation of which is that r is the radius of a circle with center (0, 0) and (x, y) is any point on the circle.

The relationship $x^2 + y^2 = r^2$ only exists in real terms when both x and y are equal to or greater than $-r$ and less than or equal to $+r$. This may be written in the form

$$-r \leq x \leq +r$$

And

$$-r \leq y \leq +r$$

or

$$|x| \leq r$$

And

$$|y| \leq r$$

where $|x|$ and $|y|$ mean the positive values of x and y. $|x|$ and $|y|$ are called the *modulus* of x and *modulus* of y or "mod x" and "mod y" or the "absolute values" of x and y.

For a circle of radius r but center at point X_c, Y_c (Circle 2 in Figure 4.10), then the general equation would take the form

$$(x - X_c)^2 + (y - Y_c)^2 = r^2$$

or

$$x^2 - 2X_c x + X_c^2 + y^2 - 2Y_c y + Y_c^2 = r^2$$

or

$$x^2 + y^2 + ax + by + c = 0$$

where

$$a, b, \text{ and } c \text{ are constants with } a = -2X_c; \quad b = -2Y_c; \quad \text{and} \quad c = X_c^2 + Y_c^2 - r^2.$$

The term *quadratic* is commonly associated with *quadratic equation*. A quadratic equation takes the form

$$ax^2 + bx + c = 0$$

Or

$$ax^2 + bx^1 + cx^0 = 0$$

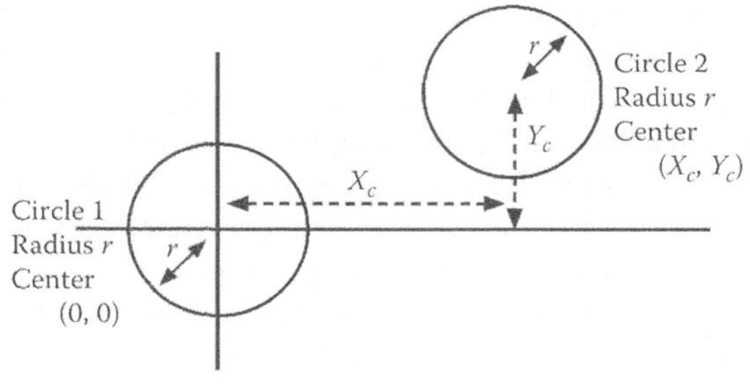

FIGURE 4.10 The equation of a circle.

In this form, the power of x takes the integer values 0, 1, and 2. Because it is of second degree, there are two possible solutions. Mathematics solves such equations in a variety of ways. For example, the equations can be rearranged by dividing every term by a (assuming a is not zero). Hence,

$$x^2 + (b / a)x + (c / a) = 0$$

or

$$x^2 + (b / a)x = -(c / a)$$

Next, we add a number n so that $x^2 + (b / a)\, x + n$ becomes a *perfect square*, that is, it is some value raised to the power of 2. Let this value be $(x + p)$ so that we have to find p such that $(x + p)^2 = x^2 + (b / a)\, x + n$.

We do not of course know n at this stage, but we do know that $(x + p)^2 = x^2 + 2px + p^2$. Hence, we need to choose p so that $2p = (b / a)$ and $n = p^2$. This means that $p = (b / 2a)$ and hence, $n = p^2 = b^2 / 4a^2$. This gives a perfect square

$$(x + b / 2a)^2 \quad \text{or} \quad x^2 + (b / a)x + b^2 / 4a^2$$

If we write our original equation $ax^2 + bx + c = 0$ as $x^2 + (b / a)\, x = -(c / a)$, then add $b^2 / 4a^2$ to both sides, we obtain:

$$x^2 + (b / a)x + b^2 / 4a^2 = b^2 / 4a^2 - (c / a)$$

Or

$$(x + b / 2a)^2 = b^2 / 4a^2 - (c / a) = b^2 / 4a^2 - (4ac / 4a^2) = (b^2 - 4ac) / 4a^2$$

Taking the square roots $(x + b / 2a) = \pm \{\sqrt{(b^2 - 4ac)}\} / 2a$
Hence,

$$x = \{-b \pm \sqrt{(b^2 - 4ac)}\} / 2a$$

("x" equals minus "b" plus or minus the square root of "b" squared minus four "a" "c" all over two "a.")

Example 4.5 illustrates how this works in practice. It is known as "solving a quadratic." Note that if $b^2 < 4ac$ then $(b^2 - 4ac)$ is negative and since when using real numbers there is no such thing as the square root of a negative number, there is no solution to the problem since there is no real value of x that satisfies $ax^2 + bx + c = 0$.

When solving simultaneous equations a similar approach is used in which we alter both sides of an equation (Example 4.6). Consider two lines with equations:

$$ax + by + c = 0 \quad \text{and} \quad dx + ey + f = 0$$

Multiply both sides of the first equation by d and both sides of the second equation by a. This gives:

$$adx + bdy + cd = 0$$

and

$$adx + aey + af = 0$$

Subtracting one from the other:

$$adx + bdy - adx - aey + cd - af = 0$$

Example 4.5 Solving a Quadratic

As shown in the main text if

$$ax^2 + bx + c = 0$$

Then,

$$x = \left\{-b \pm \sqrt{(b^2 - 4ac)}\right\} / 2a$$

Consider $3x^2 + 4x + 1 = 0$. Here,

$$a = 3,\ b = 4,\ c = 1$$

Hence,

$$x = \left\{-4 \pm \sqrt{16 - 12}\right\} / 6$$

So,

$$x = (-4 + 2) / 6 = -1/3$$

or

$$x = (-4 - 2) / 6 = -1$$

The two solutions to $3x^2 + 4x + 1 = 0$ are $x = -1/3$ or $x = -1$.

Example 4.6 Intersecting Lines Using Simultaneous Equations

As shown in the main text, if there are two equations

$$ax + by + c = 0 \quad \text{and} \quad dx + ey + f = 0$$

Then,

$$\frac{x}{(bf - ce)} = \frac{-y}{(af - cd)} = \frac{1}{(ae - bd)}$$

Consider the two lines $3x + 4y - 10 = 0$ and $5x - 2y - 8 = 0$. These will intersect where

$$\frac{x}{-52} = \frac{-y}{26} = \frac{1}{-26}. \text{ Hence, } x = 2 \quad \text{and} \quad y = 1$$

The values $x = 2$ and $y = 1$ satisfy both equations simultaneously. Hence, they must represent the point where the two lines intersect.

or

$$y(bd - ae) = (af - cd)$$

or

$$y = -(af - cd) / (ae - bd)$$

Similarly,

$$x = (bf - ce) / (ae - bd)$$

or

$$\frac{x}{(bf - ce)} = \frac{-y}{(af - cd)} = \frac{1}{(ae - bd)}$$

(In Chapter 9, we will rediscover these relationships in the context of determinants.)

4.6 FUNCTIONS AND GRAPHS

A *function* is a relationship in which, given the value of one or more variables, the overall value of the function can be calculated. For example, $y = f(x)$ can be read as "y is a function of x" with y as the value of the function and x the *argument*. The function can take any mathematical form such as

$$y = ax + b \text{(which is a line)}$$

$$y = ax^2 + bx + c \text{ (a quadratic or second-degree curve)}$$

$$y = ax^3 + bx^2 + cx + d \text{ (a cubic or third-degree curve), and so on.}$$

A particular group of functions are called *polynomials*. A polynomial is essentially an expression that contains two or more terms such as

$$x^3 + 3x^2y + 3xy^2 + y^3 + x^2 + 3y^2 + xy + x + y + 4$$

More specifically, the term is used to describe a relationship such as:

$$y = a + bx + cx^2 + dx^3 + ex^4 + fx^5 + \cdots \cdots \cdots$$

where a, b, c, d, e, f, and so on, have fixed values.

There may be a finite number of terms, for example, where

$$y = (1 + x)^4 = 1 + 4x + 6x^2 + 4x^3 + x^4$$

Alternatively, there may be an infinite number of terms in which case it will either generate an infinite number or else converge onto a specific value. The term *convergence* means that however many terms are added, the function will simply grow closer and closer to a specific value. Thus,

$$e = 1 + 1/1! + 1/2! + 1/3! + \cdots$$

has an infinite number of terms but it never gets bigger than a certain value, while the series of terms $1 + 2 + 3 + \cdots$ just gets bigger and bigger and is not convergent.

A *series* is a sequence of terms that may be finite or infinite. An example of a series is

$$S = a_0 + a_1 + a_2 + a_3 + \cdots + a_{(n-1)} + a_n$$

where n is a positive integer. It is often written as

$$\sum a_i \text{ or } \sum_{i=0}^{i=n} a_i$$

where the symbol $\sum$ (the Greek letter "sigma") means "the sum of." In practice, an infinite series will only have a finite sum if the series (a_0, $a_0 + a_1$, $a_0 + a_1 + a_2$, $a_0 + a_1 + a_2 + a_3$, …, etc.) converges.

Functions may involve one dependent and one independent variable or may have several independent variables such as: $z = f(x, y)$, meaning that z is a function of x and y where x and y are both independent variables. If z is dependent on x and y so that we have only one value of z for given values of x and y, then we have a surface. If z is linearly dependent on x and y (i.e., $z = f(x,y) = ax^1 + by^1 + c$) then, as we have seen above, this is a plane surface. If $z = ax^2 + by^2 + cx + dy + e$, then we have a second degree or quadratic surface. If $y = f(x)$ so that y is a function of x, then there is a unique

TABLE 4.1
Values of y and x for $y = 0.5x^2$

x	0	±0.2	±0.4	±0.6	±0.8	±1	±1.2	±1.4	±1.6	±1.8	±2
y	0	0.02	0.08	0.18	0.32	0.5	0.72	.98	1.28	1.62	2

value for y for every value of x. If the relationship between y and x is such that it is also possible to determine x uniquely given the value of y, then the relationship is said to have an *inverse function*, written as f^{-1}. Thus, if $y = f(x) = ax + b$, then $x = f^{-1}(y) = (y - b) / a$. This is often not possible; for instance, if $y = x^2$ then $x = +\sqrt{y}$ or $x = -\sqrt{y}$ and there are two possible relationships. Hence, there is no inverse function. In general, functions are mathematical relationships between two (or more) variables that may be in the form of one-to-one or one-to-many or many-to-many.

We have seen that the equation for a circle centered at the origin is $x^2 + y^2 = r^2$ where r = radius. For any value of x that represents a point on the circle there are two possible values of y (one positive and one negative); similarly, for any value of y there are two values of x and thus there is a one-to-two relationship between x and y and between y and x.

Relationships between two variables are often better expressed graphically than numerically since many people find a visual image easier to understand than a mathematical equation. A *graph* is a drawing showing the relationship between certain sets of quantities by means of points or lines plotted with respect to a set of coordinate axes.

As an example of a graph, consider the function $y = 0.5\,x^2$. The values of points on the graph can be calculated if we choose a series of values for x and then compute y (see Table 4.1). Every pair of (x, y) coordinates can then be plotted. Intermediate points can be calculated as necessary.

When plotted, the coordinates form the curve as shown in Figure 4.11. In general, when drawing a graph, a series of points are plotted and then are joined by straight lines. A circle may be drawn on a piece of paper using a compass, resulting in a nice smooth shape, but in most cases this is not possible and as described in Chapter 1, curves are either plotted in the form of dots (the raster method) or as a series of short vectors with the precise location of points on the curve being calculated from the algebraic formulae. In practical terms, not all points can be plotted as they may extend far beyond the limits of any piece of paper.

In Figure 4.11, if $x = 100$ then $y = 5000$ and as a result, the axes would have to be scaled down to accommodate such a range.

Figure 4.12 shows the curve for $y = 1 / x$. When $x = 1$ then $y = 1$; when $x = 0.5$ then $y = 2$; when $x = 0.1$ then $y = 10$; when $x = 0.001$ then $y = 1000$, and so on. As x gets smaller and smaller and

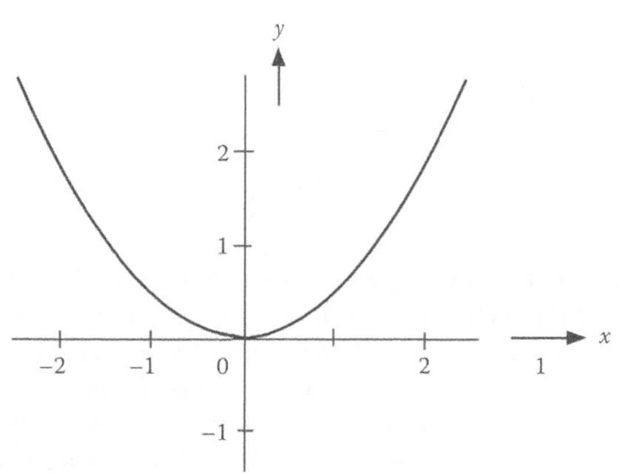

FIGURE 4.11 Graph of the function $y = 0.5x^2$.

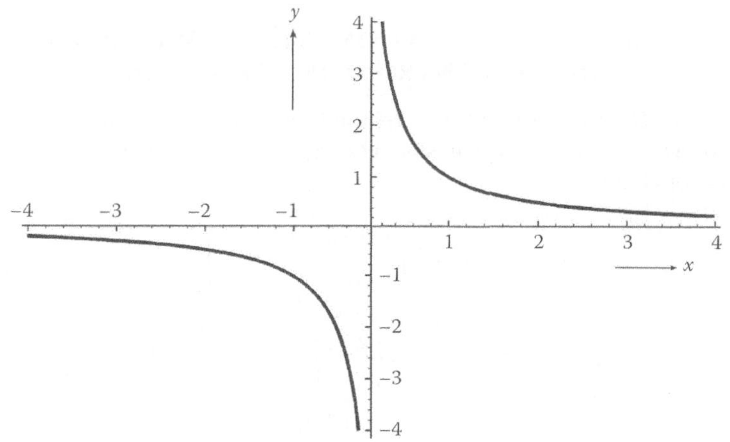

FIGURE 4.12 Graph of $y = 1 / x$.

approaches zero from the positive side, y goes off to plus infinity ($+\infty$). As x approaches zero from the negative side, y goes off to minus infinity ($-\infty$).

The value of y when $x = 0$ can only be expressed mathematically as $y = \infty$. It cannot be shown on any graph. It also has the strange property that it flips from $+\infty$ to $-\infty$ or vice versa as the value of x passes through zero.

4.7 INTERPOLATING INTERMEDIATE VALUES

Both the graphical and numerical approaches can be used to determine values in between or beyond those that have been measured. The process of determining intermediate values is known as *interpolation*. Taking a guess at what happens beyond what is already known is called *extrapolation*; for instance, we know approximately the number of people who were in the world at the end of each decade over the last century and by plotting the numbers on a graph, it is possible to predict how many people will be in the world in ad 2050 or even ad 2100 assuming that the present trends continue. For now, we will only consider simple linear interpolation and delay until later any discussion on more complex forms.

In Figure 4.13, P represents a point interpolated along the straight line between A and B; Q is an extrapolation and assumes that AB continues in a straight line.

The coordinates of intermediate points can be calculated on the basis of ratios. Thus, if in Figure 4.13, PB is one quarter of AB then it divides AB in the ratio 3 to 1 so, by interpolation:

$$X_p = (3 * X_B + X_A) / 4 \quad \text{and} \quad Y_P = (3 * Y_B + Y_A) / 4$$

If Q is the point on the line AB such that $BQ = (1 / 2)\, BA$, then B divides AQ in the ratio 2 to 1 so $B = (2Q + A) / 3$ or $Q = (3B - A) / 2$. Thus,

$$X_Q = (3 * X_B - X_A) / 2 \quad \text{and} \quad Y_Q = (3 * Y_B - Y_A) / 2$$

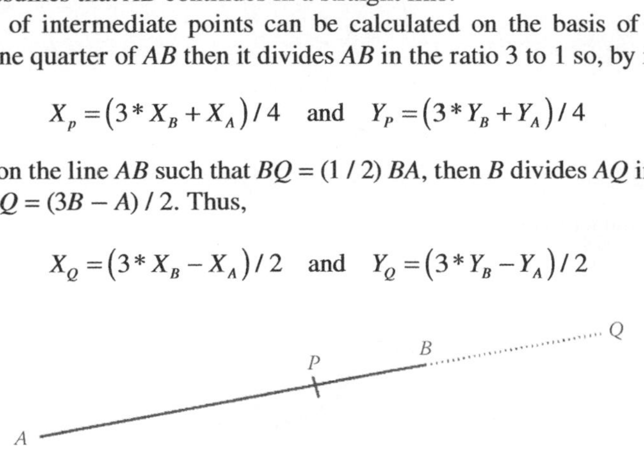

FIGURE 4.13 Linear interpolation and extrapolation.

BOX 4.3 INTERPOLATION—THE COORDINATES OF THE CENTROID OF A TRIANGLE

In Figure 4.14, let D be the midpoint of BC and let G divide AD in the ratio 2:1 (so that $GD = 1/3$rd of AD). Let the coordinates be A (x_A, y_A), B (x_B, y_B), and C (x_C, y_C). By proportion the coordinates of D are

$$(x_D, y_D) = \left\{\frac{x_B + x_C}{2}, \frac{y_B + y_C}{2}\right\}$$

$$\text{The coordinates of } G = \left\{\frac{2x_D + x_A}{3}, \frac{2y_D + y_A}{3}\right\}$$

$$= \left\{\frac{x_A + x_B + x_C}{3}, \frac{y_A + y_B + y_C}{3}\right\}$$

Note: Since this is symmetrical it would be the same for BE where E is the midpoint of AC and CF where F is the midpoint of AB. The lines joining the vertices of a triangle with the midpoints of the opposite sides pass through one point G known as the centroid. If the triangle were made of uniform density, then the centroid would be its center of mass or gravity.

We can use these types of relationships to show that in any triangle ABC the lines joining each apex to the midpoint of the opposite side all pass through one point (see Box 4.3 and Figure 4.14).

The process of reducing graphs to straight-line sections helps in interpolation, making the calculation of intermediate values relatively simple. An illustration of such a calculation was given previously in Example 4.3, which showed the cutting off of a line at a map sheet edge or window boundary. Given two points, each one being on opposite sides of the required value, the coordinates of the intermediate point can be determined. On the other hand, if as in Figure 4.15, A is the top of a slope and B the bottom, linear interpolation will overestimate lower heights and underestimate the higher ones.

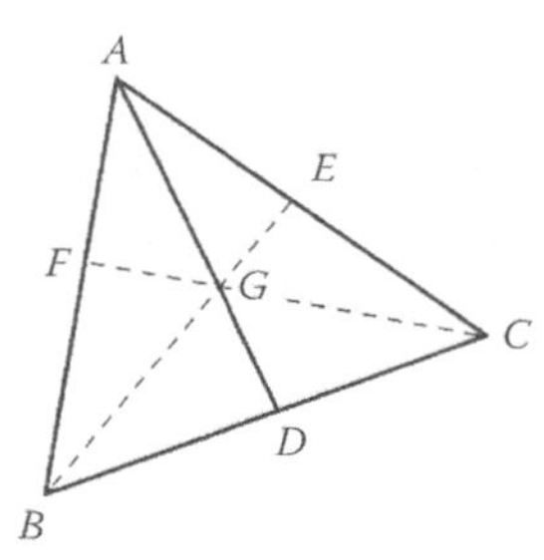

FIGURE 4.14 The midpoints of the sides of a triangle.

FIGURE 4.15 Interpolation of heights down a slope.

Intermediate values are often interpolated on the assumption that change is linear. In particular, this can happen when constructing lines of equal value such as contour lines. Neighboring points of known value (such as spot heights) are joined and the positions where a contour line crosses the lines of the resulting triangles are then interpolated. These positions in turn are joined to form the contour. The procedure can be applied to any third-dimensional value such as air pressure or temperature at a point; but here we will consider heights.

In Figure 4.16a, the five spot heights have been joined by straight lines to form triangles. In Figure 4.16b, the values equivalent to 25 meters have been linearly interpolated along each line. Thus, between the 27.5 m and 21.2 m spot heights the level drops by 6.3 m. To drop to 25 m from 27.5 means to drop 2.5 m out of the total of 6.3 m. The contour crosses the line at a distance of (2.5 / 6.3) of its length. If the end points are (x_A, y_A) and (x_B, y_B), then the coordinates of the interpolated point will be

$$\left\{\frac{3.8x_A + 2.5x_B}{6.3}, \frac{3.8y_A + 2.5y_B}{6.3}\right\}$$

The interpolated points can then be joined either (Figure 4.16c) by straight lines or (Figure 4.16d) by a curve drawn freehand or by using a computer algorithm as discussed in Chapter 13.

The assumption behind the interpolation is that the lines chosen are of uniform slope. In fact, other triangles could have been formed as in Figure 4.17b. Each will give a different contour line. Which is right?

The answer is that neither will give a perfect answer, which is why contours are generally regarded as only being reliable within a band representing about a third of the contour interval (for instance, within just over three meters for a 10-meter contour interval). The main problem is with the assumption that height changes evenly between spot heights. In some cases, the surveyor may have made sure that points are chosen to fulfill this criterion but, in general, slopes are not uniform and a linear interpolation is just a convenient guess.

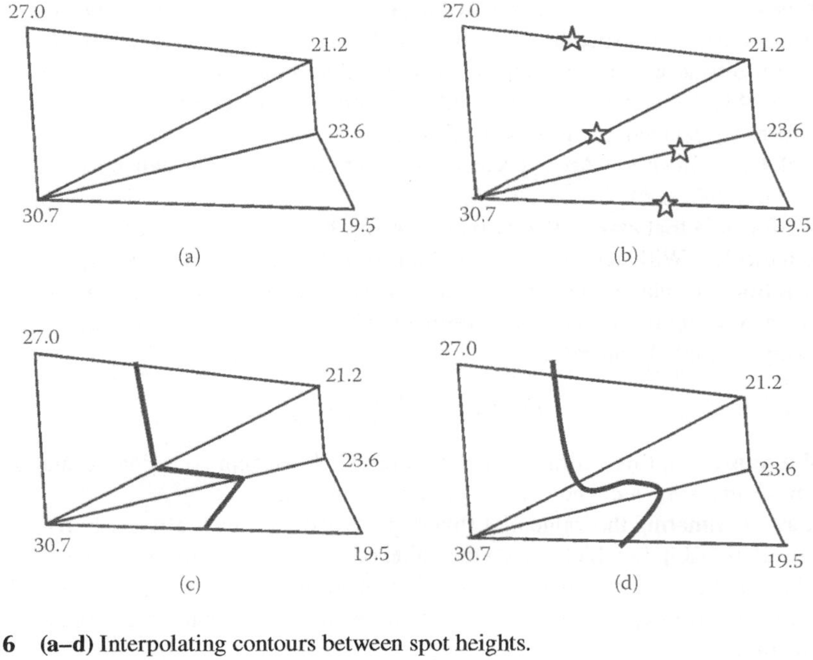

FIGURE 4.16 (a–d) Interpolating contours between spot heights.

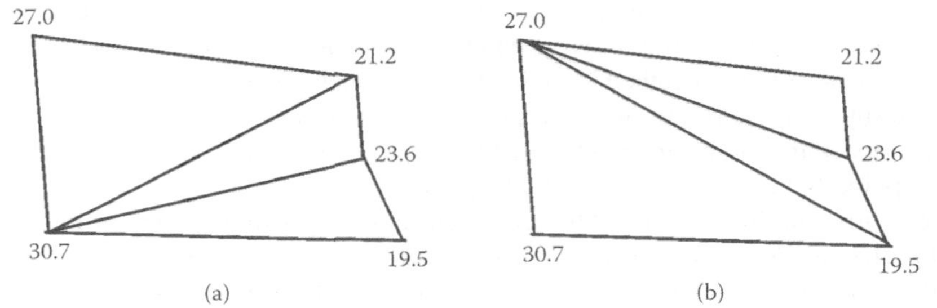

FIGURE 4.17 **(a, b)** Alternative triangulation networks.

In Chapter 5, we will describe a technique that uses what is known as the *Theissen* or *Thiessen polygon* to select good-shaped triangles. For now, it should just be noted that the accuracy of interpolation techniques depends on (i) the accuracy of the original data; (ii) assumptions behind the interpolation; and (iii) the guesswork that enters into smoothing the contour line into an acceptable curve. We will explore ways in which a smooth curve can be fitted through a series of points in Chapter 13.

GLOSSARY

Argument An element to which a mathematical operation or function is to be applied.

Centroid The point in a triangle where the lines joining each point of the triangle to the midpoint of the opposite side intersect (such lines being known as medians). For the triangle *ABC*, its coordinates are

$$\{(x_A + x_B + x_C)/3,\ (y_A + y_B + y_C)/3\}$$

Clipping The process of starting or ending a line where it reaches the edge of a map square.

Conditional equation An equation that is only true for certain values.

Convergent series A *series* in which as more and more terms are added, the value of the function gets closer and closer to a specific size. Thus, $e = 1 + 1/1! + 1/2! + 1/3! + 1/4! + 1/5! + \cdots \cdots$ is an infinite series that converges to a value of approximately 2.7182818285.

Dependent variable A quantity determined by another value such as y being dependent on x in the form $y = f(x)$ where "f" is some *function.*

Difference of two squares An expression such as $(a^2 - b^2)$ that can be factorized into the form $(a + b)(a - b)$.

Equation A formula that asserts that two expressions have the same value.

Equation of a circle With center (X_C, Y_C) and radius r then $(x - X_C)^2 + (y - Y_C)^2 = r^2$.

Equation of a line In plane rectangular Cartesian coordinates, $y = mx + c$ where m is the slope and c a constant representing the *intercept* of the line on the y-axis where $x = 0$. For a line passing through A and B then

$$m = (y_A - y_B)/(x_A - x_B).$$

Equation of a plane In three-dimensional rectangular Cartesian coordinates, an equation of the form $z = mx + ny + c$ where m, n, and c are constants.

Extrapolation Estimating the value of a function beyond the limits already known.

Function A relationship in which, given the value of one or more variables, then the overall value of the function can be calculated. For instance, the function $f(x) = ax^3 + bx^2 + cx + d$ is a cubic or third-degree curve for which the value $y = f(x)$ can be calculated for any given value of x.

Graph A visual demonstration of a function in the form of points and lines.

Identical equation or identity A relationship that is true for all values that form part of the expression, for example, all values of x and y that lie along the line $y = mx + c$.

Independent variable A variable quantity whose value determines that of a *dependent variable.*

Intercept Where a line cuts an axis.

Interpolation Estimating the value of a function within limits already known.

Inverse function If y is a *function* of x, that is, $y = f(x)$, then the inverse function written as f^{-1} is such that $x = f^{-1}(y)$. Many functions do not have an inverse.

Perfect square An expression that is the product of two identical components, for example, $x^2 + (b / a)x + b^2 / 4a^2 = (x + b / 2a)^2$ is a perfect square.

Point-in-polygon (PiP) An algorithm to determine whether a point lies inside or outside a polygon.

Polynomial An expression that is the sum of a number of terms, each consisting of a constant and one or more variables raised to a nonnegative integer power, such as $a + bx + cx^2 + cx^3 + dx^4$.

Quadratic An equation of the second degree, that is, one where no term is raised to a higher power than 2, such as $ax^2 + bx + c = 0$.

Scale The ratio between the lengths on one object with those of another, also known as the *scale factor*. The term is also used to mean a measuring instrument to determine quantities such as lengths.

Scale factor The multiplying factor applied to one measurement to obtain another.

Series The sum of a finite or infinite number of terms, expressible in the form $\sum x$ – (sigma x).

Similar triangles Triangles that are the same shape and differ only in scale.

Simultaneous equations Two or more equations that must be satisfied at the same time. For example, when finding the intersection of two lines, the equations for each line must be satisfied simultaneously.

Thiessen polygon (Also sometimes spelled *Theissen*) With a series of given points scattered around an area, a Thiessen polygon surrounds each given point in such a way that every other point within the polygon is nearer to that point than any other given point.

5 The Geometry of Common Shapes

5.1 TRIANGLES AND CIRCLES

Geometry is the study of constructible shapes. In this chapter, we will review some of the shapes that occur in geomatics and geographic information system (GIS) that occupy ordinary two- or three-dimensional (3D) "Euclidian" space. Euclid was a Greek mathematician of the 3rd century BC who worked out a series of axioms or postulations concerning points, lines, angles, surfaces, and volumes. From these, he derived 465 theorems.

His basic axioms included such statements as that for any two distinct points there is only one straight line that passes through them and if three distinct points are not on a straight line then there is only one plane that will pass through them. Euclid identified 10 axioms but subsequently a further one was added, namely that only one straight line can be drawn parallel to a given line through any point not on that line.

In Euclidian space, the shortest distance between two points is a straight line. Two lines that are either parallel or intersect form a plane—in fact parallel lines may be said to intersect at a point at infinity, an important consideration when drawing images of 3D objects in perspective on a plane (2D) surface, as discussed in Chapter 11.

The triangle is the simplest shape that is made up of straight lines. In fact, all 2D shapes can be regarded as being made up from a series of triangles, just as every curve can be thought of as a series of short straight lines. Although this can give rise to a number of errors, for example, when calculating an area enclosed by a curved line, the approximation can be adequate for many practical purposes.

Triangles come in all sorts of shapes and sizes but the basic fact is that the angles of a plane triangle add up to half of a complete turn or 180°; the angles of a spherical triangle, which is one drawn on the surface of a sphere, add up to more than 180°. For the present, we will only consider plane triangles.

In the triangle ABC in Figure 5.1a, at A face C then turn clockwise to B; at B turn clockwise to face backward at C; at C turn clockwise to face A; you will have turned through the three angles ABC and half a complete turn. It is sometimes convenient to measure this in units of radians where

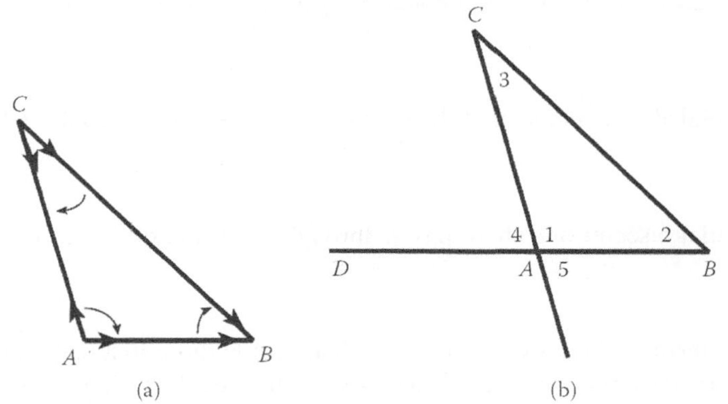

FIGURE 5.1 **(a, b)** The angles of a triangle.

DOI: 10.1201/9781003464402-7

π radians ("pi" radians) = 180°. Measures in radians are particularly important when using differential and integral calculus, as discussed in Chapter 10. For now, all angles will be considered as being in degrees, each degree being divided into 60 minutes (60′) and each minute containing sixty seconds (60″).

It is conventional for the angles of the triangle formed by points A, B, and C to be designated by capital letters, while the sides opposite the angles have lowercase letters a, b, and c. In Figure 5.1b, the internal angles 2 and 3 are both less than 90° and are *acute*, while the internal angle A is greater than 90° and is *obtuse*.

Note that since the angles $\angle 1 + \angle 2 + \angle 3 = 180°$ and also the angle marked "4" is such that $\angle 4 + \angle 1 = 180°$; hence, the external angle of the triangle is equal to the sum of the two internal opposite angles. The symbol "∠" means "angle."

This also applies to the angle marked 5 in Figure 5.1b. Thus, $\angle 4 = \angle 2 + \angle 3 = \angle 5$. In fact, whenever two lines intersect the opposite angles are equal (RULE 5.2 in Box 5.1).

BOX 5.1 ANGLES OF A TRIANGLE AND IN A CIRCLE

RULE 5.1

The external angles of a triangle obtained by extending each of the sides equal the sum of their interior opposite angles.

RULE 5.2

When two straight lines intersect, opposite angles are equal.

RULE 5.3

The angle subtended by an arc at the center of a circle equals twice the angle subtended at the circumference.

RULE 5.4

The angles subtended by an arc in the same segment of the circumference of a circle are all equal.

RULE 5.5

The opposite angles of a quadrilateral bounded by a circle add up to 180°.

RULE 5.6

The angle subtended by a diameter at the circumference of a circle equals 90°.

RULE 5.7

The perpendicular bisector of a chord passes through the center of the circle.

Triangles with unequal sides are said to be *scalene*, while those that have two angles equal are called *isosceles*. Isosceles triangles not only have two angles equal but also two sides are of the same length. If all three angles are equal, then all three sides will be of equal length and the triangle is said to be *equilateral*. The angles of an equilateral triangle are all equal to 60°.

In Figure 5.2a, a circle, known as the *circumscribing circle*, has been drawn around the triangle ABC. Its center is O, hence $OC = OA$ as both lines are radii of the circumscribing circle. Now draw the line OQ perpendicular to AC, cutting AC at P.

$\angle ACO = \angle CAO$ since the triangle AOC is isosceles because $OA = OC =$ the radius of the circumscribing circle. $\angle APO = \angle CPO = 90°$ by construction. The two triangles APO and CPO share a common side (OP) and $OC = OA$ then they must be of exactly the same size and hence point P is the midpoint of AC and $AP = CP$. Triangles that are exactly the same size and shape are said to be *congruent*. Two triangles that have the same angles but are of different size are said to be *similar*. Similar triangles differ merely in scale.

In Figure 5.2a and b, the line CO has been extended to D so that CD is a diameter of the circle. Any straight line that crosses a circle is called a *chord* (e.g., AC and CB), while the curved section that a chord cuts off is an *arc* (for example, the curved sections AQC or ADB). In Figure 5.2a, $OA = OB = OC =$ the radius of the circle. Hence, triangles AOC and BOC are isosceles and $\angle OAC = \angle OCA$. Call this angle $\varnothing$ ("phi"). Hence, $\angle AOD = 2\varnothing$ (as per RULE 5.1 in Box 5.1). Also, $\angle OCB = \angle OBC = \Omega$ ("omega") and hence $\angle BOD = 2\Omega$. Thus, $\angle ACB = \varnothing + \Omega$ while $\angle AOB = 2(\varnothing + \Omega)$. This is true wherever C is along the section of the circle $AQCB$.

In Figure 5.2b, the external angle $AOB = 360 - (2\varnothing + 2\Omega)$ will (by RULE 5.3 in Box 5.1) be twice the angle ADB, which therefore equals $180 - (\varnothing + \Omega)$. Thus, $\angle ACB + \angle ADB = 180°$ so long as C and D are in opposite segments (a *segment* being a portion of a circle bounded by an arc and a chord; any chord divides a circle into two segments). It follows, too, that angles subtended in the same segment are all equal (RULE 5.4 in Box 5.1). It also follows that as in Figure 5.2 where the line CD is a diameter, the angle at the center = 180° and hence the angle at the circumference would be 90°. Thus, $\angle DAC$ and $\angle DBC$ both equal 90°. This leads to Rules 5.5 and 5.6 in Box 5.1.

Finally, in Figure 5.2a in triangles OPA and OPC, the angle at $P = 90°$ by construction (OQ was defined as being perpendicular to AC). $\angle OAP = \angle OCP$ since triangle AOC is isosceles, OP is a common side, and $AO = CO$ as radii of the circle. Hence, triangle OAP is the mirror image of triangle OCP and thus $AP = PC$. Put another way, O lies on the perpendicular bisector of the chord. Since this applies to each of the sides of a triangle it follows that the circle that passes through each apex of a triangle has as its center the point where the three perpendicular bisectors of the sides of the triangle meet, leading to RULE 5.7 in Box 5.1.

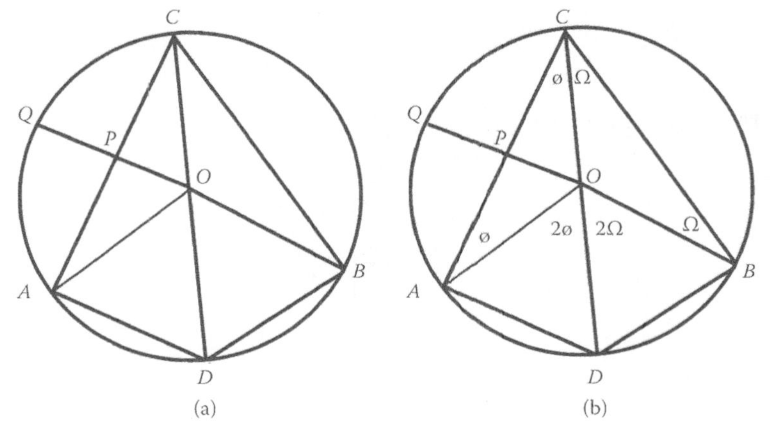

FIGURE 5.2 **(a, b)** Angles subtended by arcs.

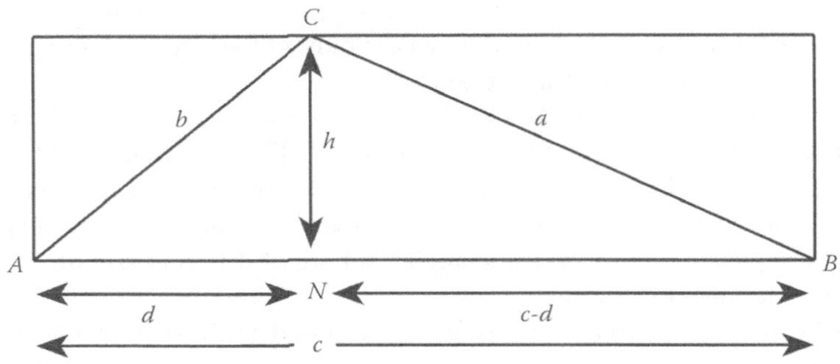

FIGURE 5.3 The area of a triangle.

5.2 AREAS OF TRIANGLES

We will return to aspects of the geometry of the circle shortly but for now, consider the triangle ABC in Figure 5.3. The triangle is half a rectangle whose area is its base times its height. Therefore, the area of a triangle = (1 / 2) * base * height = $0.5hc$. It also equals $\sqrt{s(s-a)(s-b)(s-c)}$ where s is half its perimeter as demonstrated in Box 5.2 and Example 5.1.

We can extend the calculations of the area of a triangle to the *trapezium*, which is a four-sided figure with two sides parallel as shown in Figure 5.4.

In Figure 5.4, the trapezium is made up from two triangles (ABP and QCD) and a rectangle $PBCQ$. Let $BP = CQ = h$, $AP = t$, $PQ = u$, and $QD = v$. Then,

$$area = (1/2)t*h + u*h + (1/2)v*h = (1/2)(t+u+v)*h + (1/2)u*h = (1/2)(AD+BC)h$$

= half the sum of the parallel sides times the distance between them (Box 5.3).

We can extend this simple relationship to the coordinates of A, B, and C in Figure 5.5 where the area of triangle ABC = trapeziums $\{A'ACC' + C'CBB' - A'ABB'\}$.

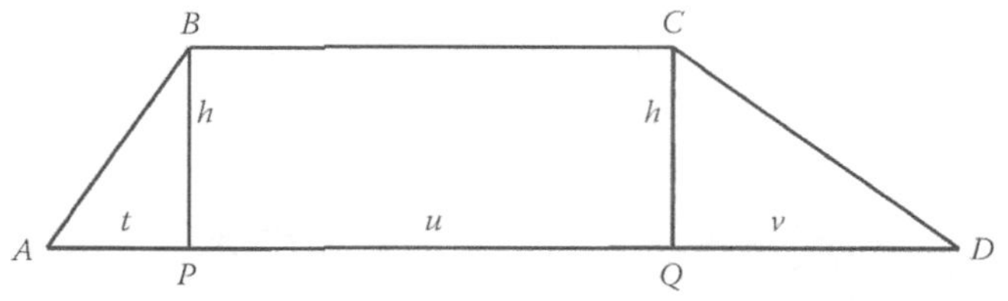

FIGURE 5.4 The area of a trapezium.

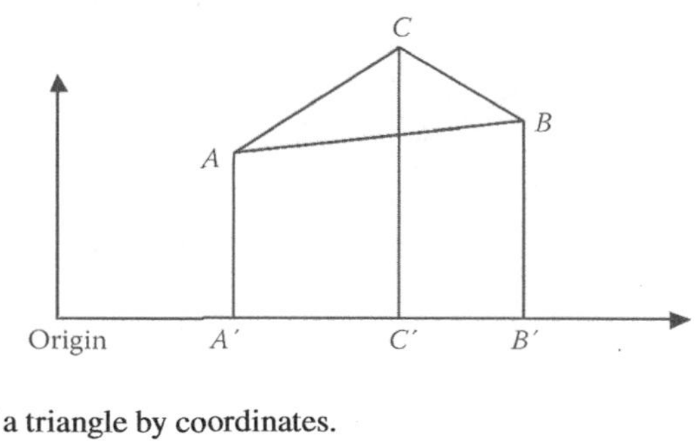

FIGURE 5.5 The area of a triangle by coordinates.

BOX 5.2 THE AREA OF A TRIANGLE FROM ITS SEMIPERIMETER

The triangle *ABC* in Figure 5.3 may be divided by the perpendicular *CN* into two areas with $AN = d$ and $NB = (c - d)$. The area of triangle $ABC = (1/2)$ base * height $= (1/2)hc$. Using Pythagoras for triangle *ANC* $b^2 = h^2 + d^2$ or $h^2 = b^2 - d^2$. Also, in triangle *BNC* $a^2 = h^2 + (c - d)^2 = h^2 + c^2 - 2cd + d^2$

$$Or\ a^2 = h^2 + d^2 + c^2 - 2cd$$

Since $b^2 = h^2 + d^2$, this means $a^2 = b^2 + c^2 - 2cd$. Rearranging,

$$2cd = b^2 + c^2 - a^2$$

Or

$$d = \left(b^2 + c^2 - a^2\right)/2c$$

Hence,

$$\begin{aligned} h^2 &= b^2 - d^2 = (b-d)(b+d)\ (\text{the difference of two squares}) \\ &= \left(b - \left(b^2 + c^2 - a^2\right)/2c\right)\left(b + \left(b^2 + c^2 - a^2\right)/2c\right) \\ &= \left\{\left(2bc - b^2 - c^2 + a^2\right)/2c\right\}\left\{\left(2bc + b^2 + c^2 - a^2\right)/2c\right\} \\ &= \left\{a^2 - (b-c)^2\right\}\left\{(b+c)^2 - a^2\right\}/4c^2 \end{aligned}$$

These again are the differences of two squares. Hence,

$$\begin{aligned} h^2 &= \{(a-b+c)(a+b-c)\}\{(b+c-a)(a+b+c)\}/4c^2,\ or \\ 4c^2h^2 &= (a+b+c-2b)*(a+b+c-2c)*(a+b+c-2a)*(a+b+c) \\ &= (2s-2b)(2s-2c)(2s-2a)(2s)\ \text{where}\ 2s = a+b+c \\ &= 16s(s-a)(s-b)(s-c) \end{aligned}$$

Or, dividing both sides by 16,

$$(1/4)c^2h^2 = s(s-a)(s-b)(s-c)$$

Taking the square roots,

$$(1/2)ch = \text{the area of } ABC = \sqrt{s(s-a)(s-b)(s-c)}$$

where s is the semiperimeter of the triangle.
Thus, the area of a triangle

$$= \sqrt{s(s-a)(s-b)(s-c)}\ \text{where } s \text{ is half its perimeter.}$$

Example 5.1 The Area of a Triangle

Consider the triangle *A, B, C* in Figure 5.3. Assume its coordinates are

$$A(1,10),\ B(9,10),\ C(4,14)$$
$$AB = c = 8;\ d = 3, \text{ and } (c-d) = 5$$

The height $h = 4$; hence its area = ½ base * height = 16.
By Pythagoras,

$$BC = a = \surd\left(h^2 + (c-d)^2\right) = \surd(41) = 6.40312$$

Likewise,

$$AC = b = \surd(h^2 + d^2) = \surd(25) = 5$$

Hence, the semiperimeter = s = (1 / 2) * (8 + 5 + 6.40312) = 9.70156. Using the formula for the area, which is

$$\surd\{s(s-a)(s-b)(s-c)\}$$

The area

$$= \surd(9.70156 * 3.29844 * 4.70156 * 1.70156) = \surd(256) = 16 \text{ as before.}$$

BOX 5.3 THE AREA OF A TRIANGLE AND A TRAPEZIUM

RULE 5.8
The area of a triangle is "half base times height." It also equals

$$\sqrt{s(s-a)(s-b)(s-c)}$$

where a, b, c are the lengths of the sides and s is the semiperimeter = (½)($a + b + c$).

RULE 5.9

The area of a trapezium equals half the sum of the length of the parallel sides times the distance between them.

For the trapezium $A'ACC'$, $A'A = y$ coordinate of A, $C'C = y$ coordinate of C, while the distance between the two sides = $x_C - x_A$. The area = ½$\{(y_A + y_C)(x_C - x_A)\}$.

Likewise, for trapezium $C'CBB'$, the area = ½$\{(y_C + y_B)(x_B - x_C)\}$ and for $A'ABB'$ the area = ½$\{(y_A + y_B)(x_B - x_A)\}$. Combining the above and simplifying, the area of the triangle ABC

$$= 1/2\{x_A y_B + x_B y_C + x_C y_A - y_A x_B - y_B x_C - y_C x_A\}$$

In fact, since every polygon can be subdivided into a series of triangles this sequence can be extended. Thus, for a six-sided polygon with points A, B, C, D, E, and F (see Example 5.2) the area would be:

$$1/2\{x_A y_B + x_B y_C + x_C y_D + x_D y_E + x_E y_F + x_F y_A$$
$$-y_A x_B - y_B x_C - y_C x_D - y_D x_E - y_E x_F - y_F x_A\}$$

For an eight-sided polygon A, B, C, D, E, F, G, and H, then the area would be (Box 5.4)

$$\text{area} = 1/2\{x_A y_B + x_B y_C + x_C y_D + x_D y_E + x_E y_F + x_F y_{G+} x_G y_H + x_H y_A$$
$$- y_A x_B - y_B x_C - y_C x_D - y_D x_E - y_E x_F - y_F x_G - y_G x_H - y_H x_A\}$$

and so on.

Example 5.2 Calculating the Area of a Polygon

Consider the six-sided polygon whose corners are A, B, C, D, E, F with A (1, 11), B (6, 13), C (10, 18), D (14, 12), E (12, 2), F (6, 7). Using the formula in Box 5.4, twice the area

$$= (1*13 + 6*18 + 10*12 + 14*2 + 12*7 + 6*11 - 11*6 - 13*10 - 18*$$
$$14 - 12*12 - 2*6 - 7*1)$$
$$= (13 + 108 + 120 + 28 + 84 + 66 - 66 - 130 - 252 - 144 - 12 - 7) = -192$$

Hence, area = 96

Note: The minus sign would not arise if the area were computed in the reverse order.

BOX 5.4 THE AREA OF A POLYGON

RULE 5.10

To obtain the area of a polygon with corners (x_A, y_A), and so forth, multiply the x value of each point by the y value of the next point and then subtract the y value of each point times the x value of the next point to give twice the area.

As an example, for an eight-sided figure the area is

$$1/2\{x_A y_B + x_B y_C + x_C y_D + x_D y_E + x_E y_F + x_F y_{G+} x_G y_H + x_H y_A$$
$$-y_A x_B - y_B x_C - y_C x_D - y_D x_E - y_E x_F - y_F x_G - y_G x_H - y_H x_A\}$$

5.3 CENTERS OF A TRIANGLE

In Chapter 4, Box 4.3, we showed that the lines joining the corners of a triangle to the middle of their opposite sides pass through a point G known as the *centroid*. On the basis that the area of a triangle is half its base times its height, then it follows that for the left-hand area in Figure 5.6, the area of ACF = area of FCB since $AF = FB$ by construction. In effect, the triangle ABC is balanced

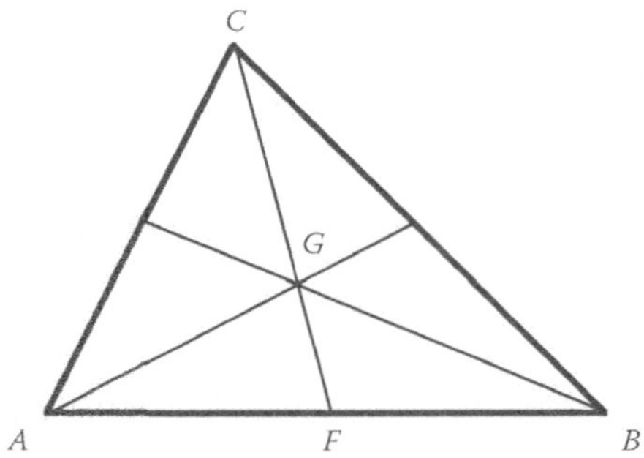

FIGURE 5.6 The centroid.

about the line CF; similarly, for the lines from A and B through G. Thus, G is the point of balance or center of gravity of the triangle. Its coordinates were shown in Box 3.3 to be $\{1/3(x_A + x_B + x_C), 1/3(y_A + y_B + y_C)\}$.

In Figure 5.7, AD is constructed so as to be perpendicular to BC, and BE to be perpendicular to AC. The two lines intersect at O and line CO meets AB at H. We will now show that CH is perpendicular to AB.

Since $\angle AEB = 90° = \angle ADB$ then a circle with AB as diameter passes through E and D (RULE 5.5), hence $\angle EDO = \angle EBA$ (angles in the same segment—RULE 5.4) $= 90 - A$. Since $\angle ODC = 90°$ by construction, $\angle EDC = A$. But E and D also lie on a circle with diameter OC since the angles subtended by a diameter are 90°. Hence, $\angle EDC = \angle EOC = A$. In triangle EOC, $\angle OEC = 90°$, hence $\angle OCE = 90 - A$. Now in triangle CAH, $\angle CAH = A$ and $\angle ACH = \angle OCE = 90 - A$, hence $\angle AHC = 90°$. Thus, CH is perpendicular to AB and the perpendiculars from each point of a triangle to its opposite side intersect at one point, known as the *orthocenter.*

In Figure 5.8, let the line CI be the bisector of the angle ACB and AI bisect CAB. Let P, Q, and R be the feet of the perpendiculars from I to the corresponding sides. Then in triangles IPC and IQC, $\angle ICP = \angle ICQ$ since IC bisects PCQ, and $\angle IPC = 90° = \angle IQC$. Thus, the two triangles have the same angles and a common side (CI) and must therefore be congruent. That means that $IP = IQ$. Similarly, triangles API and ARI are congruent, so $IP = IR$. IB is a common side between triangles IRB and IQB, and since the angles at Q and R are right angles, then by Pythagoras, QB must be the same size as RB. Thus, the two triangles IRB and IQB are congruent so that IB is the bisector of angle ABC.

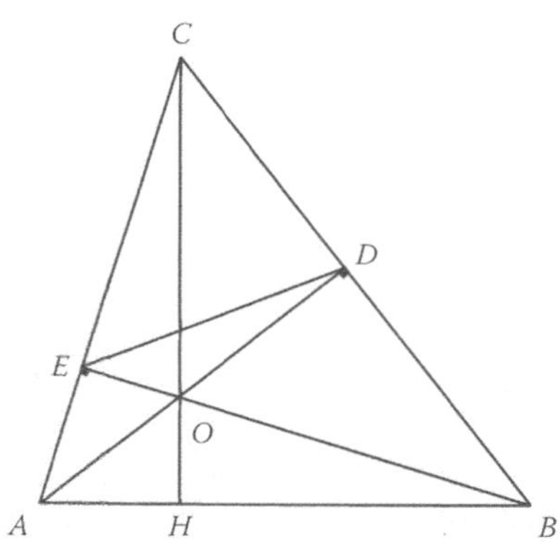

FIGURE 5.7 The orthocenter.

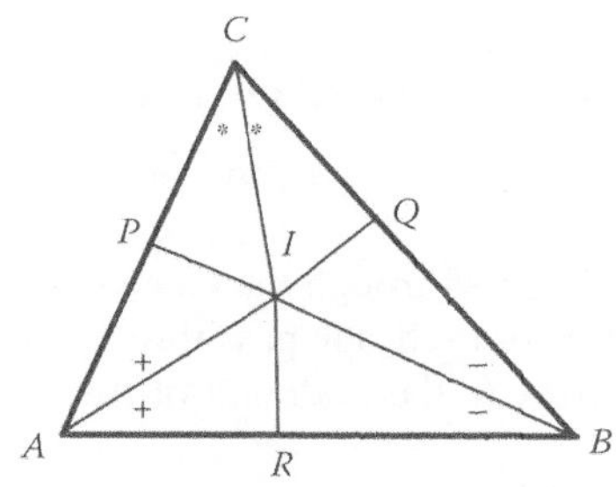

FIGURE 5.8 The incenter.

The point where the bisectors of all three angles meet is known as the *incenter.* A circle can be drawn around this point that touches each of the three sides of the triangle (Figure 5.10a).

Finally, the perpendicular bisectors of each side also meet in a point known as the *circumcenter* (*CC*) as we saw in Figure 5.2. In Figure 5.9, *P′* is the midpoint of *AC*, *Q′* is the midpoint of *CB*, and *R′* is the midpoint of *BA*. *CC* is the point from which a circle can be drawn through all three vertices *A*, *B*, *C*.

The various centers are summarized in Box 5.5. Circles have a number of well-known properties, such as the length of the circumference = $2\pi r$ where r = length of the radius. The area = πr^2. A line that just touches the circle (e.g., *AB* at point *R* in Figure 5.10a) is called a *tangent.* Any other line than a tangent will either not touch the circle at all or else will intersect it at two points. A tangent and the radius to the point where it touches the circle are at 90° and this radius is said to be *normal* to the tangent.

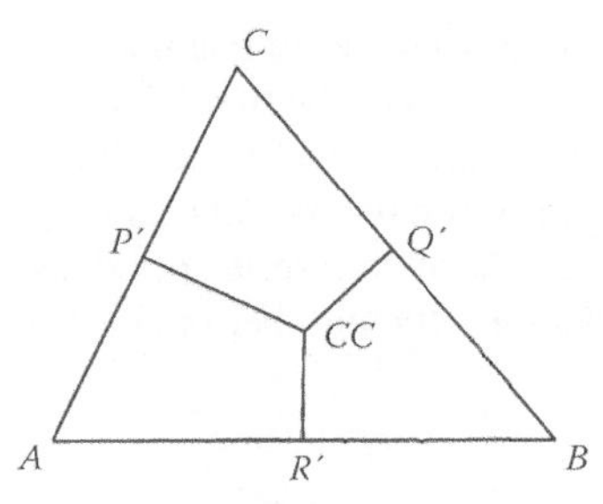

FIGURE 5.9 The circumcenter.

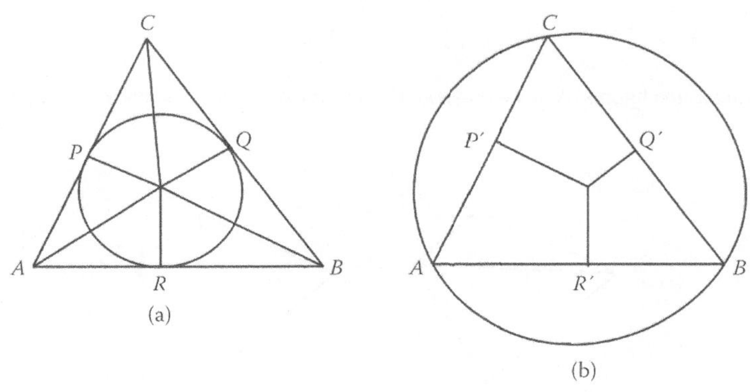

FIGURE 5.10 **(a, b)** Inscribed and circumscribed circles.

BOX 5.5 CENTERS OF A TRIANGLE

1. The perpendiculars from the vertices to the opposite sides all pass through the orthocenter.
2. The bisectors of each angle pass through the incenter.
3. The perpendicular bisectors of each side pass through the circumcenter.
4. The lines joining each vertex to the midpoint of the opposite side all pass through the centroid.

Note: Though not proven here, the three points marking the orthocenter, centroid, and circumcenter all lie on a straight line, known as the *Euler line*. The incenter does not, however, lie on this line.

5.4 POLYGONS

A *polygon* is a closed figure bounded by three or more straight lines that do not cross; it has the same number of vertices as there are sides. The polygon is said to be concave if any of its interior angles are greater than 180°, otherwise it is said to be convex.

In Figure 5.11, *A* is not a polygon; *B* is a convex polygon; *C* is a concave polygon, and *D* is essentially two polygons. The best-known examples of polygons are triangles, quadrilaterals (especially squares and rectangles), pentagons (five-sided), and hexagons (six-sided figures). A contour line that closes back on itself and is constructed out of straight-line sections would also form a polygon.

Any polygon can be broken down into a series of triangles though even with a rectangle these are not unique (see Figure 5.12). The three sides of the first triangle create internal angles that add up to 180°. A further point adds a further triangle and a further 180° to bring the total sum of the internal angles to 360°. A five-sided polygon will have its internal angles adding up to 540°, and so forth. Since the whole polygon can be thought of as a series of adjacent triangles, the sum of its interior angles will be = $(n - 2) * 180°$ where n = the number of sides (Box 5.6).

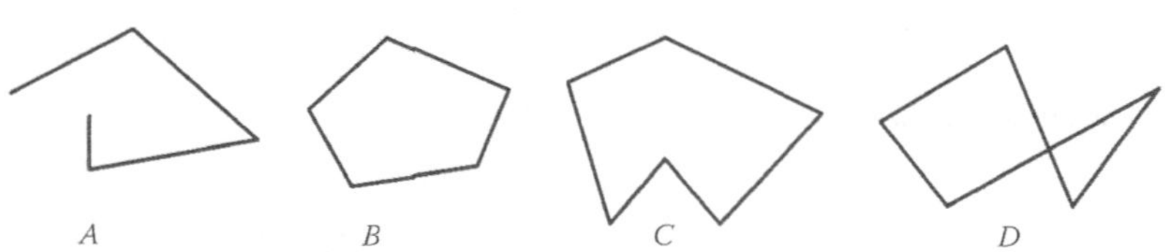

FIGURE 5.11 Straight-line figures. A: not a polygon; B: a convex polygon; C: a concave polygon; D: two polygons.

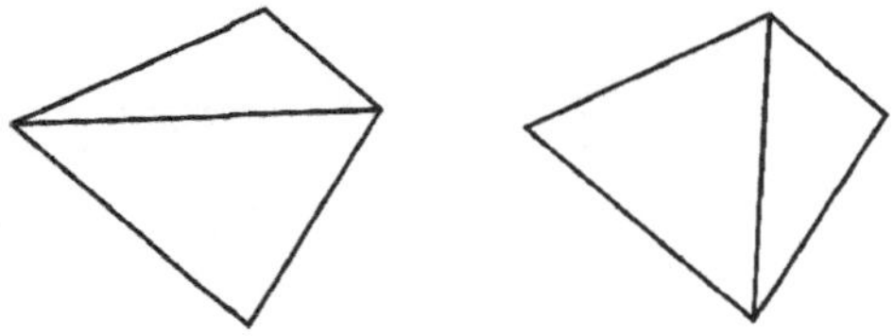

FIGURE 5.12 Two ways to divide a quadrilateral.

BOX 5.6 THE ANGLES OF A POLYGON

RULE 5.11

The sum of the interior angles of a polygon = $(n - 2) * 180°$ where n = the number of sides.

Given any set of points there are many ways in which they can be joined to form a set of triangles in a process known as *triangulation* (Figure 5.13). Until recently, the whole basis of map-making was constructed around a set of points between which the angles had been observed. The result was a network of polygons and triangles, some overlapping.

By measuring all three angles in each triangle the shape of the mesh could be rigidly determined and there would be independent checks on the accuracy of the measurements since for each triangle the measured angles had, after suitable adjustment, to add up to 180°. The overall scale was determined by measuring at least one of the sides of the network.

Although the first triangulation networks were observed during the 18th century, it was during the latter part of the 19th and most of the 20th centuries that many areas of the globe became covered by such networks. The fact that the surface of the Earth is curved meant that none of the networks of triangles were flat and hence some corrections had to be applied to the measured angles and distances in order to calculate the coordinates of the triangulation or "trig" stations. The relationships between angles and distances on a curved surface and their corresponding values on the flat are discussed further in Chapters 6 and 11. Today, positions are often determined using the global positioning systems (such as GPS) discussed in Chapter 16.

Triangulation is also used when interpolating heights from a series of irregular points in a process known as *TIN* (Triangulated Irregular Network). TIN has become the generic term that is used in GIS to describe a way in which intermediate height values can be interpolated. Although height is the most commonly used third dimension in geomatics, the technique can be extended to other characteristics at a point, such as air pressure or the strength of the magnetic field. The process involves the conversion of a set of irregularly distributed points of known third-dimensional value ("spot heights") into a network of triangles within which further height values can be interpolated, as discussed in Chapter 4 (see Figures 4.16 and 4.17).

Consider a series of points A, B, C, D, E in Figure 5.14a, for which the eastings, northings, and height or (x, y, z) values are known. The points can be joined by straight lines and the intermediate heights interpolated. The question here is "which points should be connected?" A favored method is known as the *Delaunay triangulation*. It is related to the creation of a series of polygons known variously as *Dirichlet* or *Voronoi tessellations* or *Thiessen polygons*.

In Figure 5.14a, we have a set of points of known height. We can plot these on a plane surface and then draw lines from A to all neighboring points and construct the perpendicular bisector of each of

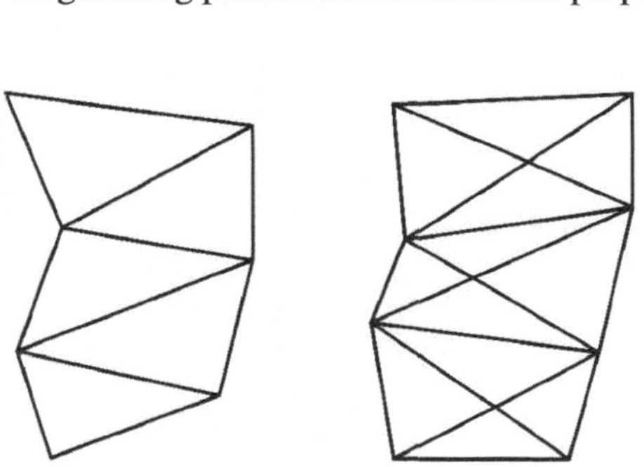

FIGURE 5.13 Triangulation networks.

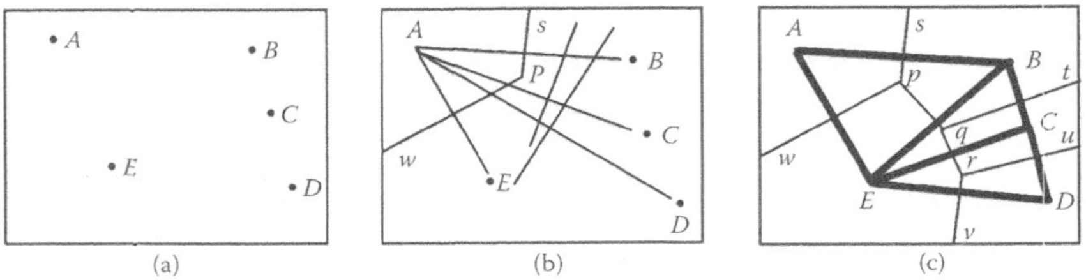

FIGURE 5.14 **(a–c)** Thiessen polygons and the Delaunay triangles.

the lines *AB*, *AC*, *AD*, and *AE*. Two of these (*WP* and *PS* in Figure 5.14b) together with the bounding area mark out a polygon in which all the points are nearer to *A* than to any of the other points *B*, *C*, *D*, and *E*. That is because all the perpendicular bisectors separate out areas one side of which is nearer *A* and the other is nearer to the second generating point.

This process can be extended to all the other points, creating a series of polygons within which all parts of the area are closer to the generating point than to any other. For example, in Figure 5.14c, the lines of the bounding rectangle together with *W* to *P* to *Q* to *R* to *V* define a polygon around *E* that is the area within which all points lie closer to *E* than to *A*, *B*, *C*, *D*, and so forth.

Such polygons are known as *Thiessen polygons* and are of particular interest in certain types of statistical mapping. The lines connecting nearby points in the final creation of these polygons form a series of triangles known as the *Delaunay triangles*. They are deemed to be the "best shaped" triangles, and are the nearest to being equilateral that can be constructed from the original points. These triangles are shown in Figure 5.14c by the darker lines.

5.5 THE SPHERE AND THE ELLIPSE

Let us now return to the geometry of the circle since it is important in two areas of geomatics—the study of positioning fixing and in the representation of the shape of the Earth, which is approximately spherical.

A plane intersects a sphere in a circle; if the plane passes through the center of the sphere, as in the case of the meridians of longitude on the Earth, the line on the surface is said to be a *great circle*. Other circles such as parallels of latitude (except the special case of the equator) are called *small circles*.

The geometry of any figure on the surface of a sphere differs from that on a plane because triangles become spherical triangles and their three angles no longer add up to 180°. Thus, in Figure 5.15 the

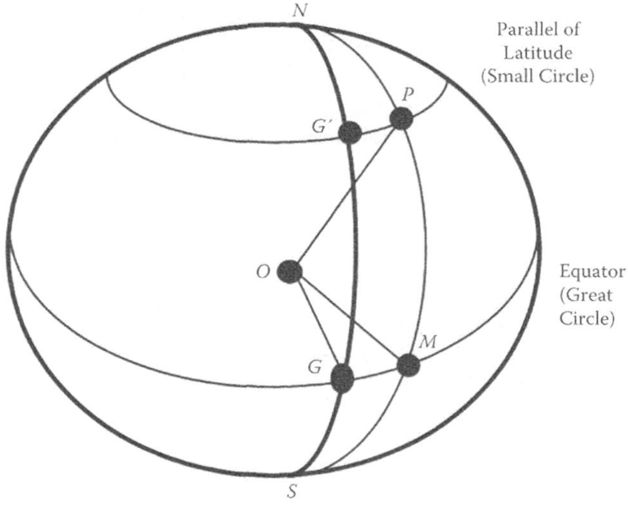

FIGURE 5.15 Great and small circles.

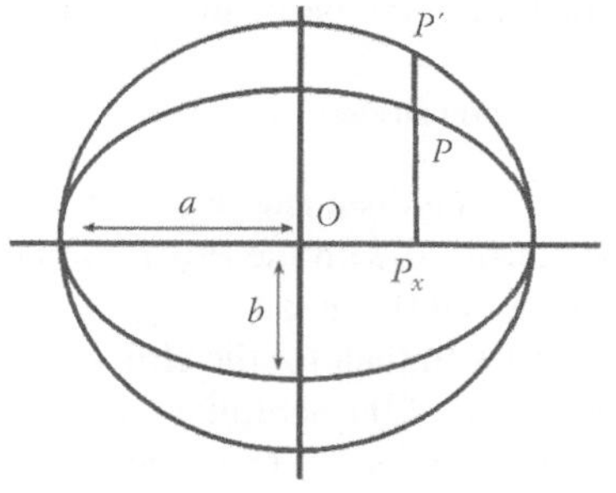

FIGURE 5.16 Ellipse and auxiliary circle.

angle $NG'P = 90° =$ angle NPG'. The angle $G'NP$ at the North Pole is the difference in longitude between the meridian or great circle through Greenwich and the meridian through P; this means that the three angles of the triangle $G'NP$ are in excess of 180°. When the sides of the triangle are defined by great circles the amount by which the sum of the three angles exceeds 180° is called the *spherical excess*. Computations involving spherical triangles will be considered in Chapter 6 on trigonometry.

A sphere is only an approximation to the shape of the earth; in reality an *ellipsoid* (obtained by rotating an ellipse about its shorter or "minor" axis) gives a better representation. An *ellipse* (Figure 5.16) is a closed figure shaped like a compressed circle that is symmetrical about two axes known as the *major axis* and the *minor axis*. The circle that is formed using the major axis as diameter is known as the *auxiliary circle*. The radius of this circle is known as the semimajor axis and is normally referred to as a, while the length of the semiminor axis is b. In the geometry of the ellipse compared with the auxiliary circle, the distances in the x-direction remain the same but the distances in the y-direction are compressed or scaled down so that P' on the circle becomes P on the ellipse.

5.6 SECTIONS OF A CONE

The ellipse and circle are two of four special second-order curves known as *conic sections*, the others being the parabola and the hyperbola (see below). They take the form of

$$ax^2 + by^2 + cxy + dx + ey + f = 0$$

where a, b, c, d, e, and f are constants. For a circle, $a = b$ and $c = 0$. Geometrically, they can be derived from a line known as the *directrix* shown as the line CD in Figure 5.17. F_1 and F_2 are two points on the major axis known as the *foci*, each *focus* being the same distance either side of the

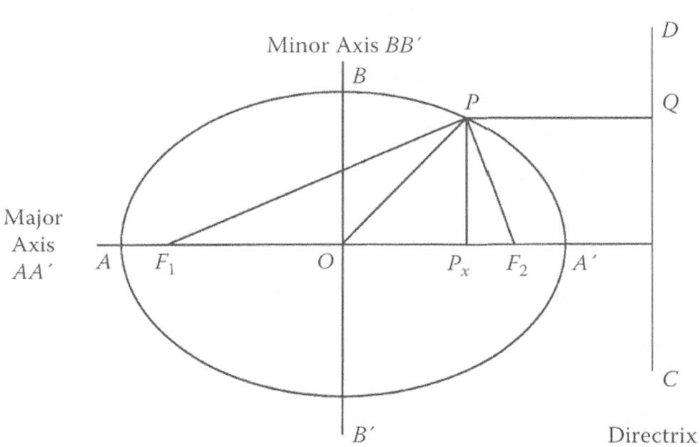

FIGURE 5.17 An ellipse and its directrix.

center O on the major axis. Q is the foot of the perpendicular from P onto the line CD so that PQ is parallel to the major axis.

The ellipse has the following characteristics:

1. For all conics, the ratio between the distances F_2P and PQ is a constant that is known as the *eccentricity* e (not to be confused with the Euler number "e" that equals 2.718…). For an ellipse, the value of e is such that $0 < e < 1$.
2. If the origin O has coordinates (0, 0) then for the ellipse F_1 is located at point $(-ae, 0)$ and F_2 at $(+ae, 0)$ where a is the length of the semimajor axis and e is the eccentricity.
3. The directrix is located at a distance (a / e) from the center of the ellipse. It thus cuts the major axis at point $(a / e, 0)$.
4. The sum of the distances $F_1P + PF_2$ is a constant for the ellipse. (This is why it is possible to draw an ellipse by placing a pin at each of the foci, a loop of string of length $F_1P + PF_2 + F_2F_1$ around the pins; then, by holding a pencil at P and keeping the string taught, the pencil can be moved around the pins to draw the ellipse).
5. $ae = \sqrt{}\left(a^2 - b^2\right)$ or $e = \sqrt{}\left(1 - b^2 / a^2\right)$.
6. The area of an ellipse is πab.
7. The length of the circumference of an ellipse is a complicated thing to calculate.
8. If P is a point on the ellipse with coordinates (x, y) and center (0, 0), then $x^2 / a^2 + y^2 / b^2 = 1$.

These relationships will be explained more fully in Chapter 13 after we have discussed trigonometric functions. Example 5.3 illustrates the quation to construct an ellipse. The circle can be regarded as a special case of an ellipse in which $a = b = r$, the radius of the circle and $x^2 / r^2 + y^2 / r^2 = 1$. The two foci are coincident at the center of the circle. The eccentricity $e = 0$ and the directrix is located a distance $(a / 0)$ from the center, which means that it is an infinite distance away.

For the ellipse $0 < e < 1$. When $e = 1$, we have $FP = PQ$ (see Figure 5.19) and instead of being gently rounded, the ellipse becomes infinite in size. The curve is known as a *parabola*. It has only one focus and its directrix is symmetrically placed on the other side of the origin, so that if F is at $(a, 0)$ the directrix passes through $(-a, 0)$. A simple example of a parabola is the function $y^2 = mx$ where m is a constant.

Example 5.3 The Equation of an Ellipse

In many graphics packages, ellipses are constructed from their enclosing rectangle. Consider the rectangle whose sides stretch from $x = 2$ to $x = 6$ and $y = 1$ to $y = 3$ (see Figure 5.18).

For an ellipse to fit within this rectangle, the major axis must have $2a$ = width of box or $a = 2$; the minor axis must have $2b$ = height of box or $b = 1$. The center of the ellipse is at $x = 4$, $y = 2$. The equation for the ellipse is

$$(x - 4)^2 / a^2 + (y - 2)^2 / b^2 = 1$$

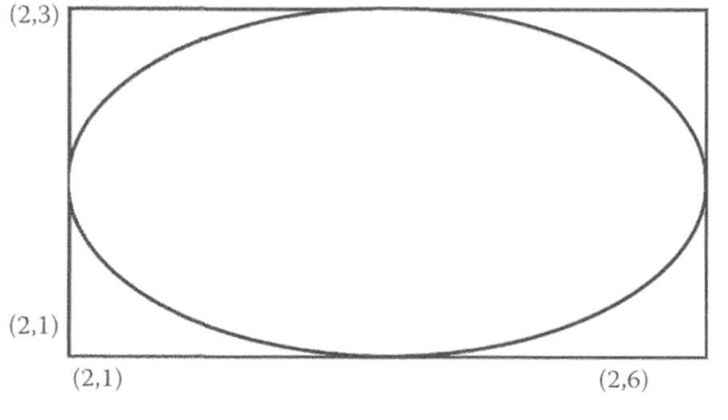

FIGURE 5.18 An enclosed ellipse.

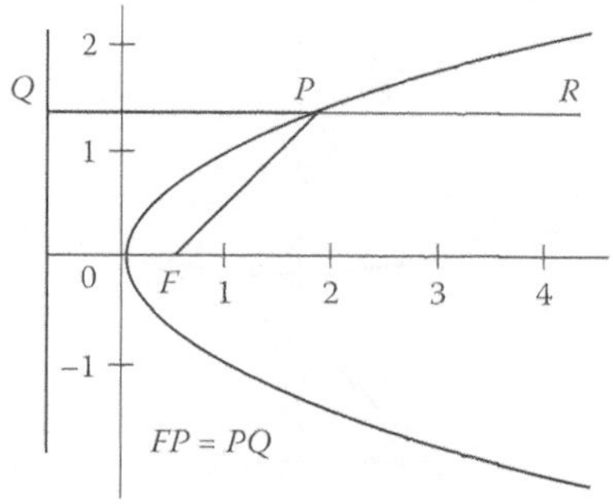

FIGURE 5.19 A parabola.

or, putting $a = 2$ and $b = 1$:

$$x^2 / 4 - 2x + 4 + y^2 - 4y + 4 = 1$$

Or

$$x^2 - 8x + 4y^2 - 16y + 28 = 0$$

To draw the ellipse we need to plot points on the curve

$$y = 2 \pm 1/2\surd\left\{4 - (x-4)^2\right\} = 2 \pm 1/2\surd\left\{8x - x^2 - 12\right\}$$

Note: In this equation there are no solutions if x or y lie outside the range, namely, $2 \le x \le 6$ or $1 \le y \le 3$.

The parabola has only one focus and the tangent to the curve at point *P* exactly bisects the lines *FP* and *QP*. This means that the line *PR* in Figure 5.19 is parallel to the axis and also at the same angle to the tangent at *P* as is the line *PF*. Hence, if a light source were placed at *F* and the inside of the parabolic surface were perfectly reflecting, then a beam of light from *F* would emerge parallel to the major axis. Conversely, a parabolic mirror can focus a parallel beam of light or other form of radiation, for instance from the sun, onto a single point (the focus, *F*).

If the eccentricity *e* has a value greater than 1, then the shape of the resulting curve is a *hyperbola*. It has an equation of the form $x^2 / a^2 - y^2 / b^2 = 1$. It consists of two branches formed by a pair of intersecting lines known as the *asymptotes* (*AB* and *CD* in Figure 5.20) to which it approaches more and more closely but only reaching them at infinity. The slope of these asymptotes is given by the equations

$$y = +(b / a)x$$

And

$$y = -(b / a)x$$

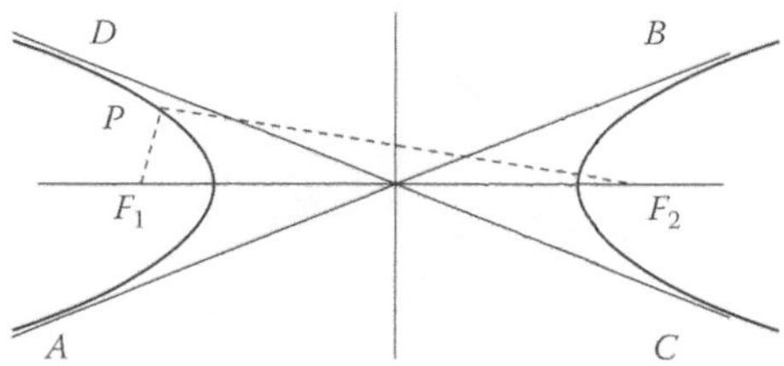

FIGURE 5.20 A hyperbola with its asymptotes.

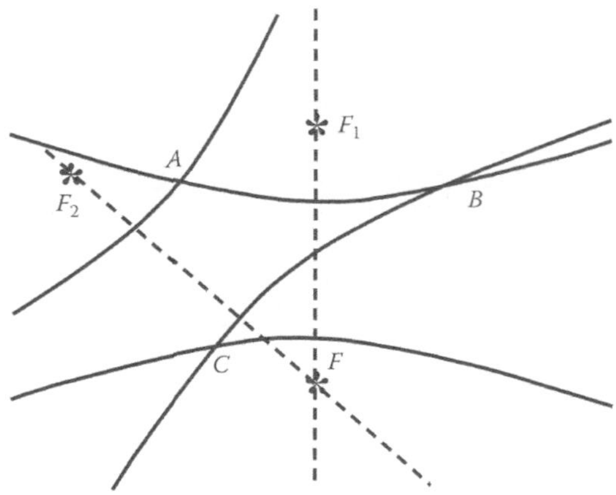

FIGURE 5.21 Two intersecting hyperbolae.

With the hyperbola there are two foci (F_1 and F_2) but whereas with the ellipse as shown in Figure 5.17, the sum of the two distances $PF_1 + PF_2$ is a constant, in the case of the hyperbola the difference between the two distances is fixed. Thus, $(PF_1 - PF_2)$ is constant in Figure 5.20. This has allowed navigation systems, especially at sea, to track a hyperbolic course using the difference in time for signals sent from two known points using what is called the *Doppler effect*. With transmitters at three known points there will be two intersecting sets of hyperbolae, allowing for the position of a vessel to be determined. In practice the two sets of hyperbolae may indicate four possible points but if the initial position is known the location of the vessel can be tracked. In Figure 5.21, the foci for the two hyperbolae are FF_1 and FF_2 where the transmitters would be located. Three possible locations are shown (*A*, *B*, and *C*). If there are more than three transmitters then the ambiguity can be resolved.

The circle, ellipse, parabola, and hyperbola all have solid equivalents (the sphere, the ellipsoid, the paraboloid, and hyperboloid). They are all quadratic forms and in their plane (i.e., 2D) form they are cross-sections of a 3D cone. Consider two intersecting lines (*AA′* and *BB′* in Figure 5.22). These form a plane. Draw one of the lines of bisection (*OO′*) and then in three dimensions rotate the original lines around this bisector to generate a double cone (one cone stuck upside down on top of the other). Cut this solid along a plane perpendicular to the axis of rotation and you have a circle; cut it at an angle less than the slope of the sides of the cone and the cross-section will be an ellipse

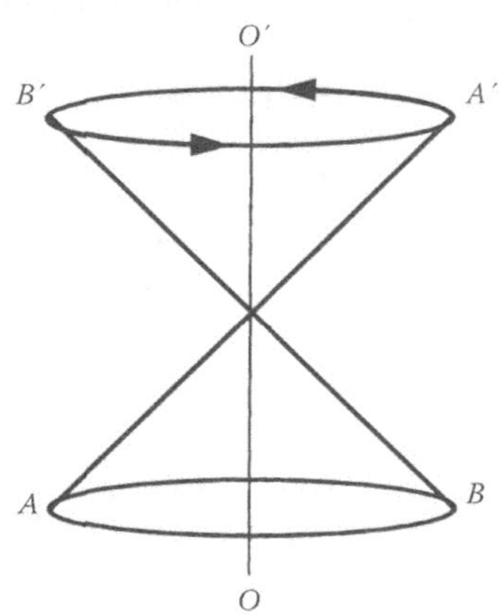

FIGURE 5.22 A double cone.

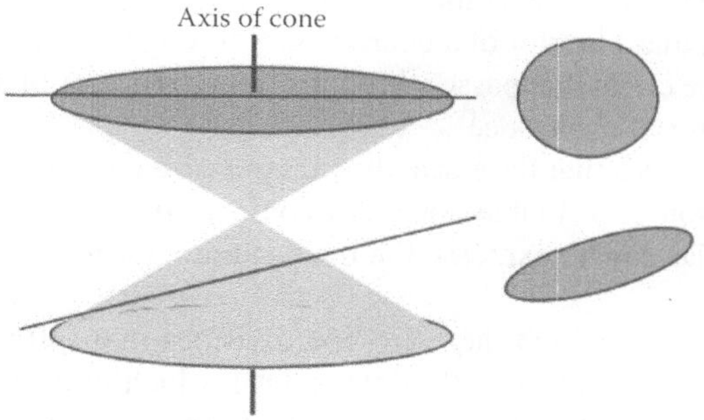

FIGURE 5.23 Sections of a cone—circle and ellipse.

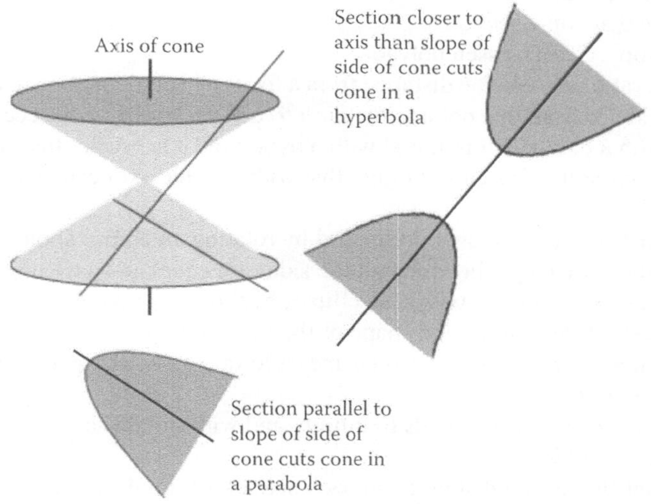

FIGURE 5.24 Sections of a cone—parabola and hyperbola.

(Figure 5.23). If the section of the double cone is cut parallel to the edges of the cone then if the cone is infinitely large then the section will be infinitely large and be a parabola. If, however, the section is greater, it will cut the double cone twice (top and bottom) in a hyperbola (Figure 5.24).

GLOSSARY

Acute angle An angle in a triangle that is less than a *right angle.*

Arc A section of the circumference of a circle. Angles subtended by an arc to any point opposite and in the same segment of the circumference of a circle are equal and half the size of the angle subtended at the center.

Area of triangle ½ Base * Height or $\sqrt{s(s-a)(s-b)(s-c)}$ where s is the *semiperimeter* or for the triangle ABC, or ½ $(x_A y_B + x_B y_C + x_C y_A - y_A x_B - y_B x_C - y_C x_A)$. The latter formula can be extended to give the area of a polygon.

Auxiliary circle The circle surrounding an ellipse based on the *major axis* as its diameter.

Centroid The point of intersection of each line joining the corner of a triangle to the midpoint of the opposite side.

Chord A straight line that divides a circle. If the division is precisely in half then the chord is a diameter. The perpendicular bisector of a chord passes through the center of a circle.

Circumcenter The center of the circle that passes through all three corners of a triangle. It lies on the perpendicular bisector of each side.

Congruent Figures such as triangles that have exactly the same size and shape.

Conic section Any plane section of a double cone. All conic sections such as circles, ellipses, parabolas, and hyperbolas can be expressed in the quadratic form $ax^2 + by^2 + cxy + dx + ey + f = 0$.

Degree The measure of an angle in which there are 360 divisions in a full turn. Each division (written as 1°) can be subdivided into 60 minutes (60′), which in turn can be subdivided into 60 seconds (60″). The angles of a plane triangle always add up to 180°. On a curved surface they add up to more than 180°.

Delaunay triangulation The formation of a network of triangles whose sides are perpendicular to the sides of a set of *Thiessen polygons.*

Directrix A line perpendicular to the major axis of an *ellipse*, *parabola*, or *hyperbola* that is used to generate that conic section.

Dirichlet tessellation *See* Thiessen polygon.

Eccentricity The ratio between the distance from a focus of a conic section to a point on the curve and the distance from that point to the *directrix.* With an ellipse the eccentricity is between 0 and 1, with a parabola it is 1, and with a hyperbola it is greater than 1.

Ellipse A *conic section* that is a closed figure that with suitable axes can be expressed in the form $x^2 / a^2 + y^2 / b^2 = 1$.

Ellipsoid A geometrical surface or solid formed by rotating an ellipse about one of its axes.

Equilateral triangle A triangle in which all the sides are equal and each angle = 60°.

Focus A fixed point on the major axis of an ellipse, hyperbola, or parabola that with the *directrix* and *eccentricity* determines the shape of the *conic section.*

Grad Also known as a *gon.* It is an angular measure in which 100 grads (or gons) equals one-quarter turn or 90°.

Hyperbola A *conic section* that extends to infinity and with suitable axes can be expressed in the form $x^2 / a^2 - y^2 / b^2 = 1$.

Incenter The point that is equidistant from each side of a triangle so that a circle can be drawn touching each side. It lies on the bisector of each angle.

Isosceles A triangle in which two sides (and hence, two angles) are equal.

Major axis The longest axis or line through the center of an *ellipse.*

Minor axis The shortest axis or line through the center of an *ellipse.*

Normal A line that is perpendicular to another line or curve.

Obtuse angle An angle in a triangle that is greater than a *right angle.*

Orthocenter The point where the perpendiculars from each corner of a triangle to the opposite side intersect.

Parabola A *conic section* that extends to infinity and with suitable axes can be expressed in the form $y^2 = 4ax$.

Polygon A closed figure bounded by three or more sides. The sum of the interior angles of a polygon with n sides add up to $(n - 2) * 180°$.

Radian An angular measurement in which there are 2π radians in a full turn where 2π is also the ratio between the circumference of a circle and its radius. $\pi \approx 3.14159265$.

Right angle An angle that is precisely a quarter turn or 90° or $\pi / 2$ *radians.*

Scalene A triangle in which all sides are of unequal length.

Segment The portion of a circle bounded by an *arc* and a *chord.*

Semiperimeter Half the sum of the length of the sides of a triangle.

Spherical excess The amount by which the sum of the three angles of a triangle on a sphere exceeds 180°.

Tangent A line that just touches a curve. Also used in trigonometry (see Chapter 6) to describe the ratio between the side opposite and the side adjacent in a right-angled triangle.

Thiessen polygon A polygon around a point such that any point within that polygon is nearer to its center than to any other point. Also known as *Dirichlet* or *Voronoi tessellations.* Often spelled as *Theissen.*

TIN (Triangulated Irregular Network) A network of irregularly shaped triangles.

Trapezium A four-sided figure with two sides parallel.

Triangulation The connection of a series of points to form a network of triangles.

Voronoi tessellation *See* Thiessen polygon.

6 Plane and Spherical Trigonometry

6.1 BASIC TRIGONOMETRIC FUNCTIONS

Trigonometry is concerned with ratios between the sides and angles of triangles. Although at its simplest it is concerned with right-angled triangles on a plane surface, its applications extend to many areas of geomatics and geographic information systems (GIS) including calculations on curved surfaces such as that of the Earth. Trigonometry is used extensively in surveying and navigation.

In Chapter 4, Figure 4.7 showed similar triangles that have a common shape but differ in scale. In Figure 6.1, the triangles ABC, $AB'C'$, and $AB''C''$ are all the same shape (they have the same angles) but they differ in size or scale. In these triangles the ratio $\frac{BC}{AB} = \frac{B'C'}{AB'} = \frac{B''C''}{AB''} =$ "side opposite over side adjacent."

Given the fixed angle A, then in any right-angled triangle, the ratio $BC\,/\,AB$ is constant. This is called the *tangent* of angle A or tan A. Similarly, the ratio $BC\,/\,AC$ is constant = "side opposite over hypotenuse." This is called the *sine* of angle A or sin A. Likewise, the ratio $AB\,/\,AC$ is a constant = "side adjacent over hypotenuse." It is called the *cosine* of angle A or cos A. (See Box 6.1 and Example 6.1.)

The ratios sin, cos, and tan are not independent. Given the theorem by Pythagoras

$$AB^2 + BC^2 = AC^2$$

Dividing both sides by AC^2 we obtain $(AB\,/\,AC)^2 + (BC\,/\,AC)^2 = 1$, or $(\sin A)^2 + (\cos A)^2 = 1$. Also,

$$(\sin\ A\,/\cos\ A) = \left(\frac{AB}{AC}\right)\Big/\left(\frac{BC}{AC}\right) = \left(\frac{AB}{AC}\right) = \tan\ A$$

Hence,

$$\tan = \frac{\sin}{\cos}$$

It is sometimes helpful to deal with the reciprocal ratios $AC\,/\,AB$ and $AC\,/\,BC$. These (and the reciprocal of tan A) have special names. 1 divided by sine is called *cosecant* or *cosec*; 1 divided by cosine is the *secant* or *sec*; and 1 divided by tangent = *cotangent* or *cot*.

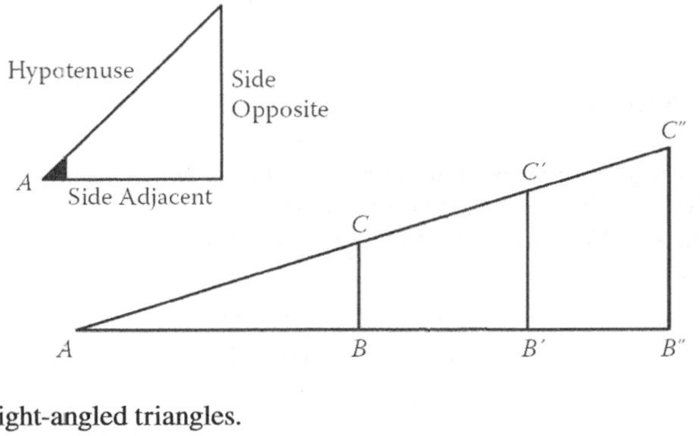

FIGURE 6.1 Similar right-angled triangles.

DOI: 10.1201/9781003464402-8

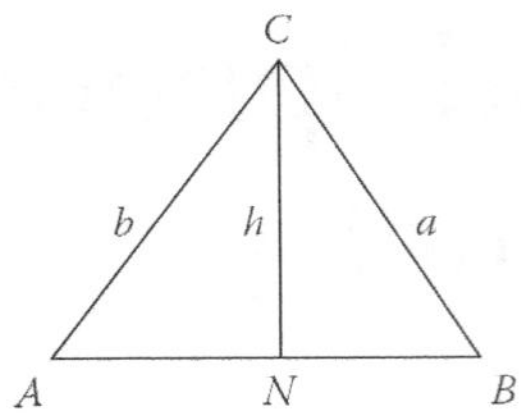

FIGURE 6.2 The height or altitude of a triangle.

By convention, just as $x * x = x^2$, so $\sin A * \sin A = (\sin A)^2$ and is written as $\sin^2 A$. Likewise, $\cos A * \cos A = \cos^2 A$, and so forth. However, although $1 / x = x^{-1}$, $1 / \sin A$ is *not* written as $\sin^{-1} A$ since this has a reserved meaning, which is "the angle whose value is sine *A*." This, in turn, is called *arc-sine A*. $\cos^{-1} A$ means "the angle whose value is cosine *A*." It is called *arccos*. Similarly, $\tan^{-1}$ = arctan, cosec^{-1} = arccosec, $\sec^{-1}$ = arcsec, and $\cot^{-1}$ = arccot.

In Figure 6.1, *ABC* is a right-angled triangle and the angle at *C* (the angle *ACB*) has as its cosine the ratio *BC* / *AC*. Thus, cos *C* is the same as sin *A*. This only happens when the angle $C = 90° - A$. For all angles *A*, $\sin A = \cos (90° - A)$; $\cos A = \sin (90° - A)$; hence, $\tan A = \cot (90 - A)$.

We can always convert non-right-angled triangles into ones with right angles as shown in Figure 6.2. Even though the triangle *ABC* has no right angle, there is still a value for "sin *A*" since this is dependent on the size of the angle *A*, not the shape or size of the triangle. We can create a right-angled triangle (Figure 6.2) by drawing the perpendicular line from *C* onto the horizontal line *AB* so that angle $ANC = 90°$. We represent the sides of the triangle as lowercase letters (with $BC = a$, $CA = b$, and $AB = c$, and the height $CN = h$, known as the *altitude*) and the angles as uppercase letters *A*, *B*, and *C*.

BOX 6.1 BASIC RELATIONSHIPS

$$\text{Sine} = \sin = \text{side opposite / hypotenuse}$$
$$\text{Cosine} = \cos = \text{side adjacent / hypotenuse}$$
$$\text{Tangent} = \tan = \text{side opposite / side adjacent}$$

$$\text{Cosecant} = \text{cosec} = 1 / \sin$$
$$\text{Secant} = \sec = 1 / \cos$$
$$\text{Cotangent} = \cot = 1 / \tan$$
$$\sin A / \cos A = \tan A$$
$$(\sin A)^2 + (\cos A)^2 = 1$$

$$\sin A = \cos(90 - A);\ \cos A = \sin(90 - A)$$

Example 6.1 Functions of Angle *A*

Sin *A*, cos *A*, and tan *A* are functions of the angle *A*. Their values can be calculated using special formulae (see Box 10.3 in Chapter 10) or be obtained from tables, pocket calculators, and so on.

As an example for $A = 35°$ from tables with seven decimal places

$$\sin 35° = 0.5735764$$
$$\cos 35° = 0.8191520$$
$$\tan 35° = 0.7002075$$

Note: (sin 35) * (sin 35) = $(\sin 35)^2 = \sin^2 35 = 0.3289899$.

$$(\cos 35) * (\cos 35) = (\cos 35)^2 = \cos^2 35 = 0.6710101$$

Thus,

$$\sin^2(35) + \cos^2(35) = 1$$

Likewise,

$$\sin 35 / \cos 35 = 0.5735764 / 0.8191520 = 0.7002075 = \tan 35$$

In Figure 6.2,

$$\sin \sin A = CN / AC = h / b$$

Or

$$h = b \sin A$$

Also,

$$\sin B = CN / BC = h / a$$

Or

$$h = a \sin B$$

Thus,

$$b \sin A = a \sin B$$

Or

$$\frac{\sin A}{a} = \frac{\sin B}{b}$$

By constructing the perpendicular from B we can also show that $(\sin A) / a = (\sin C) / c$. Thus, for any triangle ABC with sides a, b, and c

$$\frac{\sin A}{a} = \frac{\sin B}{b} = \frac{\sin C}{c}$$

This is known as the *sine formula* for triangle ABC.
Note also that the area of the triangle = (1 / 2) base * height = (1 / 2) ch

$$= (1/2) bc \sin A = (1/2) ca \sin B = (1/2) ab \sin C$$

Further, in Figure 6.2, $\cos A = AN / AC$. Thus, $AN = AC \cos A = b \cos A$. Using Pythagoras,

$$AN^2 + CN^2 = AC^2 \text{ or } CN^2 = AC^2 - AN^2 = b^2 - b^2 \cos^2 A$$

Likewise, in triangle BNC where $NB = (c - AN) = (c - b \cos A)$. Hence, we have

$$CN^2 = BC^2 - BN^2 = a^2 - (c - b\cos A)^2 = a^2 - \left\{c^2 - 2bc\cos A + (b\cos A)^2\right\}$$
$$= a^2 - c^2 + 2bc\cos A - b^2\cos^2 A$$

Also,

$$CN^2 = b^2 - b^2\cos^2 A \text{ (as shown just above)}$$

Hence,

$$a^2 - c^2 + 2bc\cos A = b^2 \text{ or } 2bc\cos A = b^2 + c^2 - a^2$$

Thus,

$$\cos A = \left(b^2 + c^2 - a^2\right) / 2bc$$

BOX 6.2 SINE AND COSINE FORMULAE FOR ANY PLANE TRIANGLE

1. $(\sin A) / a = (\sin B) / b = (\sin C) / c$
2. $\cos A = (b^2 + c^2 - a^2) / 2bc$

$$\cos B = \left(c^2 + a^2 - b^2\right) / 2ca$$

$$\cos C = \left(a^2 + b^2 - c^2\right) / 2ab$$

3. The area of the triangle = $(1 / 2)\, bc \sin A = (1 / 2)\, ca \sin B = (1 / 2)\, ab \sin C$

Similarly,

$$\cos B = \left(c^2 + a^2 - b^2\right) / 2ca$$

And

$$\cos C = \left(a^2 + b^2 - c^2\right) / 2ab$$

These relationships are known as the *cosine formulae*. Thus, if we know the lengths of the sides of a triangle, we can calculate the size of each of the angles using the formulae in Box 6.2.

6.2 OBTUSE ANGLES

So far we have assumed that the angles A, B, and C are less than or equal to a right angle. If we think of the case where ABC has B as a right angle, then the larger that A becomes, the smaller must be the angle C.

As A approaches a 90° angle, C approaches zero, while BC and AC become parallel and of similar length (Figure 6.3). In the limit sin 90° = 1 while cos 90° = 0. Since tan 90° = sin 90° / cos 90° = 1 / 0 and we cannot divide by zero, it means that as A approaches 90°, tan A gets very much bigger until it becomes infinitely large. But what happens for angles over 90°? The sine formula and the cosine

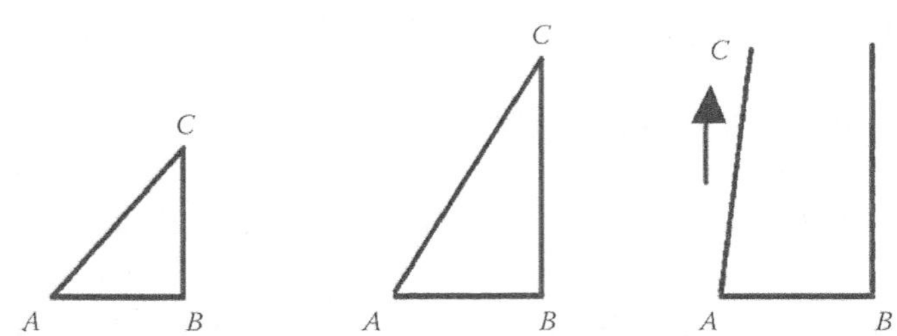

FIGURE 6.3 Toward a right angle.

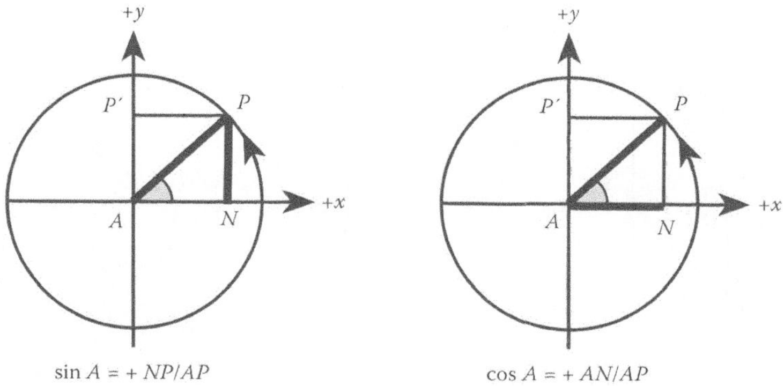

FIGURE 6.4 A circle with unit radius.

formula apply to triangles where there is a fixed relationship between the angles A, B, and C. The functions sin, cos, and tan give values that relate to the size of the angle, not to the shape or size of a triangle. We can look at them in a different way. Consider a circle of unit radius (Figure 6.4) and center A and a point P in the quadrant or quarter of the circle between the $+x$ and $+y$ axes.

If $\sin A = NP / AP$ and $AP = 1$ (unit radius), then the length of $NP = \sin A$. AP' is the projection of P onto the $+y$ axis and also $= \sin A$. P' is on the $+y$ side of the origin. Likewise, the length $AN = \cos A$. This is the projection of P onto the $+x$ axis.

As P moves anticlockwise around the circle, the angle that AP makes with the $+x$ axis increases until at 90° the projection of AP onto the $+y$ axis reaches the point where $NP = AP' = 1$ and $AN = 0$. As the angle increases further the length of NP grows shorter. It still has the value of $\sin A$. On the other hand, AN has moved from 1 down to 0 and is now on the negative side of the axis. Comparing Figures 6.4 and 6.5 with P in Figure 6.5, the mirror image of what it was in Figure 6.4, then the lengths AP' (the projection of P onto the $+y$ axis) is exactly the same but AN, the projection of P on the $+x$ axis, has in the second case become negative.

For angles where $90° \le A \le 180°$

Plane and Spherical Trigonometry

$$\sin A = +\sin(180 - A)$$
$$\cos A = -\cos(180 - A)$$

Hence, since tan = sin / cos

$$\tan A = -\tan(180 - A)$$

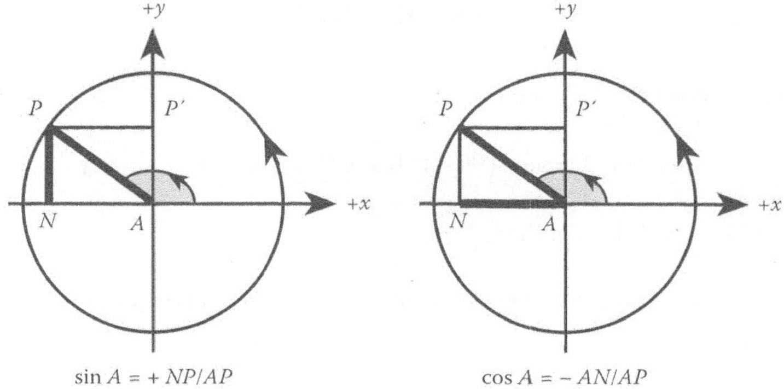

FIGURE 6.5 Angles in the second quadrant.

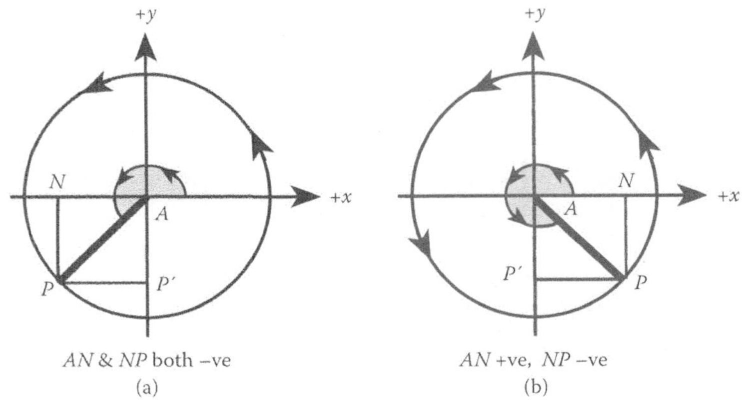

FIGURE 6.6 **(a, b)** Angles in the third and fourth quadrants.

A similar argument applies to angles greater than 180° and less than 270° in what is called the *third quadrant*. In Figure 6.6a, both *AN* and *NP* are negative

$$\sin A = -\sin(A - 180)$$
$$\cos A = -\cos(A - 180)$$
$$\tan A = +\tan(A - 180)$$

And for the fourth quadrant (Figure 6.6b) where *AN* has again become positive but *NP* is still negative

$$\sin A = -\sin(360 - A)$$
$$\cos A = +\cos(360 - A)$$
$$\tan A = -\tan(360 - A)$$

In particular, (360 – *A*) is a complete turn less *A* (= –*A*). Thus,

$$\sin(-A) = -\sin A$$
$$\cos(-A) = +\cos A$$
$$\tan(-A) = -\tan A$$

Also, since sin *A* = cos (90 – *A*), then

$$\sin(90 + A) = \cos\{90 - (90 + A)\} = \cos(-A) = +\cos A$$

And since cos (*A*) = sin (90 – *A*), then

$$\cos(90 + A) = \sin\{90 - (90 + A)\} = \sin(-A) = -\sin A$$

Hence,

$$\tan(90 + A) = -\cos A$$

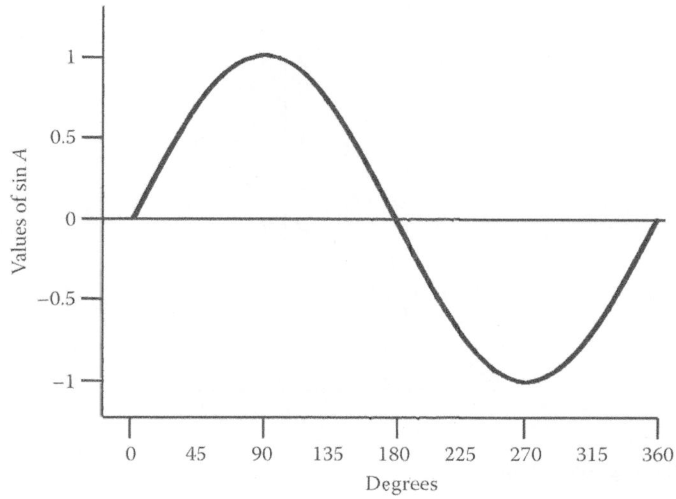

FIGURE 6.7 The cycle of values of sin A.

Figure 6.7 shows the values of sin A for angles between 0 and 360°. The cycle is repeated every 360°. The curve representing the function "cos A" has exactly the same shape but a quarter of a cycle out of phase. If 90° is subtracted from the numbers on the horizontal scale on the graph in Figure 6.7, the resulting curve would be for "cos A." The summary of all this is given in Table 6.1.

6.3 COMBINED ANGLES

In later chapters, we will use this information but for now, there are two more relationships that need to be established. Consider two adjacent angles A and B in Figure 6.8 defined by the lines OS, OR, and OQ. From point S (anywhere along the line OS—scale does not matter) draw the perpendicular from S onto OQ at P cutting OR at U. Also, choose R as the point that gives angle $ORS = 90°$ and draw the other horizontal or perpendicular lines RT and RQ. None of this alters the angles A or B.

$$\angle OUP = 90 - A; \angle OUS = 180 - (90 - A) = 90 + A$$

Hence,

$$\angle SUR = 90 - A$$

TABLE 6.1
Signs of Sine, Cosine, and Tangent

Angular Range	Sine	Cosine	Tangent
0–90	+ sin A	+ cos A	+ tan A
90–180	+ sin(180 – A)	– cos(180 – A)	– tan(180 – A)
180–270	– sin(A – 180)	– cos(A – 180)	+ tan(A – 180)
270–360	– sin(360 – A)	+ cos(360 – A)	– tan(360 – A)

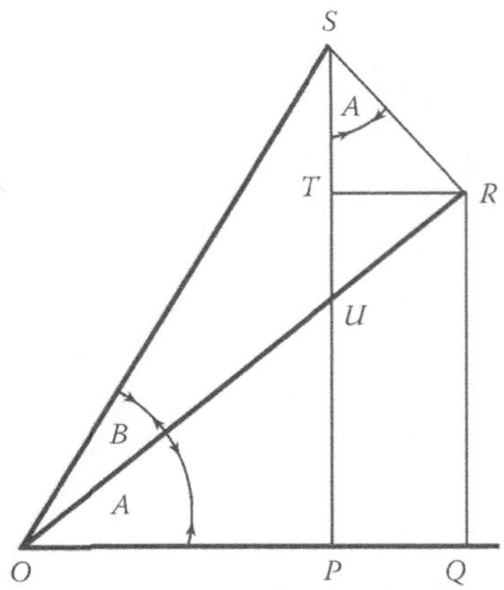

FIGURE 6.8 Combining adjacent angles.

Also,

$$\angle SUR = \angle OUP \ (\text{Rule 5.2 in Chapter 5}).$$

Since

$$\angle SRU = 90 \text{ by construction}, \angle USR = A$$

$$\sin B = \frac{RS}{OS}; \ \cos B = \frac{OR}{OS}; \ \sin A = \frac{QR}{OR}; \quad \text{and} \quad \cos A = \frac{QR}{OR}$$

$$\sin(A+B) = \frac{PS}{OS} = \frac{PT+TS}{OS} = \frac{QR+TS}{OS} (\text{since } PT = QR)$$

$$\sin(A+B) + \frac{QR}{OS} + \frac{TS}{OS} = \frac{QR}{OR} * \frac{OR}{OS} + \frac{TS}{SR} * \frac{SR}{OS}$$

(by dividing the top and bottom of the first term by *OR* and the second term by *SR*).
Hence,

$$\sin(A+B) = \sin A \cos B + \cos A \sin B$$

Similarly,

$$\cos(A+B) = \frac{OP}{OS} = \frac{OQ-PQ}{OS} = \frac{OQ-RT}{OS} = \frac{OQ}{OS} - \frac{RT}{OS}$$

$$= \frac{OQ}{OR} * \frac{OR}{OS} - \frac{RT}{SR} * \frac{SR}{OS}$$

Hence,

$$\cos(A+B) = \cos A \cos B - \sin A \sin B$$

The results are summarized in Box 6.3.

We have already shown in Box 6.2 that cos $A = (b^2 + c^2 - a^2) / 2bc$ and in Box 6.3 that cos $A = 2 \cos^2 (½A) - 1$. Hence,

$$2\cos^2(1/2A) = (b^2 + c^2 - a^2)/2bc + 1 = (b^2 + c^2 + 2bc - a^2)/2bc$$

BOX 6.3 COMBINED ANGLES

$$\sin(A+B)=\sin A\cos B+\cos A\sin B$$
$$\cos(A+B)=\cos A\cos B-\sin A\sin B$$
$$\sin 2A=\sin(A+A)=2\sin A\cos A$$

Hence,

$$\sin A=2\sin(1/2A)\cos(1/2A)$$
$$\cos 2A=\cos(A+A)=\cos^2 A-\sin^2 A;\ \text{and since}\ \cos^2 A+\sin^2 A=1,\ \text{then}$$
$$\cos 2A=1-2\sin^2 A=2\cos^2 A-1$$

Hence,

$$\cos A=\cos^2(1/2A)-\sin^2(1/2A)=1-2\sin^2(1/2A)=2\cos^2(1/2A)-1$$

Since sin $(-B)=-\sin B$ and cos $(-B)=+\cos B$, then

$$\sin(A-B)=\sin A\cos B-\cos A\sin B$$
$$\cos(A-B)=\cos A\cos B+\sin A\sin B$$

Or

$$\cos^2(1/2A)=\left(b^2+c^2+2bc-a^2\right)/4bc=\left\{(b+c)^2-a^2\right\}/4bc$$

Since the numerator is the difference of two squares

$$\cos^2(1/2A)=(b+c+a)(b+c-a)/4bc=(a+b+c)(a+b+c-2a)/4bc$$
$$=\{(a+b+c)/2\}\{(a+b+c)/2-a\}/bc=s(s-a)/bc$$

where s is the semiperimeter = ½$(a+b+c)$. Thus,

$$\cos(1/2A)=\sqrt{s(s-a)/bc}$$

Similarly,

$$\sin^2(1/2A)=(s-b)(s-c)/bc,$$

hence the results shown in Box 6.4.

6.4 BEARINGS AND DISTANCES

The formulae given in Box 6.3 are necessary and sufficient for the solution of a wide range of problems in surveying and mapping. Before considering some examples we need to clarify the difference between a mathematician's approach and a surveyor's. Traditionally in mathematics, angles

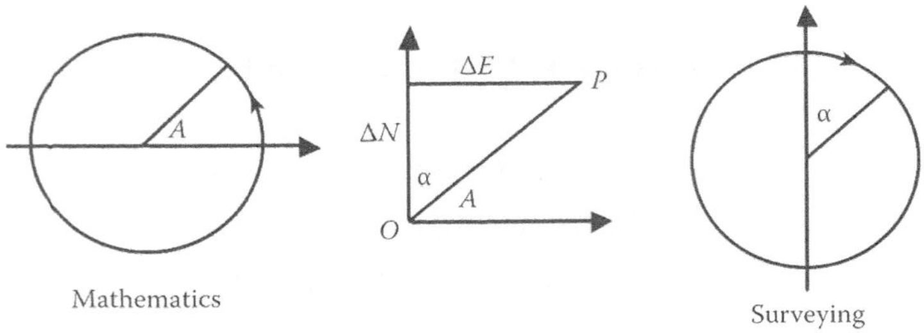

FIGURE 6.9 Angle and bearing measurements.

are measured from the horizontal anticlockwise. Surveyors and navigators start with the north and measure angles and bearings clockwise (Figure 6.9).

BOX 6.4 TRIGONOMETRICAL FUNCTIONS AND THE SEMIPERIMETER

$$\sin(1/2A) = \sqrt{(s-b)(s-c)/bc}$$
$$\cos(1/2A) = \sqrt{s(s-a)/bc}$$
$$\tan(1/2A) = \sqrt{(s-b)(s-c)/s(s-a)}$$
$$\sin(1/2B) = \sqrt{(s-c)(s-a)/ca}$$
$$\cos(1/2B) = \sqrt{s(s-b)/ca}$$
$$\tan(1/2B) = \sqrt{(s-c)(s-a)/s(s-b)}$$
$$\sin(1/2C) = \sqrt{(s-a)(s-b)/ab}$$
$$\cos(1/2C) = \sqrt{s(s-c)/ab}$$
$$\tan(1/2C) = \sqrt{(s-a)(s-b)/s(s-c)}$$

When dealing with bearings, the Greek alphabet is often used. Bearings are angles and conform to all the rules that affect the angles A, B, and C that we have been discussing. Thus, in the center triangle of Figure 6.9 the value $\tan\alpha = \Delta E / \Delta N$ while $\tan A = \Delta N / \Delta E$. In what follows we will often use the Greek alphabet for both bearings and angles measured clockwise.

The accuracy of the process of calculating bearings and distances from coordinates depends upon the number of significant figures that are used throughout the computation, as illustrated in Examples 6.2 and 6.3.

Taking the relationships a stage further, consider Figure 6.10 where point Q forms a triangle with OP. Q is to the right of the line from O to P and the measured angles are α and β. Let the coordinates of O be (E_O, N_O), P (E_P, N_P), and Q (E_Q, N_Q).

If the bearing from O to P is ϕ, then the bearing $OQ = \phi + \alpha$ and the bearing P to $O = 180 + \phi$ so that the bearing $PQ = 180 + \phi - \beta$. $\angle Q = 180 - (\alpha + \beta)$. Hence,

$$\sin Q = \sin\{180 - (\alpha + \beta)\} = \sin(\alpha + \beta) = \sin \sin \alpha \cos \beta + \cos \alpha \sin \beta$$

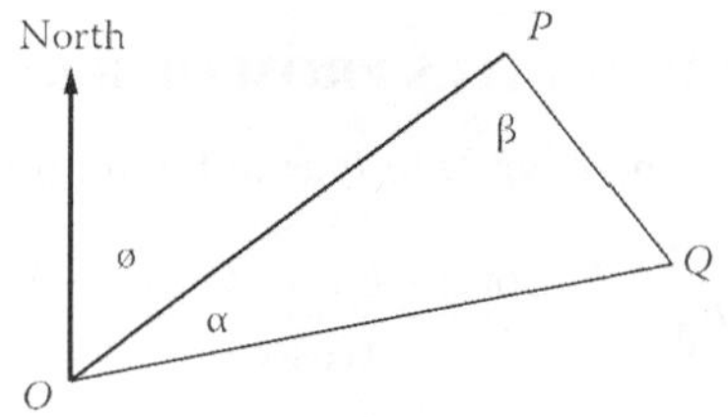

FIGURE 6.10 Fixing points from observed angles.

Example 6.2 Bearings and Distances from Coordinates

If point O in Figure 6.10 is (2624.81 E, 3427.64 N) and P is (3056.61 E, 4058.18 N), then $\Delta E = 431.80$ and $\Delta N = 630.54$. Let the bearing $OP = ø$

$$\Delta E / \Delta N = 0.6848098 \ \tan^{-1}(0.6848098) = 34.40373 \text{ degrees}$$
$$\phi = 34.40373 \text{ degrees} = 34° \ 24'13.4'' = \text{bearing from } O \text{ to } P$$
$$\sin \phi = \Delta E / OP, \text{ hence } OP = \Delta E \ \text{cosec}\phi = 431.8 * 1.76985 = 764.22$$
$$\cos \phi = \Delta N / OP, \text{ hence } OP = \Delta N \ \sec\phi = 630.54 * 1.21201 = 764.22$$

This checks the calculation.

Example 6.3 Coordinates from Bearings and Distances

If O in Figure 6.10 has coordinates (2624.81 E, 3427.64 N) and the distance $OP = 764.22$ and bearing $ø = 34° \ 24'$ then

$$\Delta E = OP \sin\phi = 764.22 * 0.5649670 = 431.76$$
$$\Delta N = OP \cos\phi = 764.22 * 0.8251135 = 630.57$$

These figures differ slightly from the values given in Example 6.2. This is because the bearing should in fact be $ø = 34° \ 24' \ 13.4''$. Using the more accurate value would give slightly different values for sin and cos, which would result in

$$\Delta E = 764.22 * 0.5650206 = 431.80$$
$$\Delta N = 764.22 * 0.8250766 = 630.54$$

Using the sine formula, $OQ = OP \sin \beta / \sin Q$. As we have seen,

$$OP = (E_P - E_O) / \sin\phi = (N_P - N_O) / \cos\phi$$
$$E_Q = E_O + OQ \sin(\phi + \alpha) = E_O + OQ \ \sin\phi \ \cos \ \alpha + OQ \ \cos\phi \ \sin \ \alpha$$
$$= E_O + \{OP \sin\beta \sin\phi \cos\alpha + OP \sin\beta \cos\phi \sin\alpha\} / [\sin\alpha \cos\beta + \cos\alpha \sin\beta]$$

By substituting for OP we obtain the relations shown in Box 6.5.

BOX 6.5 COORDINATES FROM OBSERVED ANGLES

If point Q is to the right of the line from O to P as in Figure 6.10, then

$$E_Q = \frac{E_P \cot\alpha + E_O \cot\beta + N_P - N_O}{\cot\alpha + \cot\beta}$$

$$N_Q = \frac{N_P \cot\alpha + N_O \cot\beta - E_P - E_O}{\cot\alpha + \cot\beta}$$

Example 6.4 illustrates a calculation of coordinates using the formulae in Box 6.5.

Example 6.4 Computing a Point from Two Observed Angles

Using the previous data from Example 6.2 and in Figure 6.10:

Let O have coordinates (2624.81 E, 3427.64 N) and P (3056.61 E, 4058.18 N)

$$\Delta E = 431.80$$
$$\Delta N = 630.54$$

Let $\alpha = 20°$ and $\beta = 60°$. Then

$$\cot\alpha = 2.7474774; \cot\beta = 0.5773503; \cot\alpha + \cot\beta = 3.3248277$$

$$E_Q = \frac{E_P \cot\alpha + E_O \cot\beta + N_P - N_O}{\cot\alpha + \cot\beta}$$
$$= (8397.967 + 1515.435 + 630.54) / 3.3248277 = 3171.27$$

$$N_Q = \frac{N_P \cot\alpha + N_O \cot\beta - E_P + E_O}{\cot\alpha + \cot\beta}$$
$$= (1149.758 + 1978.949 - 431.80) / 3.3248277 = 3818.82$$

Hence the coordinates of Q are (3171.27, 3818.82).

If Q were on the other side of the line OP then the two last terms in the upper line of both expressions in Box 6.5 would be of the opposite sign and references to P and O would be reversed.

As an example of the use of trigonometric functions in geomatics, consider the case of traversing, which is a technique commonly used in land surveying. It entails measuring a series of angles and distances.

The traverse would start at a known point (A in Figure 6.11) and end at another known point (B) in order to confirm the accuracy of the work. The bearings and distances from A to P to Q to B would be derived from observations so that the coordinates of the two new points P and Q could be calculated. The measurement of the distances AP, PQ, and QB would be relatively straightforward, especially with modern electronic distance measuring devices.

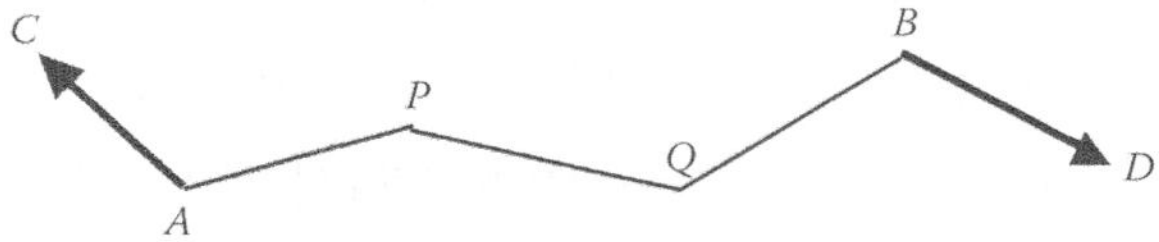

FIGURE 6.11 A traverse.

The measurement of bearings is more problematic since the bearing of *AP* is an angle relative to North; but in what direction is North? The solution is to have another known point (here *C*), calculate the bearing of *AC*, measure the angle between *AC* and *AP* clockwise and hence derive the bearing of *AP*. We can then derive the bearing *PQ* by measuring the angle at *P*.

Example 6.5 Computing a Traverse

Station	Distance	Observed Angle	Bearing	Sin Bearing	Cos Bearing	Δ East	Δ North	East	North
C								**5663.28**	**13794.22**
	497.68		**141 21 02**	**0.6245538**	**−0.7809818**	**310.83**	**−388.68**		
A		108 17 08						**5974.11**	**13405.54**
	498.93		69 38 10	0.9375015	0.3479813	467.75	173.62		
P		228 33 27						6441.86	13579.16
		318.95	118 11 37	0.8813561	−0.4724525	281.11	−150.69		
Q		108 44 11						6722.97	13428.47
	399.55		46 55 48	0.7305199	0.6828914	291.88	272.85		
B		246 54 02						**7014.85**	**13701.32**
	750.38		**113 49 50**	**0.9147443**	**−0.4040332**	**686.41**	**−303.18**		
D								**7701.26**	**13398.14**

Because all measurements and calculations are prone to error, either through human mistakes or the accumulation of small inaccuracies, it is necessary to verify the positions by some independent means. This is achieved by measuring the angle at *Q*, the distance *QB*, and a check angle at *B* to another known point (*D*).

In the calculation shown in Example 6.5, the given values are in bold, there are three observed distances and four observed angles from which the coordinates of the two new points have been derived. Note that values have been used that do not require adjustment; normally in a traverse there will be a need to apply corrections to ensure exact mathematical consistency. The theory of errors is explored from a statistical perspective in later chapters.

6.5 ANGLES ON A SPHERE

In practice, the world is not flat and we must extend some of the ideas above to what happens on a curved surface. For now, we will assume for practical purposes that the world is a sphere. As explained in Chapter 5, on a sphere any plane that passes through the center of the sphere cuts the surface along a line that is called a *great circle*. Thus, assuming that the Earth is a sphere, the equator and all meridians of longitude are great circles. The shortest distance between any two points on the surface of a sphere follows the line of a great circle.

A triangle on the surface of a sphere that has each of its arcs as part of a great circle is called a *spherical triangle*. Thus, if *O* is the center of the sphere in Figure 6.12, then the plane *OAB* cuts the surface along the arc *AB*, which is part of a great circle; similarly, for *BC* and *CA*. The length of the arc *AB* is the radius of the sphere multiplied by the angle *AOB* in radians. This is usually expressed in letters from the Greek alphabet—for example, π radians for 180°. The angle on the surface (for instance, *BAC*) is written in normal text—here as *A*. For a spherical triangle *ABC*, the angles would then be *A*, *B*, and *C*, while the side opposite *A* (i.e., *BC*) would be α (alpha), *CA* would be β (beta), and *AB* would be γ (gamma).

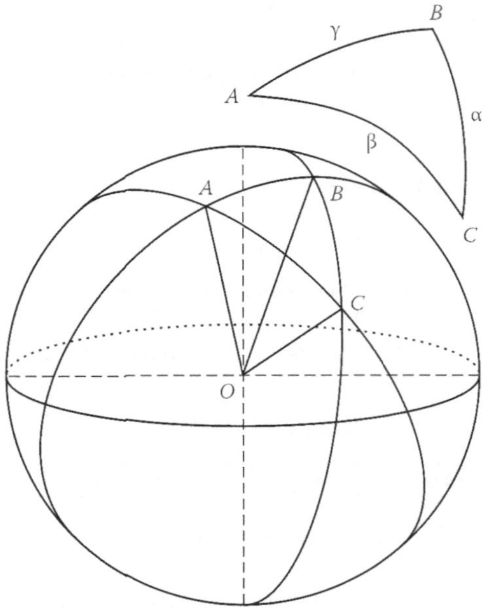

FIGURE 6.12 The spherical triangle.

Consider a sphere of radius R (Figure 6.13). Let P be the foot of the perpendicular from A onto the plane OBC; let the line PS be perpendicular to OB and PQ be perpendicular to OC. Triangles APS and APQ both have right angles at P since AP is perpendicular to the plane that contains all the points $OQCBS$.

Furthermore, since the points A, S, and P form a plane that is at right angles to the line OB it follows that the line AS (not drawn in Figure 6.13) must be at right angles to the line OB or that ASO is a right-angled triangle with the angle at $S = 90°$.

It also follows that the plane ASP is parallel to the plane that touches the sphere at B, known as the tangent plane at B since both are at right angles to the line OB, which is a radius of the sphere. The angle ASP must therefore be the same as the angle ABC at the surface.

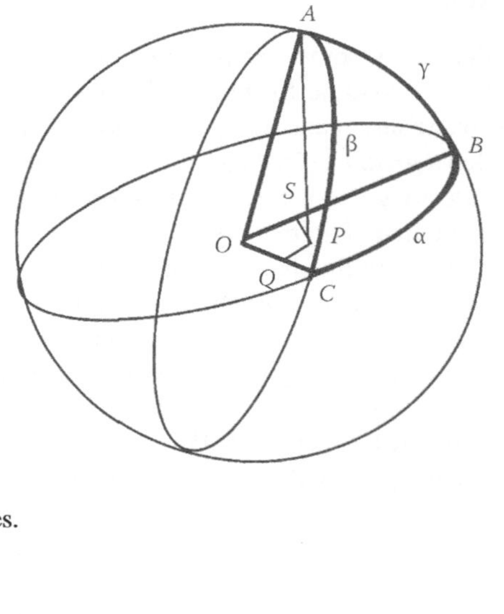

FIGURE 6.13 Spherical angles.

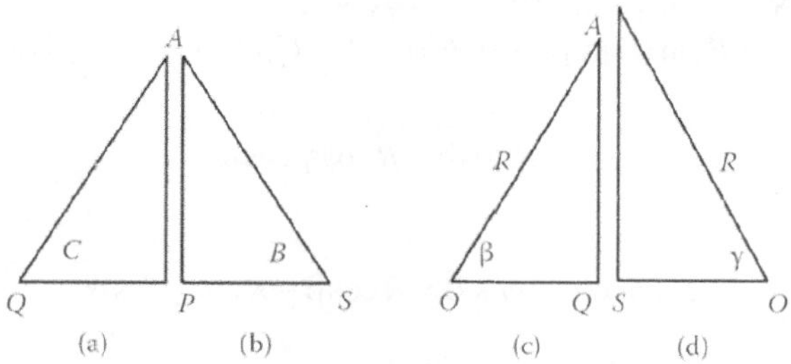

FIGURE 6.14 **(a–d)** The sine formula for spherical triangles.

Thus, $\angle ASP = B$. Finally, the $\angle AOS = \angle AOB = \gamma$. Similarly, $\angle APQ = 90° = \angle AQO$. Also, $\angle AQP = C$ and $\angle AOQ = \beta$. The relationships are shown in Figure 6.14.

All this may seem difficult to visualize until we separate out the triangles. Then we have

$$\text{From Figure 6.14}c \quad AQ = OA\sin\beta = R\sin\beta$$
$$\text{From Figure 6.14}d \quad AS = OA\ \sin\gamma = R\sin\gamma$$

Combining these with Figure 6.14a,b, we have

$$\text{From Figure 6.14}a \quad AP = AQ\sin C = R\sin\beta\sin C$$
$$\text{From Figure 6.14}b \quad AP = AS\sin B = R\sin\gamma\sin B$$

Hence,

$$\sin\beta\sin C = \sin\gamma\sin B \,\text{or}\, \sin B / \sin\beta = \sin C / \sin\gamma$$

And similarly,

$$= \sin A / \sin\alpha$$

Thus,

$$\frac{\sin A}{\sin\alpha} = \frac{\sin B}{\sin\beta} = \frac{\sin C}{\sin\gamma}$$

This is known as the *sine formula* for a spherical triangle.

Taking this a stage further, in the plane *OBC* as shown in Figure 6.13, draw *SU* perpendicular to *OC* and *PT* perpendicular to *SU*. This is shown in Figure 6.15 where $\angle BOC = \alpha$ and hence, $\angle OSU = 90° - \alpha$. Since *PS* is perpendicular to *OB*, then $\angle PST = \alpha$.

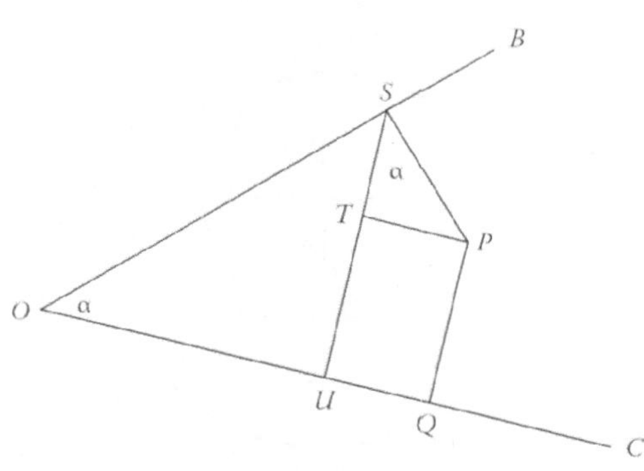

FIGURE 6.15 The cosine formula for spherical triangles.

From Figure 6.14, $AS = R\sin\gamma$ and $SP = R\sin\gamma\cos B$; also $OQ = R\cos\beta$ and $OS = R\cos\gamma$. From Figure 6.15, $TP = SP\sin\alpha = R\sin\alpha\sin\gamma\cos B$. Also, $TP = UQ$ (since $PTUQ$ is a rectangle) $= OQ - OU = R\cos\beta - OS\cos\alpha$

$$= R\cos\beta - R\cos\gamma\cos\alpha$$

Hence,

$$R\sin\alpha\sin\gamma\cos B = R\cos\beta - R\cos\gamma\cos\alpha$$

Dividing through by R and rearranging

$$\cos\beta = \cos\gamma\cos\alpha + \sin\gamma\sin\alpha\cos B$$

Similarly,

$$\cos\gamma = \cos\alpha\cos\beta + \sin\alpha\sin\beta\cos C$$

and

$$\cos\alpha = \cos\beta\cos\gamma + \sin\beta\sin\gamma\cos A$$

These are known as the *cosine formulae* for a spherical triangle. The results are summarized in Box 6.6.

BOX 6.6 SINE AND COSINE FORMULAE FOR SPHERICAL TRIANGLES

$$\frac{\sin A}{\sin\alpha} = \frac{\sin B}{\sin\beta} = \frac{\sin C}{\sin\gamma}$$

$$\cos\alpha = \cos\beta\cos\gamma + \sin\beta\sin\gamma\cos A$$

$$\cos\beta = \cos\gamma\cos\alpha + \sin\gamma\sin\alpha\cos B$$

$$\cos\gamma = \cos\alpha\cos\beta + \sin\alpha\sin\beta\cos C$$

As an example of the application for the sine and cosine formulae, consider two points A and B with the following latitude and longitude (see Figure 6.16):

$$A(40°N,\ 10°E) \text{ and } B(55°N,\ 15°E)$$

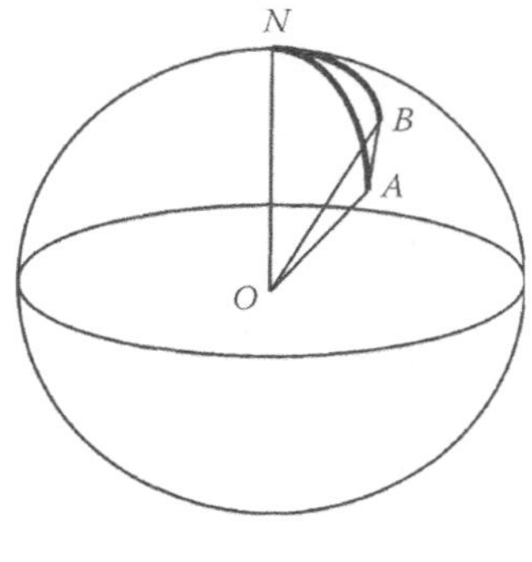

FIGURE 6.16 Colatitudes *AN*, *BN*.

The angle ANB at the North Pole = difference in longitude between A and B = 5°. The latitude of A is the angle that OA makes with the equator which means that the angle NOA = length of arc AN = 90° − latitude of A (known as the *colatitude* of A) = 50°.

Similarly, the length of arc NB = 35° (the colatitude of B). We therefore have a spherical triangle in which we know two sides (AN and BN) and an included angle (ANB). Hence, using the spherical sine and cosine formulae as shown in Box 6.6, we can calculate the bearing and distance of A to B, as shown in Example 6.6.

Spherical trigonometry underpins the conversion of coordinates on the surface of the Earth into coordinates that can be used on a flat surface such as a map. In Chapter 12, we examine map projections and some of the ways in which the data may be transformed.

Example 6.6 Bearing and Distance from Latitude and Longitude

In Figure 6.16, N is the North Pole. A (40° N, 10° E) and B (55° N, 15° E) differ in longitude by 5°. Angle AN = 50°, BN = 35°, and ANB = 5°.

From the cosine formula,

$$\begin{aligned}\cos(AB) &= \cos 50 \cos 35 + \sin 50 \sin 35 \cos 5 \\ &= 0.5265408 + 0.4377130 = 0.9642538\end{aligned}$$

Hence, the angle AB = 15° 22′ degrees or 0.2681836 radians. This is the angle AOB subtended by AB at the center.

Assuming the radius of the Earth to be 6.4 * 10^6 meters then the length of AB = 0.2681836^r * 6.4 * 10^6 = 1716.4 kilometers.

Using the sine formula,

$$(\sin ANB)/(\sin AB) = (\sin NAB)/(\sin NB), \text{ or}$$

$$\sin NAB = \sin NB * \sin ANB / \sin AB = 0.1886573$$

Hence,

$$\text{angle } NAB = 10°52' = \text{Bearing from } A \text{ to } B$$

Similarly,

$$\sin NBA = \sin NA * \sin ANB / \sin NB = 0.2519627$$

Hence,

$$\begin{aligned}\text{Angle } NBA &= 165°\ 24' \\ \text{Bearing } BA &= 194°\ 36'\end{aligned}$$

Note that in general, unlike on a plane surface, the direction from A to B on a sphere does not normally differ by 180° from the direction from B to A.

GLOSSARY

Arcsine, Arccosine, Arctangent, Arccosecant, Arcsecant, Arccotangent The inverse of the trigonometric functions *sine*, *cosine*, and so on. Written as $\sin^{-1} \phi$, $\cos^{-1} \phi$, and so on. They are the angles that give rise to the values sine ϕ, cosine ϕ, and so on.

Area of triangle In a triangle whose angles are A, B, C, and sides a, b, c, then area = ½ $bc \sin A$ = ½ $ca \sin B$ = ½ $ab \sin C$

If $s = ½(a + b + c)$ then area $= \sqrt{s(s-a)(s-b)(s-c)}$

Combined angles For two angles A, B

$$\sin(A+B)=\sin A\cos B+\cos A\sin B$$
$$\cos(A+B)=\cos A\cos B-\sin A\sin B$$

$$\sin(A-B)=\sin A\cos B-\cos A\sin B$$
$$\cos(A-B)=\cos A\cos B+\sin A\sin B$$
$$\sin(2A)=2\sin A\cos A$$
$$\cos(2A)=1-2\sin^2(A)=2\cos^2(A)-1$$
$$\sin(A)=2\sin 1/2A\cos 1/2A$$
$$\cos(A)=1-2\sin^2(1/2A)=2\cos^2(1/2A)-1$$

If $s = ½(a + b + c)$

$$\sin^2(1/2A)=(s-b)(s-c)/bc$$
$$\cos^2(1/2A)=s(s-a)/bc$$
$$\tan^2(1/2A)=\{(s-b)(s-c)\}/\{s(s-a)\}$$

Coordinates from angles If in triangle ABC, C is to the right of the line A to B and the coordinates of A are (x_A, y_A) and of B are (x_B, y_B) then point C is

$$x_C=\{x_B\cot A+x_A\cot B+y_B-y_A\}/\{\cot A+\cot B\}$$
$$y_C=\{y_B\cot A+y_A\cot B+x_B-x_A\}/\{\cot A+\cot B\}$$

Cosecant The reciprocal of *Sine*. Written as cosec ø. cosec ø = 1 / sin ø.

Cosine The ratio in a right-angled triangle between the side adjacent and the hypotenuse. Written as cos ø. $\cos^2 ø = 1 - \sin^2 ø$.

In the triangle ABC, $\cos(1/2A)=\sqrt{s(s-a)/bc}$, and so forth.

Cosine formula In a plane triangle whose angles are A, B, C, and sides a, b, c, and semiperimeter s, then

$$\cos A=(b^2+c^2-a^2)/2bc$$
$$\cos B=(c^2+a^2-b^2)/2ca$$
$$\cos C=(a^2+b^2-c^2)/2ab$$

$$\cos(1/2A)=\sqrt{s(s-a)/bc}$$
$$\cos(1/2B)=\sqrt{s(s-b)/ca}$$
$$\cos(1/2C)=\sqrt{s(s-c)/ab}$$

Cotangent The reciprocal of *Tangent*. Written as cot ø.

$$\cot\phi=1/\tan\phi$$

Obtuse angles Angles greater than 90° and less than 180°. For an angle A:

If $90° \leq A \leq 180°$: $\sin A = +\sin(180 - A)$, $\cos A = -\cos(180 - A)$, $\tan A = -\tan(180 - A)$
If $180° \leq A \leq 270°$: $\sin A = -\sin(A - 180)$, $\cos A = -\cos(A - 180)$, and $\tan A = \tan(A - 180)$
If $270° \leq A \leq 360°$: $\sin A = -\sin(360 - A)$, $\cos A = +\cos(360 - A)$, and $\tan A = -\tan(360 - A)$
Also, $\sin(90 + A) = \cos A$, $\cos(90 + A) = -\sin A$, and $\tan(90 + A) = -\cot A$
And, $\sin(-A) = -\sin A$, $\cos(-A) = \cos A$, $\tan(-A) = -\tan A$

Secant The reciprocal of *cosine*. Written as sec ø. $\sec ø = 1 / \cos ø$.

Sine The ratio in a right-angled triangle between the side opposite and the hypotenuse. Written as sin ø. $\sin^2 ø = 1 - \cos^2 ø$.

In the triangle ABC, $\sin(1/2A) = \sqrt{(s-b)(s-c)/bc}$, and so forth.

Sine formula In a plane triangle whose angles are A, B, C, and sides a, b, c, and semiperimeter s, then

$$(\sin A)/a = (\sin B)/b = (\sin C)/c$$

$$\sin(1/2A) = \sqrt{(s-b)(s-c)/bc}$$
$$\sin(1/2B) = \sqrt{(s-c)(s-a)/ca}$$
$$\sin(1/2C) = \sqrt{(s-a)(s-b)/ab}$$

Spherical triangle A triangle on the surface of a sphere formed by the arcs of three great circles. If the angles on the surface where these arcs intersect are A, B, C, and if the arcs subtend angles α, β, γ at the center of the sphere then

$$\sin A / \sin \alpha = \sin B / \sin \beta = \sin C / \sin \gamma$$
$$\cos \alpha = \cos \beta \cos \gamma + \sin \beta \sin \gamma \cos A$$
$$\cos \beta = \cos \gamma \cos \alpha + \sin \gamma \sin \alpha \cos B$$
$$\cos \gamma = \cos \alpha \cos \beta + \sin \alpha \sin \beta \cos C$$

Tangent of an angle The ratio in a right-angled triangle between the side opposite and the side adjacent. Written as tan ø.

$$\tan \phi = \sin \phi / \cos \phi.$$

In the triangle ABC, $\tan(1/2A) = \sqrt{(s-b)(s-c)/s(s-a)}$, and so forth.

The word tangent is also used to describe a line that just touches a curve or a plane that just touches a solid.

Traverse The locating of points by measuring a series of distances and associated bearings, derived from successive angular measurements.

7 Applied Spatial Analysis

7.1 SPATIAL OVERLAY

In many geographic information system (GIS) applications, where we study a set of locations, the data about these locations often come in different datasets. For example, one dataset contains the terrain model of the Earth's surface, the digital elevation model as a raster file. A separate dataset contains polygons that follow building footprints. Yet another dataset contains statistical data, such as average income and highest education level per administrative or census area. In addition, a set of lines describing the network of local roads, and the set of points describing nearby fuel stations might be also relevant to this application. Although it might seem that the datasets mentioned above have little to do with each other, they share the same spatial locations. A GIS practitioner often wants to work with all that data and understand patterns and relationships in order to solve the problem at hand.

A GIS system is able to read and interpret the data given to it in different formats. Some datasets are polygons, some are lines, some are points, and some are raster. Each dataset might use a different coordinate system and different units. A GIS system can perform appropriate coordinate transformations and data conversions, and is able to display all datasets on a computer screen in the form of a virtual map where each dataset appears as a map layer. One can control the draw order of the layers and the display transparency, as data in one layer might cover the data in other layers.

To create a visual overlay, the GIS system will refer to the description of the coordinate system (the map projection specification) used in each dataset. For more discussion about map projections, see Chapter 12. The GIS system will then convert the geographic location of a data point to a coordinate on the virtual map area. Once all data in every dataset are converted to coordinates on the virtual map area, the data can then be displayed on a computer screen as map layers.

Once the procedure is established for converting data from their native coordinate systems to the coordinate system of the virtual map area, it can be used match data spatially between different data sources. This matching process is called *Spatial Overlay*. Spatial Overlay is designed to facilitate analysis of disparate data about land areas through joining them like joining relational database tables together. Whereas relational databases require primary and foreign key fields to join tables together, Spatial Overlay uses geographic coordinates as the key field. Spatial overlay is somewhat synonymous with spatial join, but the former often implies a more rigorous process that breaks up larger geometries such as polygons into smaller pieces, each may be matched to a different data record from another map layer. For instance, one might want to find which polygons of building footprints that intersect with census areas where the average household income is in a certain range. Or, one might want to find building footprints that intersect with elevation contour lines.

Spatial overlay uses different methods depending on whether the desired output dataset is in vector or raster format. Overlaying raster data requires that all datasets have the same underlying grid structure including the same origin point location and the same grid size. Once this requirement is met, overlaying the data from different datasets in the same cell location becomes a trivial task in the computer. If different raster datasets have varying grid sizes or grid origin points, they will need to be "Re-sampled," that is, creating the same data attribute values in a different grid structure

DOI: 10.1201/9781003464402-9

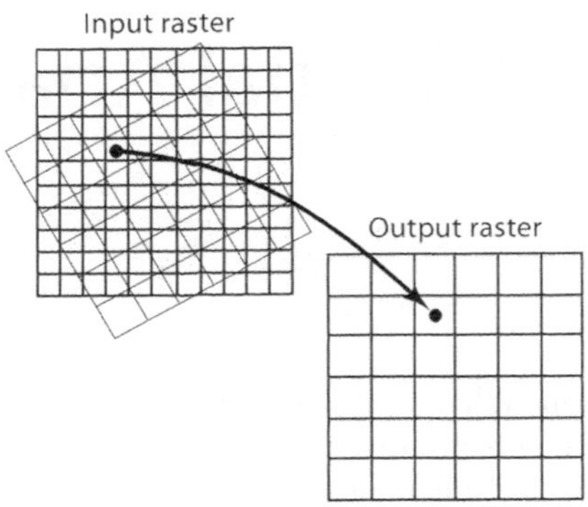

FIGURE 7.1 Resampling a raster with rotation and change of grid size.

through appropriate interpolation methods. Figure 7.1 illustrates a re-sample operation to convert a raster dataset to a different raster dataset with a rotated grid and larger grid cell size. Various interpolations methods applicable to raster data are covered in Section 7.5. Once *re-sampled*, the raster data are populated with interpolated values because of shifts in grid cell locations, but these estimated values are expected to be consistent with the original raster dataset in order to represent reality in a similar manner.

Overlaying vector data requires an entirely different methodology. The vector data must be in the same coordinate system (or Map Projection), and two-dimensional geometric calculations are necessary in the GIS system to generate output shapes of points, lines, and areas based on generating overlapping shapes between the multiple datasets. The newly generated shapes are usually different from the original shapes and tend to be smaller in extent, as each original shape can be split into multiple shapes. Figure 7.2 illustrates a simple scenario where two polygons overlaying with two different polygons results in an output of four polygons. Theoretically, the maximum number of output polygons from overlaying two polygons with two other polygons is infinite, and the minimum number is two output polygons. The minimum number of output polygons occurs when the two input polygons from one layer overlaps exactly with the two polygons of the other layer.

When performing spatial overlay processing, it is often necessary to create buffer polygons of other shapes. This operation is called *buffering*. Figure 7.3 illustrates the result of buffering various area, line, and point shapes. Buffering is required in the following situations when performing spatial overlay processing:

1. Spatial proximity logic is utilized in the analysis. This includes questions like: how many crime incidents occurred within 1000 feet of a senior housing facility? Or, which areas are within 500 meters of a proposed airport where residents would be bothered by the noise of airplanes?
2. Desire to overlay area shapes with point or line layers. Since point and line shapes do not occupy any area, when overlaying with area shapes they will disappear from the output data since geometric points or lines are abstract shapes that have size or width

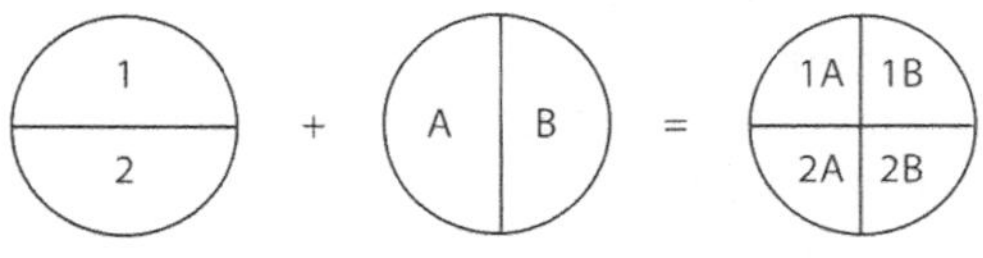

FIGURE 7.2 Overlaying two polygon layers.

FIGURE 7.3 Making buffer areas around shapes.

of 0 (i.e., infinitely small point or infinitely thin line). In order to properly consider these point and line data, one needs to estimate the influence zone of each point or line shape, for example, small streams are typically represented in GIS data as lines. When one wants to delineate the sensitive riparian zone of a stream, an estimate should be made to designate a certain proximity distance from the line where one expects to find riparian soil conditions and vegetation.

7.2 TOPOLOGY MODELS

GIS practitioners often need the computer system to help manage relationships between different spatial objects and analyze their interactions. Locations of objects and their relationships in real world can be analyzed by the mathematical concepts of topology. Topology provides a rigorous method to calculate whether certain *topological relationships* between objects exist in the spatial data, and whether these relationships match a set of pre-defined rules. There are two common types of topology models (to be illustrated by the examples below): area topology models and network topology models.

Example of an area topology model: suppose we have a land ownership management dataset that consists of privately owned land parcels, legally defined areas for public access or placement of utility infrastructure, local jurisdictional boundaries, national and regional political jurisdictions, etc. A set of implicit topology rules exist to govern the spatial relationships between these spatial entities, they can be encoded in GIS software configurations as pre-defined topology rules. Data errors that could break such pre-defined rules include locations where there are unexpected overlaps of ownership, or locations where no landowners and/or public jurisdictions are present. Figure 7.4 illustrates the four types of topological relationships between two spatial area shapes that are supposed to share a border, where three of the four types of relationships are erroneous. Having the GIS software maintain such relationships between different spatial entities ensures that instances of spatial data errors are corrected via automated detection, which notifies data administrators to initiate

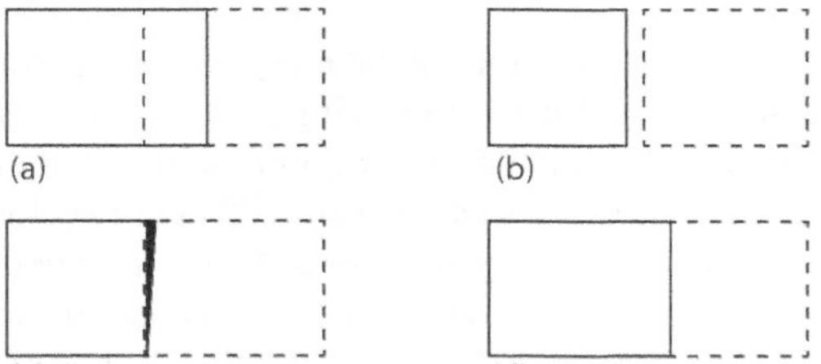

FIGURE 7.4 (a–d) Topological errors where two polygons are supposed to share a border. (a) Overlap, (b) Gap, (c) Sliver polygon, (d) No errors.

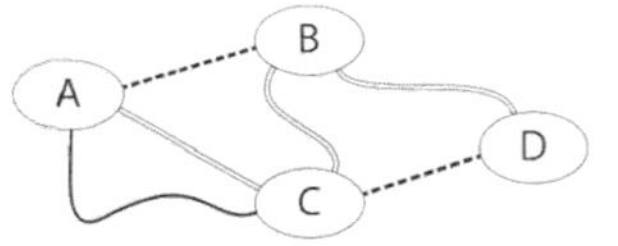

FIGURE 7.5 Cost of traveling between nodes. Solid line = by barge, dashed line = by rail, double line = by road. Node A is the starting node, node D is the Destination node. There are multiple alternative routes using different transportation modes, requiring different expenditure of time or fuel and ability to transport different loads.

data review and correction processes. Such automated error-detection and/or error-correction capabilities in the GIS software are generally called *topology models*.

Generally, topological relationships are represented by Boolean values (i.e., yes or no) rather than numeric values like map scales, distance, etc. For example, when we study parcels, we are interested in which pairs of parcels are adjacent to each other. The answer would be either yes or no. Other types of topological relationships between two objects can be inclusion (one object is inside the other), connectedness, sharing partial boundaries, and others.

Example of a network topology model (see Figure 7.5): supposed we have a transit network that comprises of different modes of transport, from buses, trains, ferry boats, airplanes, and any other modes of transport that have fixed start and stop locations. A topological structure of such a transit network can be represented as a *graph* of *nodes* connecting to *edges*, these edges are sometimes also called links. The network nodes would be point shapes or small areal shapes, which represent start or stop locations of trip segments and are usually called stations, stops, or terminals. Between different start or stop locations there are edges or links that enable travel from one to another, these can be railways, roads, bus routes, ferry routes, or airplane routes. These types of networks can be analyzed by automated computer programs to answer questions such as: what is the quickest way to travel from *point A* to *point B*, what is the most cost-saving way to travel from point A to point B, or what is the best way to travel through this network of links in order to deliver goods to multiple locations while spending the shortest travel time or the smallest total travel distance. Solving these questions above essentially optimizes resources for travel, therefore these kinds of problems are called *optimization* problems.

Edges in the graph of such network topology structures can contain data about the relationship between the nodes it connects. That can be a direction of travel, the cost of traveling between the nodes, distance or time it takes to travel, or some other information about the trip segment.

Other applications of network topology models can include managing and analyzing the flow of water through rivers and canals; estimating the flooding potential of rivers given a specific rainfall event; managing the flow of electric power through networks of wires and transformers, or the flow of fuel through networks of pipelines.

7.3 NETWORK TRACING AND RECURSION

As mentioned in the previous section, questions about navigating a topological network can be solved by computer programs. Such programs perform tracing of the connected network and identify the most optimal travel paths through the network according to user specifications. These programs are called *Network Tracing* tools. Such analyses can be run with either vector or raster data formats; the core logical processes are the same, the only difference being the logical links and nodes refer to different data elements (pixels in the case of raster vs. areas, points, or lines in the case of vector).

Network tracing analysis requires a set of constraints, that is, the links and nodes where travel is allowed in certain directions, and outside of these mutually connected links and nodes travel is forbidden. Another prerequisite of network tracing analysis is the value of cost in terms of time, financial cost, or distance (distance is often used as a proxy for other types of costs), assigned to

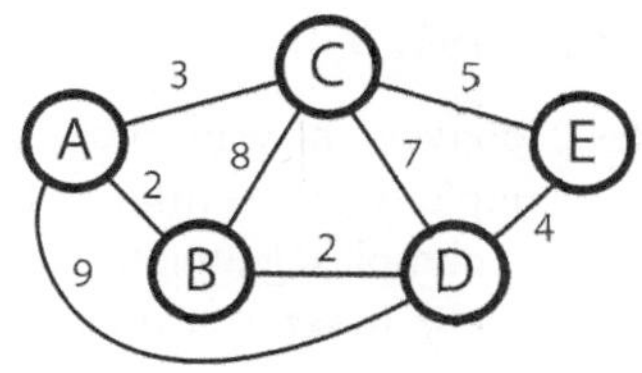

FIGURE 7.6 Solving a weighted graph. A = origin node, D = Destination node. Numbers on the links denote the cost of traveling each.

each link and node (see example in Figure 7.6). The total cost value is the overall quantity to be minimized via choosing the optimal sequence of links and nodes to connect point A with point B. The cost value assigned to each link or node is referred to as weight, therefore the graph of the network is called a *weighted graph*. The network tracing problem is then known mathematically as a problem to solve a weighted graph.

Recursive algorithms are often used to program such network tracing analyses. The core concept of *recursion* is that one can create an unlimited number of "agents" in the computer memory, each one tasked to execute the same logic but are created at different times with different initial inputs. In computing terminology these "agents" are individual *instances* of a recursive *subroutine* or *function* that is created in the computer memory when needed, and removed after its results are returned. Such recursive algorithms are usually run by a single CPU (Central Processing Unit) so that there will not be *competition* between different "agents" working on the same input data at the same time, which may lead to unintended results. Below are the logical steps to solve this optimization problem:

1. The first node in the network is point A, the programmer starts the tracing program by placing agent #001 at this node.
2. Agent #001 looks around for possible paths to another connected node and found two. It creates agent #002 and sends it to node X, and creates agent #003 and sends it to node Y. However, agent #001 waits until agent #002 returns from its assignment before sending out agent #003.
3. Agents #002 and #003 does the same thing as agent #001, but at their individual node locations. This happens in a chain of events and eventually creates hundreds or even more agents. Basically, one agent is sent to every node on each possible path outward from point A.
4. If an agent finds that there are no more ways forward, and the location it's at is not point B, it will report to its parent agent that its own path does not reach the destination.
5. If an agent finds that another agent had previously been here and had already reached point B, it compares the distance it has accumulated starting from point A with the accumulated distance of the previous agent that was here. If its own accumulated distance is the smaller of the two, the agent reports back to its parent agent that it found a path that's better than the previous agent.
6. If an agent finds that another agent had previously been here but did not reach the destination, then it will report to its parent that it cannot reach the destination.
7. If an agent receives more than one report from its child agents who had reached point B, it will compare the cumulative distance traveled by each child agent and report to its parent the shorter distance, and also mark the child agent's shorter path as the better path. All parent agents do the same thing after their child agents report back.
8. When agent #001 receives reports from all its child agents and determines which child agent had the shorter distance, it has essentially determined the optimal path from point A to point B.

This type of path finding algorithm is called the *Dijkstra algorithm*, it was invented by Dutch computer scientist Edsger Dijkstra in 1956 and is the classic path finding algorithm for weighted graphs used by most GIS software.

7.4 SPATIAL ANALYSIS USING RASTER

The raster GIS data format provides a convenient way to apply numerical methodology to break down complex spatial operations into simpler ones, similar to how a *Finite Element* model is used in engineering. Using this methodology, complex geometries such as curves and surfaces can be reduced to a collection of pixels or grid cells in a raster dataset. By dealing with individual pixels or grid cells instead of complex shapes, one can more easily apply approximation methods using simpler mathematics, such as using arithmetic calculations to replace calculus formulas. Figure 7.7 shows a raster representation of a circle. Using raster arithmetic, one can calculate area of the circle by counting the grid cells the circle encloses. Even though this result will be less accurate than using the geometric formula, if one uses smaller grid sizes the raster calculation results will get closer to the correct result. When the shape is not a circle but a more complicated geometric shape where a closed-form area formula is difficult to find, using raster calculations can require much lower effort and meet the accuracy requirements of the application at hand.

For example, the *gravity model* is an attraction model based on the formula of gravitational attraction forces in physics. The force is inversely proportional to the square of distance between the two objects, and it is also proportional to the mass, or attraction strength, of each of the objects that attract the other. This means as distance is reduced by half, the gravitational attraction is strengthened by four times. When modeling various attraction phenomena in natural and human geography, the inverse distance squared function can be made more generic by changing the –2 exponent to any negative exponent.

In urban planning such an attraction force can be used describe the economic attraction of larger cities or major political/economic centers toward nearby smaller towns or settlements. People from these smaller towns are attracted to travel to these larger centers to conduct business. Such attraction forces generate commercial or civic activities as well as demand for transportation. In Chapter 6, we discussed Thiessen polygons, also called Voronoi tessellations, which are graphs that divide the influence areas of points in a two-dimensional space. We suppose these influencing points are larger cities. The Thiessen polygons are essentially a division of the two-dimensional space into sectors based on nearest-neighbor proximity relationships. If we use the Thiessen polygon to model

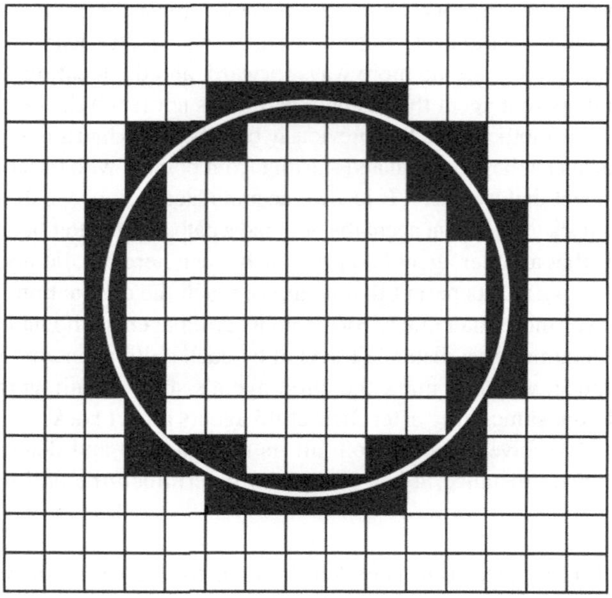

FIGURE 7.7 Modeling a circle with raster.

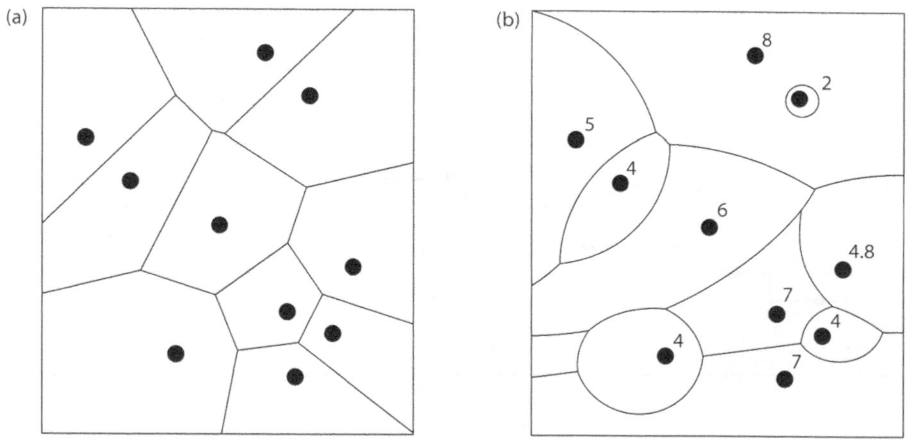

FIGURE 7.8 Weighted Voronoi diagram compared with unweighted. (a) Unweighted Thiessen/Voronoi diagram, (b) weighted Voronoi diagram using the same set of points.

attraction phenomena, it means each location will be attracted by the nearest larger city regardless of the city's attraction strength, as shown in Figure 7.8(a). As a result, Thiessen/Voronoi graphs are not suitable when we want to model attraction to larger cities with varying attraction strengths, where the attraction strength may be based on population size of these cities. The bigger the city the more likely it will attract consumers and business activity from further away, and potentially leapfrogging smaller cities nearer to them. The aim of this example is to generate modified Thiessen/Voronoi graphs that gives larger cities more territory compared to smaller cities nearby.

Using the raster data format it is relatively easy to compute this gravity model attraction using the distance from each pixel in the two-dimensional space to each large city in the map. Then the attraction forces from different cities to the pixel can be compared, and the city that has the biggest attraction force to the pixel can be discovered. The result of the analysis would color each pixel according to the large city that has the biggest attraction, as in Figure 7.8(b). essentially generating a Voronoi diagram weighted by the size of attraction points.

7.5 INTERPOLATION METHODS

As described in Chapter 4, interpolation methods are important for creating estimates of data that are not exactly at the locations where data have been collected. It can be used to add granularity to the collected data, and to provide a fully populated dataset that is expected by analytical algorithms. For example, in Figure 7.9 one attempts to create estimates at each cell in a raster grid from irregularly located raw data. Besides *linear interpolation* discussed in Chapter 4, there are more

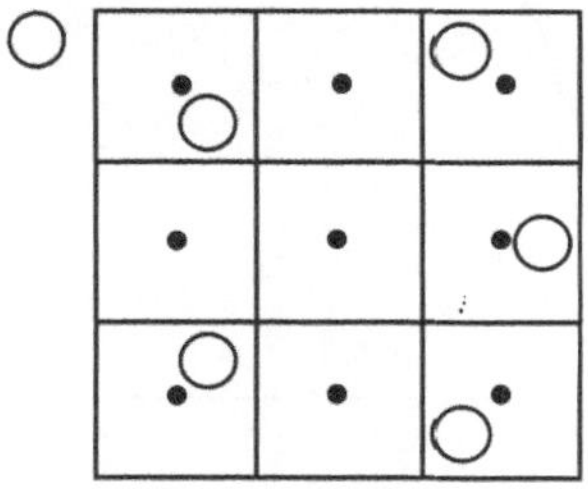

FIGURE 7.9 Interpolating values for raster grid cells (small dots) from observations at nearby locations with irregular spatial distribution (circles).

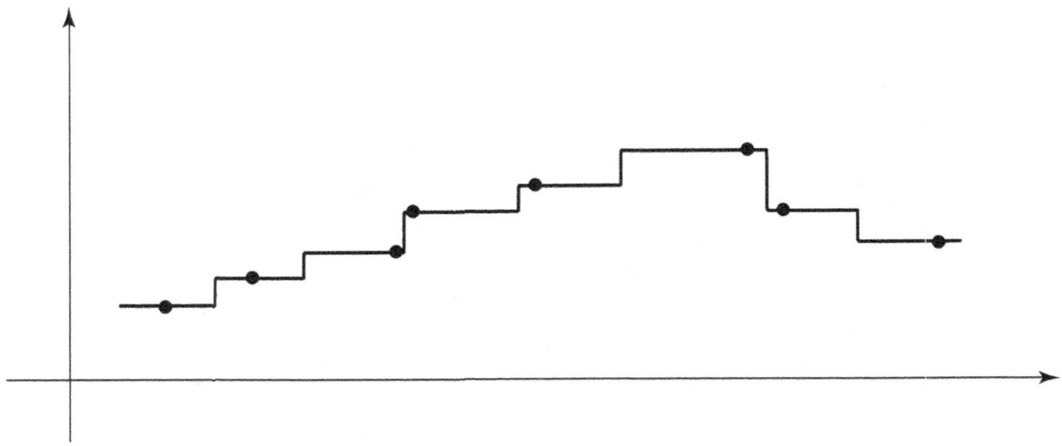

FIGURE 7.10 Nearest neighbor interpolation.

sophisticated interpolation algorithms utilizing higher order polynomials or a combination of multiple algorithms.

For example, before the advent of aerial surveying technology, elevation values in the terrain are measured at certain point locations by surveyors. These locations must be physically accessible by the surveyor, therefore they do not necessarily conform to regular spatial patterns in the two-dimensional space. Even with aerial surveying technology such as *LiDAR*, the laser beams that are emitted by the LiDAR instrument from the airplane will measure locations on the ground that do not form a regular spatial pattern. In order to construct a raster dataset of elevation values in a raster grid format, the choice of interpolation methods will determine the accuracy and resolution of the output data. The following are a few commonly used interpolation methods for GIS data interpolation (examples of some of these methods are illustrated in Figures 7.10–7.13):

1. *Nearest neighbor interpolation*: each estimated data point takes the value of one measured data point closest to itself.
2. *Bilinear interpolation*: each estimated data point uses a linear-based algorithm to interpolate between four measured data points located closest to itself.

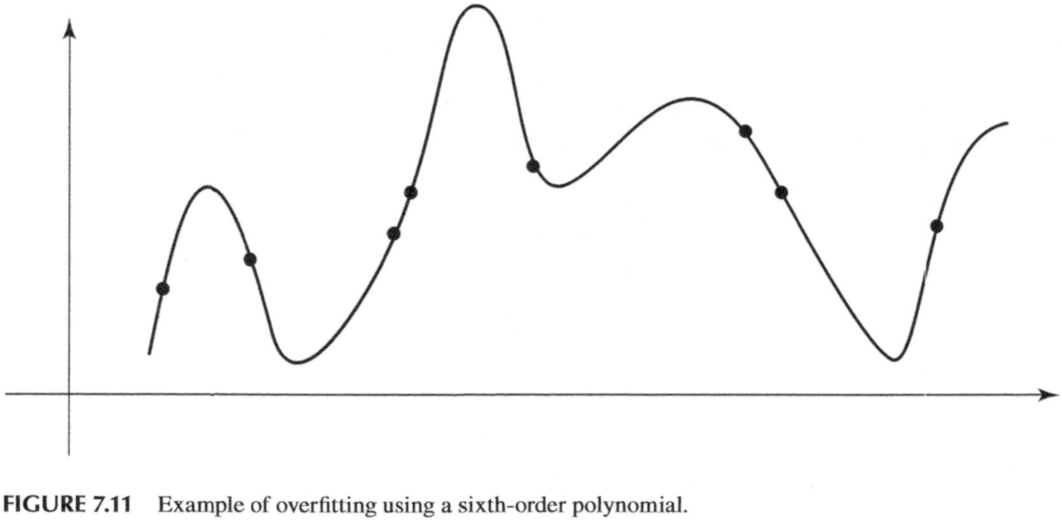

FIGURE 7.11 Example of overfitting using a sixth-order polynomial.

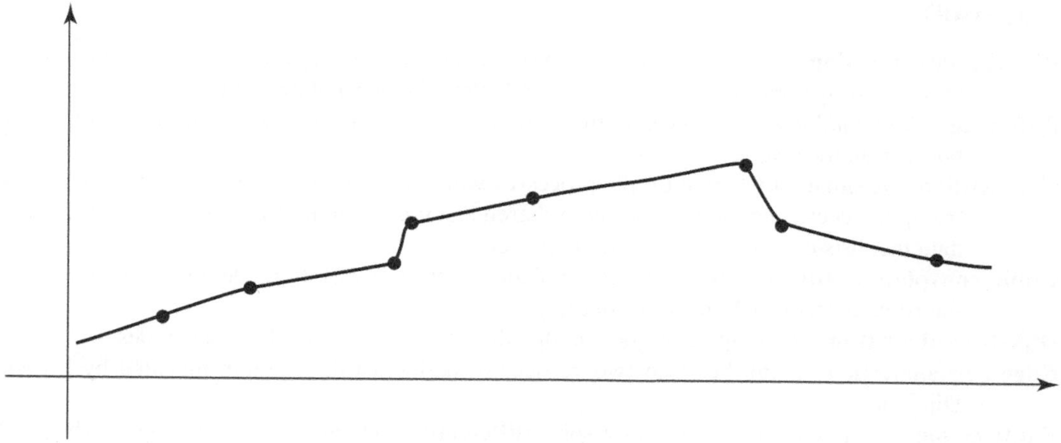

FIGURE 7.12 Kriging.

3. *Cubic convolution interpolation*: each estimated data point uses a third order polynomial-based algorithm to interpolate between sixteen measured data points located closest to itself.
4. Inverse-distance-weighted (*IDW*) interpolation: a method similar to applying a form of gravity model, which allows data points located closer to have greater influence on the result.
5. *Kriging* interpolation: also called Gaussian process regression, Kriging is a geostatistical interpolation method.
6. *Majority* or *median* values: chooses among values of nearby data points based on a statistical method such as majority, or median.

A number of the above interpolation methods above touch on statistics, which is covered in Chapter 14. For categorical, or discrete, data types that are often represented by text strings, numerical interpolation methods that computes a new value from existing values are not applicable, therefore the nearest neighbor, majority, or median value interpolation methods are the only appropriate methods among the methods discussed above.

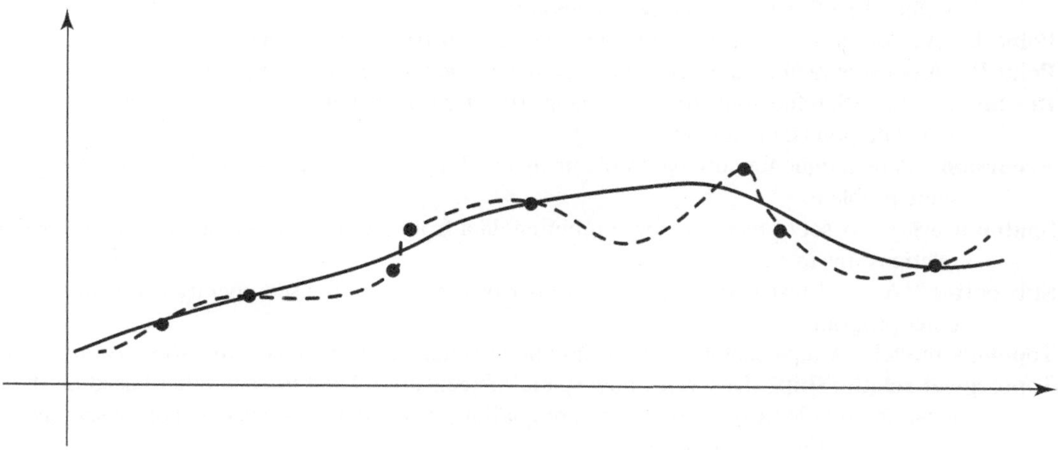

FIGURE 7.13 Kriging using a different set of parameters (solid curve). This result is more plausible for interpolating terrain because it seems to reduce noise from likely measurement error.

GLOSSARY

Bilinear interpolation An interpolation method using linear interpolation in two-dimensional space. Four nearby samples are used to calculate the interpolated value.

Buffering A method to create zones or areas around a spatial shape that is within a specific distance from the shape.

Competition A situation where multiple processes within a computer system are simultaneously trying to access and modify the same shared resource, potentially leading to conflicts and data inconsistencies if not properly managed.

Cubic convolution interpolation An interpolation method using third degree polynomials to interpolate between 16 nearby samples.

Dijkstra's algorithm A computer algorithm that finds the shortest path between nodes in a graph.

Edge A connection or link between two vertices or nodes within a graph, invented by Edsger Dijkstra.

Finite element Methods to numerically solve differential equations arising in engineering and mathematical modeling.

Function A block of computer programming code designed to perform a specific task.

Graph Visual representation depicting a set of objects and the relationships between them.

Gravity model A tool used in various social sciences like geography and economics to predict interactions between locations much like Newton's law of gravity predicts interactions between physical objects.

IDW Stands for Inverse-Distance Weighting, is an interpolation method using a weighted mean of nearby data points, where it is assumed that the closer the distance, the more alike two data points will be.

Kriging Also known as Gaussian process regression, it is a geostatistical interpolation method.

Linear interpolation An interpolation method that interpolates by fitting nearby data values into a linear function.

Nearest neighbor interpolation A simple interpolation method using one nearest existing data point to calculate an identical value as the neighboring data point.

Network tracing A group of methods used to analyze activities (typically transport activities) in a connected network.

Node Represents entities within a graph, which is created to show interconnections between such entities.

Optimization Methodology of making something (such as design, system, or decision) as fully perfect, functional, or effective as possible.

Point A A conventional label to designate the origin, or from- point of a trip.

Point B A conventional label to designate the destination, or to- point of a trip.

Re-sample In GIS, refers to methods to change the grid dimensions of a raster dataset by creating new data points, or samples.

Recursion A technique for solving problems by breaking them down into smaller versions of the same problem.

Spatial overlay A set of methods for combining data from multiple layers based on their spatial relationships to each other.

Subroutine A set of instructions designed to perform a frequently used operation within a computer program.

Topology model A representation of spatial relationships between objects in GIS software.

Topological relationship Describes how spatial objects are related to each other based on their location. Topological relationships are qualitative relationships, such as adjacency, enclosure, connectivity, overlap, etc.

Weighted graph A graph where each edge has a numerical value, or weight. Weight can represent costs, lengths, capacities, etc.

Unit III

Transforming Space

The Earth's surface area consists of many types of highly varied geography, much of which cannot be easily broken down into simpler geometric shapes like those described in Unit II. Besides relatively flat plains that cover a small percentage of the earth's continental surface, much more areas are covered by more rugged landforms such as mountains, valleys, hills, canyons, coastlines, river channels, ocean bottoms, etc. Unit III is about establishing the mathematical foundations that enable geographic information system (GIS) practitioners to understand and analyze more complex spatial geometries like these, using mathematical tools invented to handle three-dimensional irregular shapes. The mathematical concepts of vectors, matrices, and calculus encompass more intensive calculation methods that model movement and transformation of spatial objects. Because the complexity of irregular shapes can be understood as the results of transformations of individual simpler geometric objects, the mathematical methods covered in Unit III are well suited to model the Earth's surface. These methods are often used to develop computational models of dynamic real-life processes including hydrological change, light reflection, etc.

The more intensive calculation methods covered in Unit III are usually part of GIS software packages or algorithm libraries. As a result, many GIS practitioners do not implement these methods on a daily basis. Mastering the mathematical methods involving vectors, matrices, and calculus is required for GIS practitioners who build or improve GIS software libraries, but understanding the fundamental concepts behind these methods will help the practitioner better utilize the available computational tools. In Chapter 8, we introduce the concept of vectors which provides the mathematical representation of spatial movements. In Chapter 9, we introduce the tool of matrices which provides a concise framework to streamline complex calculations involving vectors. In Chapter 10, we introduce calculus, the mathematical concepts and methods that model change. Although closed-form calculus formulae exist for simpler curves, performing calculus on complex geometries often require numerical methods. With the computing speed of modern-day computers, it is feasible for most GIS practitioners to utilize numerical computation methods.

DOI: 10.1201/9781003464402-10

8 Vectors

8.1 THE NATURE OF VECTORS

A vector is a multi-dimensional quantity often used to represent movement in space. Its mathematical form resembles a one-dimensional array. In the context of this chapter, a *vector* is a quantity such as velocity that has both magnitude and direction, while a *scalar* is a quantity that has magnitude but no direction.

The magnitude may represent a true distance such as a length or a quantity such as force or speed. In the study of geographic information systems there is a tendency to think of vectors as the lengths and directions of lines and compare them with raster representation. In practice they may represent the strength and direction of a variety of phenomena such as gravity. A bearing combined with a distance is only one example of a vector.

OP in Figure 8.1 represents a vector based on orthogonal axes that could be called x, y, and z but which, here, we will call the directions **i, j, k**. They are written in bold letters to identify them as *unit vectors*, that is, vectors of unit length. If the actual length of the sides of a rectangular block of which OP is a diagonal are a_p, b_p, and c_p, then we can express the vector **OP** as:

$$\mathbf{OP} = a_p\mathbf{i} + b_p\mathbf{j} + c_p\mathbf{k}$$

What this means is that P is a distance from O of a_p in the x-direction, b_p in the y-direction, and c_p in the z-direction. If we have another vector **PQ** starting at P as in Figure 8.2 such that

$$\mathbf{PQ} = a_q\mathbf{i} + b_q\mathbf{j} + c_q\mathbf{k}$$

where a_q is the distance from P to Q in the x-direction, and so forth, then

$$\mathbf{OQ} = \mathbf{OP} + \mathbf{PQ} = (a_p + a_q)\mathbf{i} + (b_p + b_q)\mathbf{j} + (c_p + c_q)\mathbf{k}$$

From Q to O one needs to travel negatively in each of the directions **i, j, k** so that

$$\mathbf{OP} + \mathbf{PQ} + \mathbf{QO} = 0$$

Vectors may also be written with an arrow on top so that $\overrightarrow{OP} + \overrightarrow{PQ} = \overrightarrow{OQ}$. Thus, $\overrightarrow{OP} + \overrightarrow{PQ} + \overrightarrow{QO} = 0$. We can also describe a vector in two parts—its length and its direction in the form

$$\mathbf{a} = |\mathbf{a}|\,\hat{\mathbf{a}}$$

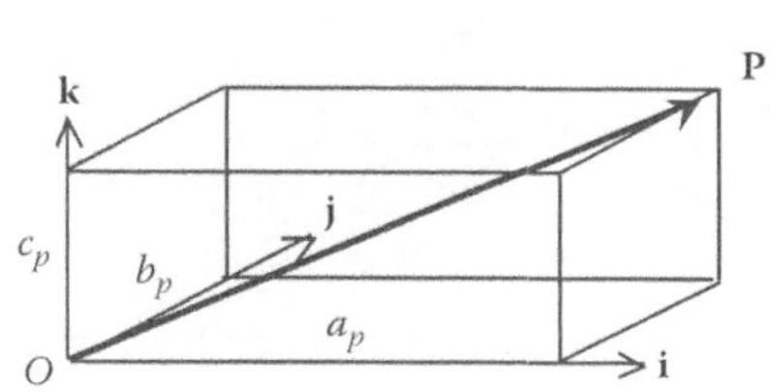

FIGURE 8.1 The axes i, j, k for vector P.

DOI: 10.1201/9781003464402-11

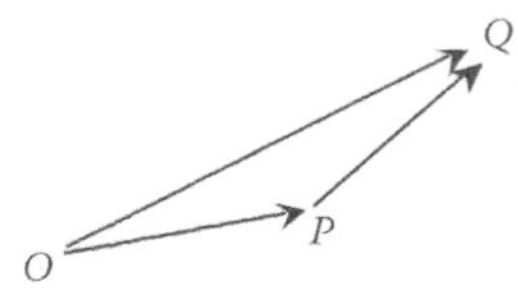

FIGURE 8.2 Vector addition.

This means that the vector **a** has length determined by its *modulus* |**a**| and the direction expressed as $\hat{\mathbf{a}}$, where $\hat{\mathbf{a}}$ is known as the *unit vector.* When

$$\mathbf{OP} = a_p\mathbf{i} + b_p\mathbf{j} + c_p\mathbf{k}$$

then the length of *OP* can be calculated by Pythagoras to be

$$|\mathbf{OP}| = \sqrt{\left(a_p^2 + b_p^2 + c_p^2\right)}$$

Note that |**OP**| (the modulus of vector **OP**) is a scalar quantity and has no direction (see Example 8.1), while $\hat{\mathbf{a}}$ has direction but is of unit length.

The cosines of the angles that **OP** makes with the reference vectors **i**, **j**, and **k** are called the *direction cosines* and are marked as α (angle **POi**), β (angle **POj**), and γ (angle **POk**) in Figure 8.3 and Example 8.2.

$$a_p = |\mathbf{OP}|\cos\alpha \quad b_p = |\mathbf{OP}|\cos\beta \quad c_p = |\mathbf{OP}|\cos\gamma$$

A three-dimensional vector can be defined by its length and its three direction cosines. The length of a vector can be changed simply by applying a scale factor so that, for example,

$$3\mathbf{a} = 3|\mathbf{a}|\hat{\mathbf{a}} \text{ so if } \mathbf{a} = 2\mathbf{i} + 3\mathbf{j} + 4\mathbf{k} \text{ then } 3\mathbf{a} = 6\mathbf{i} + 9\mathbf{j} + 12\mathbf{k}$$

In two dimensions we only need **i** and **j**. In Example 8.3 we calculate the intersection of two lines on a plane surface, showing the relationship between Cartesian coordinates and vectors. The procedure explained in the first part of the example could, however, be extended to three (or more dimensions).

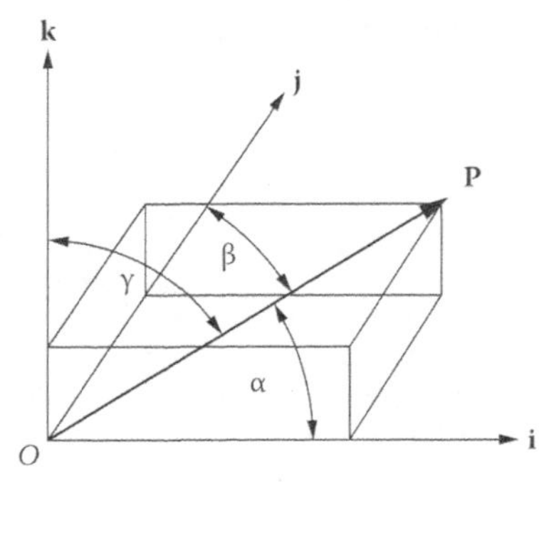

FIGURE 8.3 Direction cosines.

Example 8.1 A Triangle in Space

Let P (5, 9, 2), Q (3, 7, 4), and R (9, 6, 8) be the coordinates of three points in space.

If O(0, 0, 0) is the origin then

$$\overrightarrow{OP} = 5\mathbf{i} + 9\mathbf{j} + 2\mathbf{k}; \overrightarrow{OQ} = 3\mathbf{i} + 7\mathbf{j} + 4\mathbf{k}; \overrightarrow{OR} = 9\mathbf{i} + 6\mathbf{j} + 8\mathbf{k}$$

Also,

$$\overrightarrow{PQ} = -2\mathbf{i} - 2\mathbf{j} + 2\mathbf{k}: \overrightarrow{QR} = 6\mathbf{i} - \mathbf{j} + 4\mathbf{k}; \overrightarrow{RP} = -4\mathbf{i} + 3\mathbf{j} - 6\mathbf{k}$$

Note that

$$\overrightarrow{PQ} + \overrightarrow{QR} + \overrightarrow{RP} = 0$$

The length

$$PQ = \sqrt{\left\{(3-5)^2 + (7-9)^2 + (4-2)^2\right\}} = \sqrt{(2^2 + 2^2 + 2^2}; \; QR = \sqrt{(6^2 + 1^2 + 4^2)}; \; RP = \sqrt{(4^2 + 3^2 + 6^2)}$$

Thus, the length $PQ = |\mathbf{PQ}| = 2\sqrt{3}$; $|\mathbf{QR}| = \sqrt{(53)}$; $|\mathbf{RP}| = \sqrt{(61)}$.

Example 8.2 Direction Cosines

Let P (5, 9, 2) be the coordinates of a point in space (Figure 8.3).

If O is the origin then

$$\mathbf{OP} = 5\mathbf{i} + 9\mathbf{j} + 2\mathbf{k}$$

$$|\mathbf{OP}| = \text{the length } OP = \sqrt{(5^2 + 9^2 + 2^2)} = \sqrt{(110)} = 10.488$$

The direction cosines for **OP** are

$$\cos\alpha = 5/10.488 = 0.4767; \cos\beta = 9/10.488 = 0.8581; \cos\gamma = 2/10.488$$
$$= 0.1907(\alpha \approx 61.5°; \beta \approx 30.9°; \gamma \approx 79.0°)$$

Example 8.3 Intersection of Two Lines in 2D

In two dimensions, consider an origin O and four points P_1, P_2, P_3, P_4 with coordinates (x_1, y_1), and so forth. Any point P_n on the line P_1 to P_2 can be represented by the vector $\mathbf{p}_n = \mathbf{p}_1 + s(\mathbf{p}_2 - \mathbf{p}_1)$ where s is a variable quantity that equals 0 at P_1 and 1 at P_2. If P_n is also on the line P_3 to P_4 then the vector $\mathbf{p}_n = \mathbf{p}_3 + t(\mathbf{p}_4 - \mathbf{p}_3)$ where t is a variable quantity. Thus, at the point of intersection of these two lines,

$$\mathbf{p}_1 + s(\mathbf{p}_2 - \mathbf{p}_1) = \mathbf{p}_3 + t(\mathbf{p}_4 - \mathbf{p}_3)$$

Hence,

$$s = \{(\mathbf{p}_3 - \mathbf{p}_1) + t(\mathbf{p}_4 - \mathbf{p}_3)\} / (\mathbf{p}_2 - \mathbf{p}_1) \text{ or } t = \{(\mathbf{p}_1 - \mathbf{p}_3) + s(\mathbf{p}_2 - \mathbf{p}_1)\} / (\mathbf{p}_4 - \mathbf{p}_3)$$

This is true for both the set of x values and the set of y values (and z values if we were considering three dimensions). Hence,

$$t = \{(x_1 - x_3) + s(x_2 - x_1)\} / (x_4 - x_3)$$
$$= \{(y_1 - y_3) + s(y_2 - y_1)\} / (y_4 - y_3)$$

Hence,

$$s = \frac{x_1(y_4 - y_3) + x_3(y_1 - y_4) + x_4(y_3 - y_1)}{(x_4 - x_3)(y_2 - y_1) - (x_2 - x_1)(y_4 - y_3)}$$

and

$$t = \frac{x_1(y_3 - y_2) + x_2(y_3 - y_1) + x_3(y_2 - y_1)}{(x_2 - x_1)(y_4 - y_3) - (x_4 - x_3)(y_2 - y_1)}$$

From $\mathbf{p}_n = \mathbf{p}_1 + s(\mathbf{p}_2 - \mathbf{p}_1)$

$$x_n = x_1 + s(x_2 - x_1) \quad y_n = y_1 + s(y_2 - y_1)$$

For instance, using the data in Examples 4.1 and 4.2 in Chapter 4,

$$A = P_1 = (1234.56, 2345.67) \text{ and } B = P_2 = (1296.32, 2417.38)$$
$$C = P_3 = (1300.24, 2351.77) \text{ and } D = P_4 = (1212.45, 2431.78)$$
$$s = (-5790.5758) / (-11236.8385) = 0.51532073$$
$$x_n = 1266.3 \quad y_n = 2382.62$$

These are the same values as derived in Chapter 4.

8.2 DOT AND CROSS PRODUCTS

Vectors can be added and subtracted. They can also be multiplied or divided by a scalar quantity. Two vectors can also be multiplied together but the concept has a different meaning from ordinary multiplication. Consider two vectors **a** and **b** such that

$$\mathbf{a} = a_x\mathbf{i} + a_y\mathbf{j} + a_z\mathbf{k} \text{ and } \mathbf{b} = b_x\mathbf{i} + b_y\mathbf{j} + b_z\mathbf{k}$$

There are two forms of product, known as the *dot product* and the *cross product*. The dot product results in a scalar quantity defined as "a dot b"

$$\mathbf{a} \cdot \mathbf{b} = |\mathbf{a}||\mathbf{b}| \cos \varnothing$$

where $\varnothing$ is the difference in direction or angle between the two vectors. In particular, since cos(0) = 1 and cos (90°) = 0 and by definition $|\mathbf{i}| = |\mathbf{j}| = |\mathbf{k}| = 1$ since they are unit vectors

$$\mathbf{i} \cdot \mathbf{i} = \mathbf{j} \cdot \mathbf{j} = \mathbf{k} \cdot \mathbf{k} = 1 \quad \text{while} \quad \mathbf{i} \cdot \mathbf{j} = \mathbf{j} \cdot \mathbf{k} = \mathbf{k} \cdot \mathbf{i} = 0$$

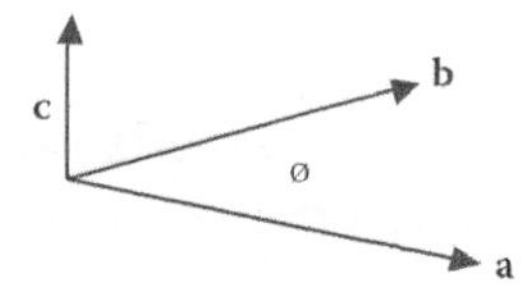

FIGURE 8.4 Dot and cross products.

Thus, $\mathbf{a} \cdot \mathbf{b} = (a_x\mathbf{i} + a_y\mathbf{j} + a_z\mathbf{k}) . (b_x\mathbf{i} + b_y\mathbf{j} + b_z\mathbf{k})$ giving 3 * 3 = 9 relationships that contain terms such as $\mathbf{i} \cdot \mathbf{i}$ (which = 1) or $\mathbf{i} \cdot \mathbf{j}$ (which = 0) or $\mathbf{j} \cdot \mathbf{i}$ (which also = 0). Working this through,

$$\mathbf{a} \cdot \mathbf{b} = a_x b_x + a_y b_y + a_z b_z$$

which is simply a number or scalar quantity (see Example 8.4).

The cross product ("a cross b") creates a new vector perpendicular to the plane containing **a** and **b** (shown by the vector **c** in Figure 8.4). Its origins lie in the study of electromagnetism although it has relevance to a number of problems in geomatics and geographic information system (GIS), for instance, when determining whether any surface element of a feature, such as a building facade, is facing toward or away from the observer. It also allows levels of shading to be calculated or hidden surfaces removed. The cross product

$$\mathbf{a} \times \mathbf{b} = |\mathbf{a}||\mathbf{b}| \sin \varnothing \ \mathbf{c}$$

The magnitude = |**a**||**b**| sin ø is a scalar quantity, while the direction is determined by convention to be a right-handed system. It follows that if we reverse the order of multiplication, the vector **c** will be pointing in the opposite direction, that is,

$$\mathbf{a} \times \mathbf{b} = -\mathbf{b} \times \mathbf{a}$$

Example 8.4 The Angle between Two Vectors

Using the data in Example 8.1 where P (5, 9, 2) and Q (3, 7, 4) were the coordinates of two points in space and $\overrightarrow{OP} = 5\mathbf{i} + 9\mathbf{j} + 2\mathbf{k}; \overrightarrow{OQ} = 3\mathbf{i} + 7\mathbf{j} + 4\mathbf{k}$.

$$\begin{aligned} OP \cdot OQ &= 5*3 + 9*7 + 2*4 = 15 + 63 + 8 = 86 \\ &= |OP| \cdot |OQ| \cos \varnothing \text{ where } \varnothing = \text{the angle between } OP \text{ and } OQ \\ &= \sqrt{(110)} * \sqrt{(74)} \cos \varnothing \\ &= 90.22 \cos \varnothing \end{aligned}$$

Hence,

$$\cos \varnothing = 86 / 90.22 = 0.953 \text{ or } \varnothing = 17.6°$$

$$\text{Note: Using the formula } \cos \varnothing = \frac{a^2 + b^2 - c^2}{2ab} \text{ from Chapter 6, Box 6.2}$$
$$= (OP^2 + OQ^2 - PQ^2) / (2 * OP * OQ)$$

Then

$$PQ^2 = 12, OP^2 = 110, OQ^2 = 74$$
$$\cos \varnothing = (110 + 74 - 12) / 2\sqrt{(110 * 74)} = 172 / 180.44 = 0.953$$

which confirms the calculation.

For unit vectors, the angles between them are 90° and sin (90°) = 1 and cos (90°) = sin (0°) = 0. Hence,

$$\mathbf{i} \times \mathbf{i} = \mathbf{j} \times \mathbf{j} = \mathbf{k} \times \mathbf{k} = 0$$

Also,

$$\mathbf{i} \times \mathbf{j} = \mathbf{k}; \mathbf{j} \times \mathbf{k} = \mathbf{i}; \mathbf{k} \times \mathbf{i} = \mathbf{j} \quad \text{and} \quad \mathbf{j} \times \mathbf{i} = -\mathbf{k}; \mathbf{k} \times \mathbf{j} = -\mathbf{i} \text{ and } \mathbf{i} \times \mathbf{k} = -\mathbf{j}$$

If, as before, we have

$$\mathbf{a} = a_x\mathbf{i} + a_y\mathbf{j} + a_z\mathbf{k} \text{ and } \mathbf{b} = b_x\mathbf{i} + b_y\mathbf{j} + b_z\mathbf{k}$$

then

$$\mathbf{a} \times \mathbf{b} = \left(a_x\mathbf{i} + a_y\mathbf{j} + a_z\mathbf{k}\right) \times \left(b_x\mathbf{i} + b_y\mathbf{j} + b_z\mathbf{k}\right)$$

This gives 3 * 3 = 9 relationships that contain terms such as $\mathbf{i} \times \mathbf{i}$ (which = 0) or $\mathbf{i} \times \mathbf{j}$ (which = $\mathbf{k}$) or $\mathbf{j} \times \mathbf{i}$ (which = $-\mathbf{k}$). Working this through,

$$\mathbf{a} \times \mathbf{b} = \left(a_y b_z - a_z b_y\right)\mathbf{i} + \left(a_z b_x - a_x b_z\right)\mathbf{j} + \left(a_x b_y - a_y b_x\right)\mathbf{k}$$

Example 8.5 The Area of a Triangle

Again using the data from Example 8.1, where *P* (5, 9, 2), *Q* (3, 7, 4), and *R* (9, 6, 8) are the coordinates of three points in space.

$$|\mathbf{PQ}| = \sqrt{(4 + 4 + 4)} = 2\sqrt{3} \approx 3.46; |\mathbf{PR}| = \sqrt{(16 + 9 + 36)} = \sqrt{(61)} \approx 7.81$$
$$\mathbf{PQ} = -2\mathbf{i} - 2\mathbf{j} + 2\mathbf{k}\text{: } \mathbf{PR} = 4\mathbf{i} - 3\mathbf{j} + 6\mathbf{k}; \mathbf{QR} = 6\mathbf{i} - \mathbf{j} + 4\mathbf{k}$$

The cross product

$$\mathbf{PQ} \times \mathbf{PR} = \begin{vmatrix} i & j & k \\ -2 & -2 & 2 \\ 4 & -3 & 6 \end{vmatrix} = -6\mathbf{i} + 20\mathbf{j} + 14\mathbf{k} \text{ the modulus of which} \approx 25.14$$

Since the area of a triangle = ½ *ab* sin *C* and the modulus of the cross product = |**PQ**| × |**PR**| * sin (angle *QPR*) then the modulus of **PQ** × **PR** = twice the area of the triangle. The area of triangle *PQR* is 1 / 2 * 25.14 ≈ 12.57.

Since *QR* = $\sqrt{(53)}$ ≈ 7.28, this can be checked using the semiperimeter formula (see Box 5.2) where *s* ≈ 9.275 giving the area as

$$\sqrt{(9.275 * 5.815 * 1.465 * 1.995)} \approx 12.56$$

The difference is due to rounding errors. If more decimal places were used, the two values would be identical.

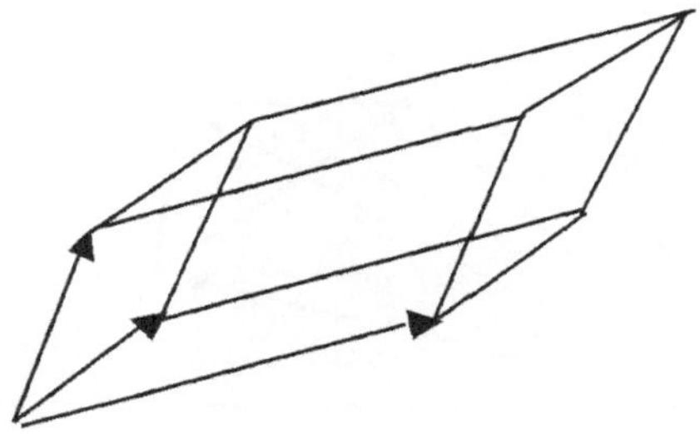

FIGURE 8.5 A parallelepiped.

Or, to express this in determinant form (see Example 8.5)

$$\mathbf{a}\times\mathbf{b}=\begin{vmatrix}\mathbf{i} & \mathbf{j} & \mathbf{k}\\ a_x & a_y & a_z\\ b_x & b_y & b_z\end{vmatrix}=\begin{vmatrix}a_y & a_z\\ b_y & b_z\end{vmatrix}\mathbf{i}-\begin{vmatrix}a_x & a_z\\ b_x & b_z\end{vmatrix}\mathbf{j}+\begin{vmatrix}a_x & a_y\\ b_x & b_y\end{vmatrix}\mathbf{k}$$

The relation $\mathbf{a}\cdot(\mathbf{b}\times\mathbf{c})$ represents the scalar product between two vectors, the vector **a** and the vector $\mathbf{b}\times\mathbf{c}$. It is called the *scalar triple product* and gives the volume of the solid whose edges are represented by the vectors **a**, **b**, and **c**. This solid has parallel sides and is in effect a squashed brick and is known as a *parallelepiped* (Figure 8.5).

If

$$\mathbf{a}=a_x\mathbf{i}+a_y\mathbf{j}+a_z\mathbf{k};\ \mathbf{b}=b_x\mathbf{i}+b_y\mathbf{j}+b_z\mathbf{k};\text{ and }\mathbf{c}=c_x\mathbf{i}+c_y\mathbf{j}+c_z\mathbf{k}$$

then

$$\mathbf{b}\times\mathbf{c}=\left(b_yc_z-b_zc_y\right)\mathbf{i}+\left(b_zc_x-b_xc_z\right)\mathbf{j}+\left(b_xc_y-b_yc_x\right)\mathbf{k}$$

Hence,

$$\begin{aligned}\mathbf{a}\cdot(\mathbf{b}\times\mathbf{c}) &=\left\{a_x\mathbf{i}+a_y\mathbf{j}+a_z\mathbf{k}\right\}\cdot\left\{\left(b_yc_z-b_zc_y\right)\mathbf{i}+\left(b_zc_x-b_xc_z\right)\mathbf{j}+\left(b_xc_y-b_yc_x\right)\mathbf{k}\right\}\\ &=a_x\left(b_yc_z-b_zc_y\right)+a_y\left(b_zc_x-b_xc_z\right)+a_z\left(b_xc_y-b_yc_x\right)\\ &=a_x\left(b_yc_z-b_zc_y\right)-a_y\left(b_xc_z-b_zc_x\right)+a_z\left(b_xc_y-b_yc_x\right)\end{aligned}$$

Thus,

$$\mathbf{a}\cdot(\mathbf{b}\times\mathbf{c})=\begin{vmatrix}a_x & a_y & a_z\\ b_x & b_y & b_z\\ c_x & c_y & c_z\end{vmatrix}$$

In Box 8.1, we summarize vector multiplication.

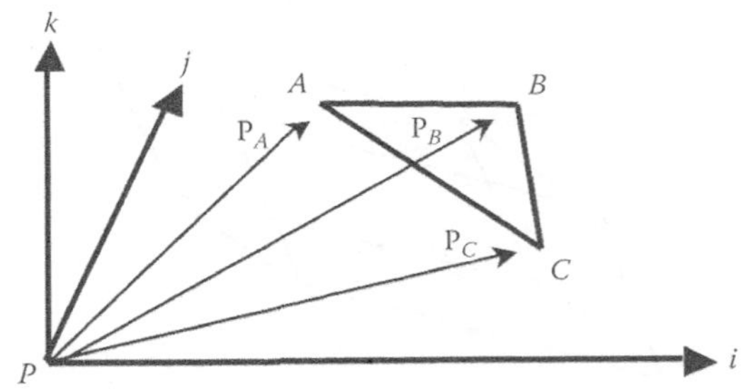

FIGURE 8.6 Vectors and a plane.

8.3 VECTORS AND PLANES

If we have three points A (x_A, y_A, z_A), B (x_B, y_B, z_B), and C (x_C, y_C, z_C), then we can represent these with three vectors (Figure 8.6) from the origin P (0, 0, 0).

$$\mathbf{P}_A = (x_A\mathbf{i}, y_A\mathbf{j}, z_A\mathbf{k}), \mathbf{P}_B = (x_B\mathbf{i}, y_B\mathbf{j}, z_B\mathbf{k}), \text{ and } \mathbf{P}_C = (x_C\mathbf{i}, y_C\mathbf{j}, z_C\mathbf{k})$$

The lines CA and CB are represented by $(\mathbf{P}_A - \mathbf{P}_C)$ and $(\mathbf{P}_B - \mathbf{P}_C)$. The cross product between these two lines is given by

$$(\mathbf{P}_A - \mathbf{P}_C) \times (\mathbf{P}_B - \mathbf{P}_C)$$

BOX 8.1 VECTOR MULTIPLICATION

If $\mathbf{a} = a_x\mathbf{i} + a_y\mathbf{j} + a_z\mathbf{k}$; $\mathbf{b} = b_x\mathbf{i} + b_y\mathbf{j} + b_z\mathbf{k}$; and $\mathbf{c} = c_x\mathbf{i} + c_y\mathbf{j} + c_z\mathbf{k}$, then

$$\mathbf{a} \cdot \mathbf{b} = a_x b_x + a_y b_y + a_z b_z$$

$$\mathbf{a} \times \mathbf{b} = (a_y b_z - a_z b_y)\mathbf{i} + (a_z b_x - a_x b_z)\mathbf{j} + (a_x b_y - a_y b_x)\mathbf{k}$$

$$\mathbf{a} \cdot (\mathbf{b} \times \mathbf{c}) = a_x(b_y c_z - b_z c_y) - a_y(b_x c_z - b_z c_x) + a_z(b_x c_y - b_y c_x) = \begin{vmatrix} a_x & a_y & a_z \\ b_x & b_y & b_z \\ c_x & c_y & c_z \end{vmatrix}$$

$$\begin{aligned}
\text{This} &= \{(x_A - x_C)\boldsymbol{i} + (y_A - y_C)\boldsymbol{j} + (z_A - z_C)\boldsymbol{k}\} \times \{(x_B - x_C)\boldsymbol{i} + (y_B - y_C)\boldsymbol{j} + (z_B - z_C)\boldsymbol{k}\} \\
&= \{(x_A - x_C)(y_B - y_C)\boldsymbol{k} - (x_A - x_C)(z_B - z_C)\boldsymbol{j} - (y_A - y_C)(x_B - x_C)\boldsymbol{k} \\
&\quad + (y_A - y_C)(z_B - z_C)\boldsymbol{i} + (z_A - z_C)(x_B - x_C)\boldsymbol{j} - (z_A - z_C)(y_B - y_C)\boldsymbol{i}\} \\
&= \{(y_A(z_B - z_C) + y_B(z_C - z_A) + y_C(z_A - z_B)\}\boldsymbol{i} \\
&\quad + \{z_A(x_B - x_C) + z_B(x_C - x_A) + z_C(x_A - x_B)\}\boldsymbol{j} \\
&\quad + \{x_A(y_B - y_C) + x_B(y_C - y_A) + x_C(y_A - y_B)\}\boldsymbol{k}
\end{aligned}$$

This cross product represents a vector that is perpendicular to both the lines CA and CB in Figure 8.6, which means that it is normal to the plane of CAB. This direction can be used, for instance, to calculate the level of illumination of a surface. Note that by reversing the sequence so that we have the cross product for *CB* and *CA* (rather than *CA* and *CB*), then the direction of the normal will be reversed, that is, it will be in the opposite direction.

If we join any general point *Q* (*x*, *y*, *z*) to *C* then the vector for the line *CQ* will be ($\mathbf{P}_Q - \mathbf{P}_C$). If point *Q* lies in the plane *ABC* then the dot product between this vector and the normal to the plane will be zero. That is, for *Q* to lie in the plane *ABC*

$$(\mathbf{P}_A - \mathbf{P}_C)\times(\mathbf{P}_B - \mathbf{P}_C)\cdot(\mathbf{P}_Q - \mathbf{P}_C) = 0$$

or

$$(\mathbf{P}_A - \mathbf{P}_C)\times(\mathbf{P}_B - \mathbf{P}_C)\cdot\{(x - x_C)\mathbf{i} + (y - y_C)\mathbf{j} + (z - z_C)\mathbf{k}\} = 0$$

This gives

$$\begin{aligned}&(x - x_C)\{y_A(z_B - z_C) + y_B(z_C - z_A) + y_C(z_A - z_B)\}\\&+(y - y_C)\{z_A(x_B - x_C) + z_B(x_C - x_A) + z_C(x_A - x_B)\}\\&+(z - z_C)\{x_A(y_B - y_C) + x_B(y_C - y_A) + x_C(y_A - y_B)\} = 0\end{aligned}$$

This is the equation of the plane *ABC* (see Box 8.2 and Example 8.6).

BOX 8.2 THE EQUATION OF A PLANE

The plane that contains three points:

$$A(x_A, y_A, z_A), B(x_B, y_B, z_B), \text{ and } C(x_C, y_C, z_C)$$

is given by

$$\begin{aligned}&(x - x_C)\{y_A(z_B - z_C) + y_B(z_C - z_A) + y_C(z_A - z_B)\}\\&+(y - y_C)\{z_A(x_B - x_C) + z_B(x_C - x_A) + z_C(x_A - x_B)\}\\&+(z - z_C)\{x_A(y_B - y_C) + x_B(y_C - y_A) + x_C(y_A - y_B)\} = 0\end{aligned}$$

which reduces to $ax + by + cz + d = 0$ where in determinant form

$$a = \begin{vmatrix}(y_A - y_C) & (z_A - z_C)\\(y_B - y_C) & (z_B - z_C)\end{vmatrix};\ b = \begin{vmatrix}(z_A - z_C) & (x_A - x_C)\\(z_B - z_C) & (x_B - x_C)\end{vmatrix};\ c = \begin{vmatrix}(x_A - x_C) & (y_A - y_C)\\(x_B - x_C) & (y_B - y_C)\end{vmatrix};\ \text{and}$$

$$d = -(ax_C + by_C + cy_C)$$

This equation reduces to the form $ax + by + cz + d = 0$ where a, b, c, d are constants that define the plane through the points ABC. The relationships given for the reduced equation of the plane in Box 8.2 (see Example 8.6) can also be expressed as:

$$a = \begin{vmatrix} 1 & y_C & z_C \\ 1 & y_A & z_A \\ 1 & y_B & z_B \end{vmatrix}; b = \begin{vmatrix} x_C & 1 & z_C \\ x_A & 1 & z_A \\ x_B & 1 & z_B \end{vmatrix}; c = \begin{vmatrix} x_C & y_C & 1 \\ x_A & y_A & 1 \\ x_B & y_B & 1 \end{vmatrix}; d = -(ax_C + by_C + cy_C)$$

Let us designate the vector normal to the plane as $\mathbf{n}$, that is, the vector derived from the cross product $(\mathbf{P}_A - \mathbf{P}_C) \times (\mathbf{P}_B - \mathbf{P}_C) = \mathbf{n}$.

If the equations of the plane is given by $ax + by + cz + d = 0$, then

$$\mathbf{n} = a\mathbf{i} + b\mathbf{j} + c\mathbf{k}$$

and

$$|\mathbf{n}| = \sqrt{a^2 + b^2 + c^2}$$

If we have two planes defined by $a_1x + b_1y + c_1z + d_1 = 0$ and $a_2x + b_2y + c_2z + d_2 = 0$, then the angle between the planes is given by

$$\mathbf{n}_1 \cdot \mathbf{n}_2 = |\mathbf{n}_1| * |\mathbf{n}_2| * \cos(\varnothing)$$

Example 8.6 The Equation of a Plane

Once again, using the values in Example 8.1 where $P(5, 9, 2)$, $Q(3, 7, 4)$, and $R(9, 6, 8)$ and using the formulae for a, b, c, d

$$a = \begin{vmatrix} 1 & 6 & 8 \\ 1 & 9 & 2 \\ 1 & 7 & 4 \end{vmatrix} = -6; b = \begin{vmatrix} 9 & 1 & 8 \\ 5 & 1 & 2 \\ 3 & 1 & 4 \end{vmatrix} = 20; c = \begin{vmatrix} 9 & 6 & 1 \\ 5 & 9 & 1 \\ 3 & 7 & 1 \end{vmatrix} = 14; d = -178$$

The equation of the plane PQR is therefore

$$-6x + 20y + 14z - 178 = 0$$

The vector for the normal to the plane is

$$-6\mathbf{i} + 20\mathbf{j} + 14\mathbf{k}$$

where $\mathbf{n}_1$ and $\mathbf{n}_2$ are the vectors of the normal to each plane, that is,

$$\mathbf{n}_1 = a_1\mathbf{i} + b_1\mathbf{j} + c_1\mathbf{k} \text{ and } \mathbf{n}_2 = a_2\mathbf{i} + b_2\mathbf{j} + c_2\mathbf{k}$$

Thus, ϕ, the angle between the planes is such that

$$\phi = \cos^{-1}\left(\frac{\mathbf{n}_1 \cdot \mathbf{n}_2}{|\mathbf{n}_1||\mathbf{n}_2|}\right)$$

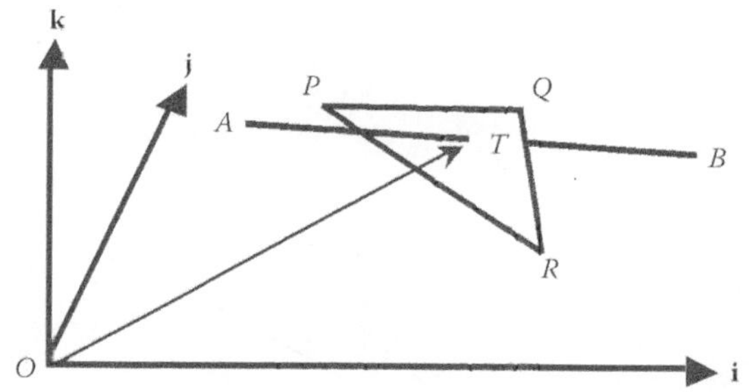

FIGURE 8.7 Angle of incidence.

8.4 ANGLES OF INCIDENCE

We can extend this idea further by considering a vector **v** that represents a line such as a ray of light striking the plane surface defined by PQR in Figure 8.7. The angle between **n**, the normal to the surface (here the plane PQR) and the vector **v** is given by the dot product

$$\mathbf{n}\cdot\mathbf{v} = |\mathbf{n}| * |\mathbf{v}| * \cos(\varnothing)$$

where $\varnothing$ is the angle that the line makes with the normal to the plane. Thus, the angle with the normal to the plane is given by $\phi = \cos^{-1}\left(\frac{\mathbf{n}\cdot\mathbf{v}}{|\mathbf{n}||\mathbf{v}|}\right)$.

Now consider the line from A to B in Figure 8.7. A is point (x_A, y_A, z_A) and B is point (x_B, y_B, z_B), and the line AB cuts the plane containing the triangle PQR at T. Let T be (x_T, y_T, z_T) and the vector $\mathbf{OT} = \mathbf{t} = x_T\mathbf{i} + y_T\mathbf{j} + z_T\mathbf{k}$. Let the vector **OA** be $\mathbf{a} = x_A\mathbf{i} + y_A\mathbf{j} + z_A\mathbf{k}$ and the vector **OB** be $\mathbf{b} = x_B\mathbf{i} + y_B\mathbf{j} + z_B\mathbf{k}$. Finally, let the vector **AB** be $\mathbf{v} = (x_B - x_A)\mathbf{i} + (y_B - y_A)\,\mathbf{j} + (z_B - z_A)\,\mathbf{k}$.

T must satisfy the condition that it is both on the line AB and the plane PQR, which is defined by $ax + by + cz + d = 0$. The normal to this plane is $\mathbf{n} = a\mathbf{i} + b\mathbf{j} + c\mathbf{k}$.

$$\mathbf{n}\cdot\mathbf{t} = (a\mathbf{i}+b\mathbf{j}+c\mathbf{k})\cdot(x_T\mathbf{i}+y_T\mathbf{j}+z_T\mathbf{k}) = ax_T + by_T + cz_T$$

since $\mathbf{i}\cdot\mathbf{i} = \mathbf{j}\cdot\mathbf{j} = \mathbf{k}\cdot\mathbf{k} = 1$ and $\mathbf{i}\cdot\mathbf{j} = \mathbf{j}\cdot\mathbf{k} = \mathbf{k}\cdot\mathbf{i} = 0$.

Since T lies on the plane, $ax_T + by_T + cz_T + d = 0$, then

$$\mathbf{n}\cdot\mathbf{t} + d = 0$$

We can also express T as the vector **OA** plus **AT** which, in turn, is a proportion of the vector **AB** in the form $\mathbf{t} = \mathbf{a} + \lambda\mathbf{v}$, where λ is some unknown scalar quantity. Thus,

$$\mathbf{n}\cdot(\mathbf{a}+\lambda\mathbf{v})+d=0 \quad \text{or} \quad \mathbf{n}\cdot\mathbf{a}+\lambda\mathbf{n}\cdot\mathbf{v}+d=0 \quad \text{or} \quad \lambda = \frac{-(\mathbf{n}\cdot\mathbf{a}+d)}{\mathbf{n}\cdot\mathbf{v}}$$

This means that we can determine the coordinates of point T based on the coordinates of A and B. We also have the value of $\mathbf{n}\cdot\mathbf{t}$ from which we have the cosine of the angle between the vector **OT** and the normal to the plane and hence the angle of incidence. See Example 8.7.

Note that if $\mathbf{n}\cdot\mathbf{v}$ is zero there is no solution as the line AB will be parallel to the plane PQR.

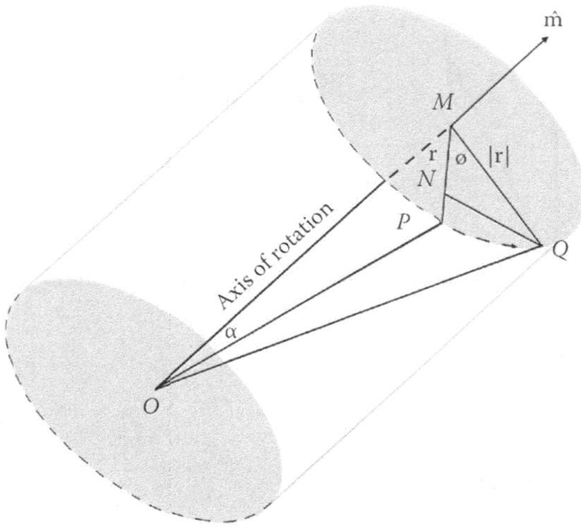

FIGURE 8.8 Rotation using vectors.

8.5 VECTORS AND ROTATIONS

Matrices are a mathematical form of shorthand to manage multiple similar calculations, often in different dimensions such as x, y, or z. For in-depth explanations of matrices, see Chapter 9. To illustrate how vectors can be used to find the same answer as might be achieved through the use of matrices, consider an arbitrary axis *OM* about which we wish to rotate a point *P* through an angle ø to a new position *Q* (Figure 8.8).

The center for the rotation of *P* is *M* so that $MP = MQ = r =$ radius of the circle center *M* as *P* is rotated. The plane *MPQ* is at right angles to the axis of rotation *OM*.

Let the vector **OM** = **m** and let the axis of rotation be given by the unit vector $\widehat{\mathbf{m}}$ where $\widehat{\mathbf{m}} = a\mathbf{i} + b\mathbf{j} + c\mathbf{k}$. The notation $\widehat{\mathbf{m}}$ showing the "hat" over the **m** indicates that it is a unit vector, which means that *a*, *b*, and *c* have been scaled so that $\sqrt{a^2 + b^2 + c^2} = 1$.

Example 8.7 The Intersection of a Line and a Plane

In Example 8.6, we derived the equation of the plane *PQR* as

$$-6x + 20y + 14z - 178 = 0$$

In Figure 8.7, let *A* be point (1, 1, 2) and *B* be (5, 7, 10).
The line *AB* is the vector

$$\mathbf{v} = 4\mathbf{i} + 6\mathbf{j} + 8\mathbf{k}$$

If **n** is the normal to the plane,

$$\mathbf{n} = -6\mathbf{i} + 20\mathbf{j} + 14\mathbf{k}$$

The dot product $\mathbf{n} \cdot \mathbf{v} = -24\mathbf{i}^2 + 120\mathbf{j}^2 + 112\mathbf{k}^2 = -24 + 120 + 112 = 208$;

$$|\mathbf{n}| = \sqrt{(632)} \approx 25.14; |\mathbf{v}| = \sqrt{(116)} \approx 10.77$$

The line AB will strike the plane at an angle ø with the normal to the plane where

$$\varnothing = \cos^{-1}\left(\frac{\mathbf{n} \cdot \mathbf{v}}{|\mathbf{n}||\mathbf{v}|}\right) \quad \text{or} \quad \varnothing \approx \cos^{-1}\{208 / (25.14 * 10.77)\} \approx 39.8°$$

Let the line AB intersect the plane PQR at T.

Since T lies on the line AB its vector **OT** = **OA** + λ **AB** where λ is a scalar quantity. If **OT** = **t** and **OA** = **a** then **t** = **a** + λ**v**. Also, **n** • **t** + d = 0 since T lies on the plane.

Thus,

$$\mathbf{n} \cdot \mathbf{a} + \lambda\, \mathbf{n} \cdot \mathbf{v} + d = 0 \quad \text{or} \quad \lambda = \frac{-(\mathbf{n} \cdot \mathbf{a} + d)}{\mathbf{n} \cdot \mathbf{v}}$$

Since **a** = 1**i** + 1**j** + 2**k**, then **n** • **a** = −6 + 20 + 28 = 42. Hence,

$$\lambda = 136 / 208 = 0.653846$$
$$\mathbf{t} = (1\mathbf{i} + 1\mathbf{j} + 2\mathbf{k}) + 0.653846 * (4\mathbf{i} + 6\mathbf{j} + 8\mathbf{k})$$
$$= 3.6\mathbf{i} + 4.9\mathbf{j} + 7.2\mathbf{k} \text{ rounding to one decimal place.}$$

Thus, T has coordinates (3.6, 4.9, 7.2).

Let point P be (x_P, y_P, z_P) and its new location after rotation about OM through an angle ø be (x_Q, y_Q, z_Q). Also, in the plane MPQ, let QN be such that N is the foot of the perpendicular from Q to the line PM.

By the rule of vectors, the vector **OQ** = vector **OM** + vector **MN** + vector **NQ**, or

$$\overrightarrow{OQ} = \overrightarrow{OM} + \overrightarrow{MN} + \overrightarrow{NQ}$$

The length OM = length OP cos α where α is the angle MOP.

If **OP** is the vector **p**, then the length OP = |**p**| and the length OM = |**p**| cos α or since the angle α is built into the dot product,

$$\text{Length of } OM = \hat{\mathbf{m}} \cdot \mathbf{p}$$

The vector **OM** is therefore $\hat{\mathbf{m}}$ multiplied by its length, namely,

$$\overrightarrow{OM} = \mathbf{m} = (\hat{\mathbf{m}} \cdot \mathbf{p})\hat{\mathbf{m}}$$

The length of MP = OP sin α = |**p**| sin α

If we call the vector **MP** = **r**, then its length = |**r**| = |**p**| sin α, which, of course, is also the length of MQ as P is rotated about M. Hence, the length MN = MQ cos ø = |**r**| cos ø and the vector for **MN** is

$$\overrightarrow{MN} = (\cos \varnothing)\mathbf{r}$$

Point P is such that the vector $\mathbf{p} = \overrightarrow{OP} = \overrightarrow{OM} + \overrightarrow{MP} = \mathbf{m} + \mathbf{r}$, or

$$\mathbf{r} = \mathbf{p} - \mathbf{m} = \mathbf{p} - (\hat{\mathbf{m}} \cdot \mathbf{p})\hat{\mathbf{m}}$$

Hence,

$$\overrightarrow{MN} = (\cos \varnothing)\left[\mathbf{p} - (\hat{\mathbf{m}} \cdot \mathbf{p})\hat{\mathbf{m}}\right]$$

To find the vector $\overrightarrow{NQ}$ we need to find a line at right angles to MN and in the plane that is at right angles to OM. This will be given by the cross product between the two vectors and **r**, giving us a vector **s** for the direction $\overrightarrow{NQ}$.

$$\mathbf{s} = \hat{\mathbf{m}} \times \mathbf{r} = \hat{\mathbf{m}} \times \left[\mathbf{p} - (\hat{\mathbf{m}} \cdot \mathbf{p})\hat{\mathbf{m}}\right]$$

Or since $(\hat{\mathbf{m}} \cdot \mathbf{p})$ is a scalar and $\hat{\mathbf{m}} \times \hat{\mathbf{m}} = 0$, then

$$\mathbf{s} = \hat{\mathbf{m}} \times \mathbf{p}$$

We also know from triangle MNQ that the length $NQ = MQ \sin ø = |\mathbf{r}| \sin ø$.
Since $\mathbf{S} = \hat{\mathbf{m}} \times \mathbf{r}$ and $\hat{\mathbf{m}}$ is a unit vector then $|\mathbf{s}| = |\mathbf{r}|$. Hence, $\overrightarrow{NQ} = \mathbf{s}\ \sin\varnothing = (\hat{\mathbf{m}} \times \mathbf{p})\sin\ \varnothing$

$$\overrightarrow{OQ} = \overrightarrow{OM} + \overrightarrow{MN} + \overrightarrow{NQ}$$

hence,

$$\begin{aligned}\overrightarrow{OQ} &= (\hat{\mathbf{m}} \cdot \mathbf{p})\hat{\mathbf{m}} + \left[\mathbf{p} - (\hat{\mathbf{m}} \cdot \mathbf{p})\hat{\mathbf{m}}\right]\cos\ \varnothing + (\hat{\mathbf{m}} \times \mathbf{p})\sin\ \varnothing \\ &= \mathbf{p}\ \cos\ \varnothing + (\hat{\mathbf{m}} \cdot \mathbf{p})\hat{\mathbf{m}}(1 - \cos\ \varnothing) + (\hat{\mathbf{m}} \times \mathbf{p})\sin\ \varnothing\end{aligned}$$

The dot product

$$(\hat{\mathbf{m}} \cdot \mathbf{p}) = (a\mathbf{i} + b\mathbf{j} + c\mathbf{k}) \cdot (x_p\mathbf{i} + y_p\mathbf{j} + z_p\mathbf{k}) = (ax_p + by_p + cz_p)$$

Since $\mathbf{i} \times \mathbf{i} = 0$, $\mathbf{i} \times \mathbf{j} = 1$, and $\mathbf{j} \times \mathbf{i} = -1$, then the cross product

$$\begin{aligned}(\hat{\mathbf{m}} \times \mathbf{p}) &= (a\mathbf{i} + b\mathbf{j} + c\mathbf{k}) \times (x_p\mathbf{i} + y_p\mathbf{j} + z_p\mathbf{k}) \\ &= \left[(bz_p - cy_p)\mathbf{i} + (cx_p - az_p)\mathbf{j} + (ay_p - bx_p)\mathbf{k}\right]\end{aligned}$$

Hence,

$$\begin{aligned}\overrightarrow{OQ} = &(x_p\mathbf{i} + y_p\mathbf{j} + z_p\mathbf{k})\cos\ \varnothing + (ax_p + by_p + cz_p)(a\mathbf{i} + b\mathbf{j} + c\mathbf{k})(1 - \cos\ \varnothing) \\ &+ \left[(bz_p - cy_p)\mathbf{i} + (cx_p - az_p)\mathbf{j} + (ay_p - bx_p)\mathbf{k}\right]\sin\ \varnothing\end{aligned}$$

Rearranging terms,

$$\begin{aligned}\overrightarrow{OQ} = &\left[x_p\left\{a^2(1 - \cos\varnothing) + \cos\varnothing\right\} + y_p\left\{ab(1 - \cos\varnothing) - c\ \sin\varnothing\right\} + z_p\left\{ac(1 - \cos\varnothing) + b\ \sin\varnothing\right\}\right]\mathbf{i} \\ &+ \left[x_p\left\{ab(1 - \cos\varnothing) + c\ \sin\varnothing\right\} + y_p\left\{b^2(1 - \cos\varnothing) + \cos\varnothing\right\} + z_p\left\{bc(1 - \cos\varnothing) - a\ \sin\varnothing\right\}\right]\mathbf{j} \\ &+ \left[x_p\left\{ab(1 - \cos\varnothing) - b\ \sin\varnothing\right\} + y_p\left\{bc(1 - \cos\varnothing) + a\ \sin\varnothing\right\} + z_p\left\{c^2(1 - \cos\varnothing) + \cos\varnothing\right\}\right]\mathbf{k}\end{aligned}$$

Example 8.8 Rotating a Cylinder

Consider a cylinder centered on the z-axis (i.e., in Figure 8.8, call OM the z-axis).

Let P be a point on the cylinder with coordinates $(2, 0, z)$ so that the cylinder has a diameter of 4 units. Let us rotate the cylinder through $ø = 45°$ so that P becomes Q.

For the z-axis, the unit vector = **k** and if $\hat{\mathbf{m}}$, we can write $a = b = 0$ and $c = 1$. Also, since we are considering $ø = 45°$, then $\cos ø = \sin ø = 1 / \sqrt{2}$.

Hence, the relationship

$$\begin{bmatrix} X_Q \\ Y_Q \\ Z_Q \end{bmatrix} = \begin{bmatrix} a^2(1-\cos\phi)+\cos\phi & ab(1-\cos\phi)-c\sin\phi & ac(1-\cos\phi)+b\sin\phi \\ ab(1-\cos\phi)+c\sin\phi & b^2(1-\cos\phi)+\cos\phi & bc(1-\cos\phi)-a\sin\phi \\ ac(1-\cos\phi)-b\sin\phi & bc(1-\cos\phi)+a\sin\phi & c^2(1-\cos\phi)+\cos\phi \end{bmatrix} * \begin{bmatrix} X_P \\ Y_P \\ Z_P \end{bmatrix}$$

reduces to

$$\begin{bmatrix} X_Q \\ Y_Q \\ Z_Q \end{bmatrix} = \begin{bmatrix} \{1/\sqrt{2}\} & \{-1/\sqrt{2}\} & \{0\} \\ \{1/\sqrt{2}\} & \{1/\sqrt{2}\} & \{0\} \\ \{0\} & \{0\} & \{1\} \end{bmatrix} \begin{bmatrix} X_P \\ Y_P \\ Z_P \end{bmatrix}$$

or

$$x_Q = x_P/\sqrt{2} - y_P/\sqrt{2} = \sqrt{2}; y_Q = x_P/\sqrt{2} + y_P/\sqrt{2} = \sqrt{2}; z_Q = z_P$$

Repeating through a further 45° to move P on to R, $x_R = 0$; $y_R = 2$; $z_R = z_Q = z_P$, and so on. Expressed in matrix form (see Example 8.8)

$$\begin{bmatrix} X_Q \\ Y_Q \\ Z_Q \end{bmatrix} = \begin{bmatrix} a^2(1-\cos\phi)+\cos\phi & ab(1-\cos\phi)-c\sin\phi & ac(1-\cos\phi)+b\sin\phi \\ ab(1-\cos\phi)+c\sin\phi & b^2(1-\cos\phi)+\cos\phi & bc(1-\cos\phi)-a\sin\phi \\ ac(1-\cos\phi)-b\sin\phi & bc(1-\cos\phi)+a\sin\phi & c^2(1-\cos\phi)+\cos\phi \end{bmatrix} * \begin{bmatrix} X_P \\ Y_P \\ Z_P \end{bmatrix}$$

8.6 MODELING TERRAIN

GIS practitioners need to derive information from terrain data for many kinds of applications. Before the advent of aerial surveying technology, terrain data were generally collected via land surveying, and this is still the most accurate method to collect/measure high resolution terrain data. Land surveying work always start from a known *landmark* location and measure the subsequent points in the survey scope via bearing and distance, as discussed in Chapter 1. Using this approach, the boundaries of a land parcel or the alignment of a street centerline can be drawn on the map as a series of connected survey points. This method of gathering spatial data is essentially a vector-based method. Note that the bearing angle is an angle constrained within the two-dimensional horizontal plane.

Besides the horizontal plane, land measurement also deals with interaction of the vertical with the horizontal. Slope, aspect, contours, and volumetric calculations are needed. Figure 8.9 illustrates where the slope and aspect angles are measured from the three-dimensional terrain.

1. *Slope*: the rate of vertical change per horizontal distance. This can be understood as a derivative dz/ds (s is the horizontal distance). See Chapter 10 for in-depth discussions on derivatives. At any point of terrain that is not flat and is not on the surface of a water body, a drop of water will flow from this point potentially in multiple directions, but the greatest component of the gravitational force along the surface of the terrain will pull the water toward the direction where the magnitude of dz/ds is the greatest. The greatest dz/ds value is the slope and the vector along this downward direction is called the gradient vector. Slope is usually measured in units of percent or degree. The percentage format has two versions depending on the application: one is the sine of the slope, and the other version is the tangent of the slope. There can be a big difference in the percent slope if the wrong version is calculated.
2. *Aspect*: the horizontal or azimuthal bearing angle of the gradient vector as measured from the North direction.

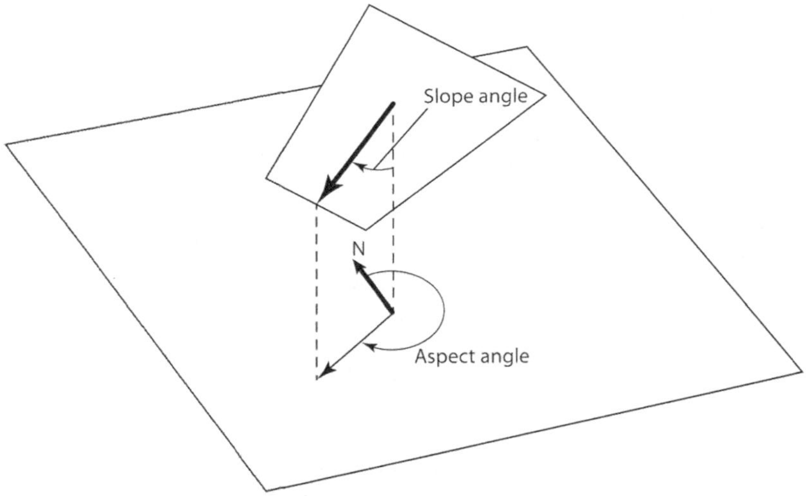

FIGURE 8.9 Slope and aspect angle measurements.

3. *Contour line*: isometric curves of the Z (elevation) coordinate projected onto the terrain surface at regular intervals of Z values. The contour curves facilitate convenient depiction of the slope and aspect data of the terrain surface on a two-dimensional map, and this was the standard way to store elevation data before computers. The shortest horizontal distance measured between successive contour lines can be used to estimate the slope—the shorter the distance, the greater the slope. Aspect can be measured from the bearing angle of the vector representing the shortest horizontal distance between successive contour lines.
4. Volumetric calculations: calculating the volume between two different elevation surfaces in the same two-dimensional area is useful to estimate the amount of water in a reservoir, or the amount of earth or dirt to be excavated or filled by a construction project. This is a three-dimensional integral that is most easily done using raster formatted elevation data. See Chapter 10 for more in-depth discussions of integrals.

Terrain data in GIS are commonly stored in raster format which makes it easier to calculate properties of the terrain described above. In order to calculate slope and aspect, more grid cells than the nearest-neighbor cell values will be needed, since the nearest-neighbor cells (8 cells around any given cell, see Figure 8.10) would only provide 8 distinct aspect angles (0, 45, 90, 135, 180, 225, 270, 315 degrees). As a result, interpolation algorithms are involved in GIS software tools that calculate slope, aspect, or generate contour curves from raster terrain data.

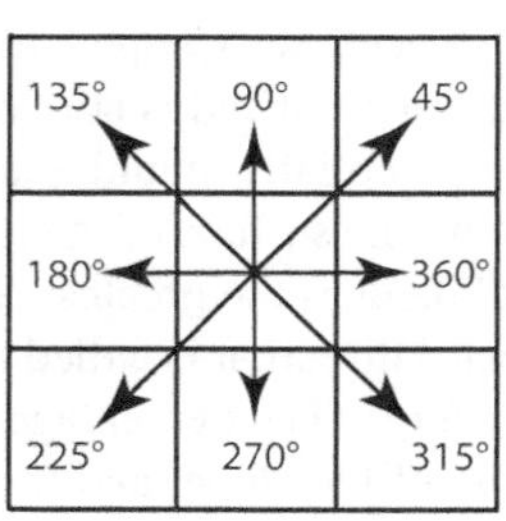

FIGURE 8.10 Eight possible aspect angles toward nearest neighbor cells in a raster dataset. Calculating more precise aspect angles (and therefore slope angles) requires accessing values from cells further away.

8.7 THREE-DIMENSIONAL RAY TRACING

Certain applications such as 3D computer graphics require a better depiction of terrain data than the standard raster data format, because raster data requires interpolation to calculate slope and aspect, both crucial data for computer graphics and *ray tracing*. A vector data format invented for this purpose is the Triangulated Irregular Network (*TIN*). This data format allows slope and aspect to be easily derived from the simplest geometric unit - the triangle. It also has the advantage of reducing the number of data points where the terrain approximates a three-dimensional plane over a larger area, and therefore has uniform slope and aspect values. This saves data storage space, and also translates into faster computations when generating 3D renderings.

Now we describe in more detail a simple 3D ray-tracing scenario, as illustrated in Figure 8.11. In such an application, the source of illumination can be the sun or artificial light sources. Typically, the sun is assumed to be infinitely far away, therefore only needs to be specified by the bearing (azimuth angle) and the inclination (i.e. the slope of the sun's rays). Rays of the sun are simulated to travel to each triangle in the TIN network, then reflected to the observer who is located at a precise 3D coordinate location. The optimal reflection angle is calculated via a vector calculation between the triangle's perpendicular vector and the vector representing the sun ray. The further away the observer's angle is from the optimal reflection angle, the less intensity of light will travel to the observer.

There are many algorithms that generate this triangulation network with certain preferences, such as to optimize orientation or angles of the resulting triangles. We can create more precise triangulation by subdividing the sides of existing triangles into multiple piecewise linear line segments that conform better to terrain elevation, then new nodes are created and the resulting triangles will be smaller. Since the triangles are of arbitrary size this is the reason the word "irregular" is in the name of triangulated irregular network (TIN). When a TIN is used to model a surface, the desired triangulation is the one that has a minimum possible number of triangles while keeping the error between the triangulated surface and the real curved surface of the terrain under a given limit.

Suppose a terrain under study was examined and the elevation values of a few surface points are known. If these points are connected by straight lines, they will serve as vertices of triangles, which can be used to build a TIN. This is a simpler process than building a TIN from a raster terrain model. Compared to building TINs from raster terrain data, this alternative method also avoids introducing the errors that inevitably come with interpolation methods.

In computer graphics a model of a surface is called by a more general term—*mesh*. Meshes are not limited to be made exclusively of triangles and can contain polygons with the number of sides greater than three, which could also be called "quads" (4 sides) or n-gons (n sides). In GIS, modeling surfaces can be useful not only to represent Earth surfaces, but also to represent man-made structures like buildings, bridges, roads, even vegetation, movable objects, and people. Polygons in a mesh can be given colors or photographic images to drape onto its surfaces when shown on the computer screen. They are used to represent surfaces of three-dimensional objects and to simulate how these objects look in the real world.

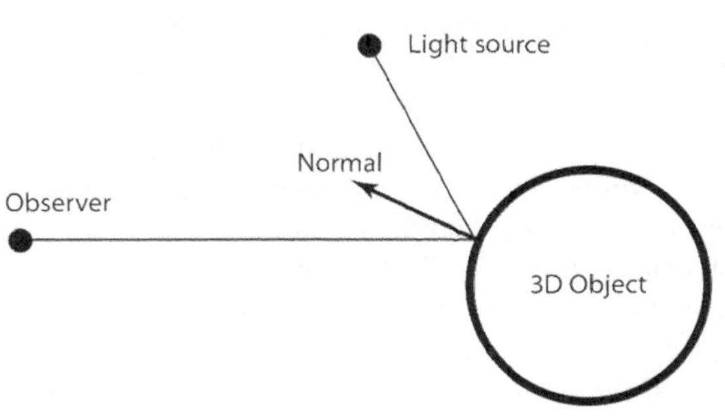

FIGURE 8.11 Ray tracing.

A similar application is using 3D ray tracing to simulate visibility of things from a view point. If one places the view point at the location of a light source such as in Figure 8.11, then any locations on the terrain or building surfaces that do not receive the rays of light would also be locations that are not visible from the view point. This method can also calculate where shadows form because shadows are locations that do not receive the rays of light. This type of methods are often called *viewshed analysis* in GIS, and have applications outside the computer graphics field. Viewshed analyses are used to inform decision making regarding visibility of landmark features when planning construction projects. In an urban setting, blockage of pre-existing views of certain landmark features by the neighbors can be considered a disadvantage of a proposed construction project.

GLOSSARY

Aspect The compass orientation of a downward slope, represented by an azimuth angle.

Contour line In GIS, contour lines are lines on a map that connect points of equal land elevation. Typically contour lines are drawn at regular intervals of elevation values, for example, one contour line for every 100 feet rise of elevation.

Cross product A new vector **c** at right angles to the two generating vectors **a** and **b** such that "**a** cross **b**" or $\mathbf{a} \times \mathbf{b} = (|\mathbf{a}| * |\mathbf{b}| * \sin \phi)\,\mathbf{c}$ where ϕ is the angle between the generating vectors. By convention the direction of **c** is determined by the right-hand rule (see Chapter 9). Since the two vectors **a** and **b** define a plane, **c** is normal to the plane.

Since sin (0°) = 0 and sin (90°) = 1, then for the unit vectors **i, j, k** of three orthogonal axes:

$$\mathbf{i}\times\mathbf{i} = \mathbf{j}\times\mathbf{j} = \mathbf{k}\times\mathbf{k} = \vec{0} \quad \text{and} \quad \mathbf{i}\times\mathbf{j} = \mathbf{j}\times\mathbf{k} = \mathbf{k}\times\mathbf{i} = \vec{1}$$

Also,

$$\mathbf{j}\times\mathbf{i} = \mathbf{k}\times\mathbf{j} = \mathbf{i}\times\mathbf{k} = -\vec{1}$$

Direction cosines The cosines of angles that a vector makes with the axes used to describe the vector.

Dot product If ϕ is the angle between two vectors **a** and **b**, then the dot product is a scalar quantity given by $\mathbf{a} \cdot \mathbf{b} = |\mathbf{a}| * |\mathbf{b}| * \cos \phi$.

Since cos (0°) = 1 and cos (90°) = 0, then for the unit vectors **i, j, k** of three orthogonal axes

$$\mathbf{i}\cdot\mathbf{i} = \mathbf{j}\cdot\mathbf{j} = \mathbf{k}\cdot\mathbf{k} = 1$$

and

$$\mathbf{i}\cdot\mathbf{j} = \mathbf{j}\cdot\mathbf{k} = \mathbf{k}\cdot\mathbf{i} = 0$$

Landmark Also called geodetic monument, are permanent physical monuments installed by government surveyors to establish horizontal positioning of specific locations in order to anchor land surveying work.

Matrix A set of numbers aligned in rows and columns forming a rectangular *array*. Often written in bold as **M**. The number of rows is not necessarily the same as the number of columns but each row must be the same length, as must each column.

Mesh Also called polygon mesh, is a collection of vertices, edges and faces that represent the spatial location and shape of 3D objects in computer graphics.

Modulus The length of a *position vector*. If the vector is **v** then the modulus is written as |**v**|—not to be confused with the symbol for a determinant. If vector $\mathbf{OP} = a\mathbf{i} + b\mathbf{j} + c\mathbf{k}$ where **i, j, k** are the unit vectors for the reference axes, then $|\mathbf{OP}| = \sqrt{a^2 + b^2 + c^2}$

Position vector A *vector* representing the location of a point in a coordinate system.

Scalar A quantity that has magnitude but no direction, such as pressure or the number of people in the world.

Ray tracing A computer graphics technique that simulates how light moves through a three-dimensional space, including reflecting off of surfaces.

Scalar triple product A scalar quantity determined by combining a *dot product* with a *cross product* in the form $\mathbf{a} \cdot \mathbf{b} \times \mathbf{c}$. It gives the volume of a solid whose orthogonal edges are formed by the vectors **a**, **b**, and **c**. If the scalar triple product is zero, the volume must be zero and hence the three points represented by **a**, **b**, and **c** must all lie in a plane.

Slope Also called gradient, it is the rise or fall of land surface derived from the steepest angle a drop of water on the land surface would flow.

TIN Stands for Triangulated Irregular Network, is a representation of a continuous surface utilizing a mesh of triangular facets.

Unit vector A vector of unit length.

Vector A quantity that has both magnitude and direction, such as velocity or a ray of light from the sun. The term is also applied to a straight line that has length and direction.

Viewshed analysis A process of identifying what areas or objects are visible from a specific location.

9 Matrices and Determinants

9.1 BASIC MATRIX OPERATIONS

Matrices are a mathematical form of shorthand. At its simplest level, a *matrix* is a set of numbers aligned in rows and columns in the form of a rectangle and then enclosed in brackets. Each number within the matrix is an *element* and the whole set of numbers may be referred to as an *array*. Rather than talk about each of the elements within the array, we can simply refer to the matrix as a whole and call it **M**.

For example,

$$\mathbf{M} = \begin{pmatrix} 1 & 2 & 3 \\ 4 & 5 & 6 \end{pmatrix}$$

is a matrix with the first six integers arranged with two rows and three columns and is called a 2 * 3 *matrix*. Think of **M** as six boxes or cells, each containing a number, rather like a spreadsheet:

1	2	3
4	5	6

Likewise, we may have a single column of cells with boxes stacked vertically, as in

$$\begin{pmatrix} 7 \\ 8 \\ 9 \end{pmatrix}$$

This is a column matrix with three rows and one column or 3 * 1. A matrix with one row or one column is often used as a convenient way to handle vector calculations. If the number of rows equals the number of columns then the matrix is said to be *square*. Thus,

$$\begin{pmatrix} -1 & 0 & 0 \\ 0 & 2 & 0 \\ 0 & 0 & -3 \end{pmatrix}$$

is a 3 * 3 square matrix. It has nine cells and since in this example the only numbers other than zero lie along the diagonal, it is called a *diagonal matrix*. The diagonal from top left to bottom right is called the *leading diagonal*. If all the numbers in the leading diagonal are 1 and all the other elements are zero, then the matrix is called an *identity matrix*. Thus,

$$\begin{pmatrix} 1 & 0 \\ 0 & 1 \end{pmatrix} \text{and} \begin{pmatrix} 1 & 0 & 0 \\ 0 & 1 & 0 \\ 0 & 0 & 1 \end{pmatrix} \text{and} \begin{pmatrix} 1 & 0 & 0 & 0 \\ 0 & 1 & 0 & 0 \\ 0 & 0 & 1 & 0 \\ 0 & 0 & 0 & 1 \end{pmatrix}$$

are all identity matrices and are often written as **I**.

The way that matrices are manipulated follows certain rules that are ideally suited to handling in a computer as the operations are in general repetitive. Two matrices can be added or

DOI: 10.1201/9781003464402-12

subtracted if they are the same size. This is done by adding or subtracting the corresponding elements (Box 9.1). Thus,

$$\begin{pmatrix} 1 & 2 & 3 & 4 \\ 5 & 6 & 7 & 8 \end{pmatrix}$$

is a 2 * 4 matrix. If we add another 2 * 4 matrix

$$\begin{pmatrix} 6 & -5 & 4 & -3 \\ 9 & -8 & 7 & 6 \end{pmatrix}$$

we obtain

$$\begin{pmatrix} 1+6 & 2-5 & 3+4 & 4-3 \\ 5+9 & 6-8 & 7+7 & 8+6 \end{pmatrix} = \begin{pmatrix} 7 & -3 & 7 & 1 \\ 14 & -2 & 14 & 14 \end{pmatrix}$$

The matrix $\begin{pmatrix} 1 & 2 & 3 \\ 4 & 5 & 6 \end{pmatrix}$ cannot, however, be added to or subtracted from $\begin{pmatrix} 7 \\ 8 \\ 9 \end{pmatrix}$ because the two matrices are not of the same size.

BOX 9.1 MATRIX ADDITION AND SUBTRACTION

If **A** is a matrix with the number in the *i*th row and *j*th column = $a_{ij,}$ and **B** is the same size with the *i*th row and *j*th column = $b_{ij,}$ then

the sum of $\mathbf{A}+\mathbf{B}=\mathbf{C}$ where $c_{ij}=a_{ij}+b_{ij}$

Likewise, if $\mathbf{A}-\mathbf{B}=\mathbf{D}$ then $d_{ij}=a_{ij}-b_{ij}$

If we multiply the whole of a matrix by a number then each element must be multiplied by that number. Thus, if

$$M = \begin{pmatrix} 1 & 2 & 3 & 4 \\ 5 & 6 & 7 & 8 \end{pmatrix} \text{ then } 4M = \begin{pmatrix} 4 & 8 & 12 & 16 \\ 20 & 24 & 28 & 32 \end{pmatrix}$$

Note that to refer to a matrix we write it in **bold** lettering (e.g., **M** above).

To multiply one matrix by another matrix then the number of columns in the first matrix must equal the number of rows in the second. Under such an arrangement the matrices are said to be *conformable.* This is necessary because the multiplication is carried out in a special way, which creates a new matrix with the same number of rows as the first and the same number of columns as the second. Put another way, if the first matrix has p rows and q columns and the second has q rows and r columns then the product of the two will have p rows and r columns.

Consider a 2 * 2 matrix

$$\mathbf{A} = \begin{pmatrix} a & b \\ d & e \end{pmatrix} \text{ with numbers } a, b, d, \text{ and } e$$

Let **B** be a 2 * 1 matrix with numbers u and w such that

$$\mathbf{B} = \begin{pmatrix} u \\ w \end{pmatrix}$$

then

$$\mathbf{A} * B = \begin{pmatrix} a & b \\ d & e \end{pmatrix} * \begin{pmatrix} u \\ w \end{pmatrix} = \begin{pmatrix} au + bw \\ du + ew \end{pmatrix}$$

For the first row, multiply each element across the row by the element that is in the same position down the column of the second matrix. Repeat for the second row of the first matrix and first column of the second. The result of multiplying a 2 * 2 matrix times a 2 * 1 matrix is a matrix with 2 rows and 1 column. Conversely, we cannot multiply a 2 * 1 matrix times a 2 * 2 since the columns in the first matrix are not as many as the rows in the second. In this case **B** * **A** does not exist.

The same procedure is followed with a 3 * 3 multiplying a 3 * 2

$$\begin{pmatrix} a & b & c \\ d & e & f \\ g & h & i \end{pmatrix} * \begin{pmatrix} u & x \\ v & y \\ w & z \end{pmatrix} = \begin{pmatrix} au + bv + cw & ax + by + cz \\ du + ev + fw & dx + ey + fz \\ gu + hv + iw & gx + hy + iz \end{pmatrix} = C$$

C is a 3 * 2 matrix.

In the above example, if in **A** the numbers $a = e = i = 1$ (i.e., the elements in the leading diagonal are all unity) and all the other elements of **A** are zero, which means that **A** is the identity matrix **I**, then

$$\mathbf{A} = \mathbf{I} = \begin{pmatrix} 1 & 0 & 0 \\ 0 & 1 & 0 \\ 0 & 0 & 1 \end{pmatrix} \text{ and } \quad \text{if } B = \begin{pmatrix} ux \\ vy \\ wz \end{pmatrix}$$

then when we multiply **I** * **B** to obtain **C**, all the elements of **C** will be the same as those in **B**, hence,

$$\mathbf{I} * \mathbf{B} = \mathbf{B}.$$

Once again, if **A** is the 3 * 3 matrix and **B** is the 3 * 2, we can multiply **A** times **B** to obtain a new matrix **C**. The order of the matrices is important for we cannot multiply **B** times **A** since this time the number of columns in the first (**B**) would not equal the number of rows in the second (now **A**). If, however, the number of rows in **A** equals the number of columns in **B**, then we can obtain both **A** * **B** and **B** * **A**. Thus, if **A** is 4 * 2 and **B** is 2 * 4, then **A** * **B** is 4 * 4, but **B** * **A** is 2 * 2 and **A** * **B** ≠ **B** * **A**. Thus, unlike ordinary arithmetic, the order of multiplication is of critical importance (Box 9.2 and Example 9.1).

BOX 9.2 MATRIX MULTIPLICATION (1)

If **A** is a matrix with the element in the ith row and jth column = a_{ij} and **B** has the same number of rows as **A** has columns and has its ith row and jth column = b_{ij}, then

$$\mathbf{A} * \mathbf{B} = \mathbf{C}$$

where c_{ij} = the sum of the elements of row i in **A** multiplied by the corresponding elements of column j in **B**

$$c_{ij} = \left\{ a_{i1}b_{1j} + a_{i2}b_{2j} + a_{i3}b_{3j} + \cdots + a_{in}b_{nj} \right\}$$

Example 9.1 Matrix Multiplication

If **A** is a 4 * 2 matrix and **B** is a 2 * 4 matrix and

$$\mathbf{A}=\begin{pmatrix}2&5\\5&7\\7&3\\8&6\end{pmatrix},\mathbf{B}=\begin{pmatrix}7&9&3&1\\5&8&4&3\end{pmatrix};\text{ then }\mathbf{A}*\mathbf{B}=\begin{pmatrix}39&58&26&17\\70&101&43&26\\64&87&33&16\\86&120&48&26\end{pmatrix}$$

$$\mathbf{B}=\begin{pmatrix}7&9&3&1\\5&8&4&3\end{pmatrix},\mathbf{A}=\begin{pmatrix}2&5\\5&7\\7&3\\8&6\end{pmatrix};\text{ then }\mathbf{B}*\mathbf{A}=\begin{pmatrix}88&106\\102&111\end{pmatrix}$$

There are, however, exceptions to the rule that $\mathbf{A} * \mathbf{B} \neq \mathbf{B} * \mathbf{A}$, for example, when **B** is such that $\mathbf{A} * \mathbf{B} = \mathbf{I}$ (the identity matrix). Consider the special case where

$$\mathbf{A}=\begin{pmatrix}a&b\\d&e\end{pmatrix}$$

and

$$\mathbf{B}=\frac{1}{ae-bd}*\begin{pmatrix}e-b\\-d\ a\end{pmatrix}$$

If we multiply **A** * **B**, we obtain

$$\begin{pmatrix}1&0\\0&1\end{pmatrix}$$

which is the identity matrix. We obtain the same result if we multiply **B** times **A** (= **B** * **A**). The matrix **B** that has this special property is said to be the *inverse* of **A** and is written as $\mathbf{A}^{-1}$. It has the property that $\mathbf{AA}^{-1} = \mathbf{A}^{-1}\mathbf{A} = \mathbf{I}$. For an illustration, see Example 9.2.

Example 9.2 Matrix Inverse

If **A** is a 3 * 3 matrix $\begin{pmatrix}1&4&3\\2&5&4\\7&6&7\end{pmatrix}$ and $\mathbf{B}=\begin{pmatrix}-5.5&5&-0.5\\-7&7&-1\\11.5&-11&1.5\end{pmatrix}$ then

$$\boldsymbol{A}*\boldsymbol{B}=\begin{pmatrix}1&0&0\\0&1&0\\0&0&1\end{pmatrix}=\boldsymbol{B}*\boldsymbol{A}$$

$\mathbf{A} * \mathbf{B} = \mathbf{I}$ the identity matrix and hence $\mathbf{B} = \mathbf{A}^{-1}$, the inverse of **A**.

9.2 DETERMINANTS

The number $(ae - bd)$ that was used above to scale the whole multiplication to make the diagonal numbers all equal to 1 is a special number for the 2 * 2 matrix **A** that is called its *determinant*. This is usually written with straight-line brackets so that:

$$\text{The determinant of } \boldsymbol{A}=|\boldsymbol{A}|=\begin{vmatrix}a&b\\d&e\end{vmatrix}=ae-bd.$$

The notation should not be confused with the idea of a *modulus*, which also uses straight-line brackets. The modulus (or "mod") of a value is a positive real number that is the absolute value of a quantity, that is, regardless of whether mathematically it is positive or negative. Thus, mod (–3) = |–3| = +3.

A determinant is always a square matrix with *n* rows and *n* columns, *n* being an integer such as 1, 2, 3, 4, and so on. If it happens in a 2 * 2 matrix that $ae = bd$, then 1 / |**A**| = 1 / 0 which is infinite, hence the matrix **A** has no inverse. The matrix is then said to be *singular.*

For a 3 * 3 matrix the process is slightly more complicated. The calculation is broken down into three stages where

$$\begin{vmatrix} a & b & c \\ d & e & f \\ g & h & i \end{vmatrix} = \text{three parts} \begin{vmatrix} a & . & . \\ . & e & f \\ . & h & i \end{vmatrix} \text{and} \begin{vmatrix} . & b & . \\ d & . & f \\ g & . & i \end{vmatrix} \text{and} \begin{vmatrix} . & . & c \\ d & e & . \\ g & h & . \end{vmatrix}$$

Each of these subcomponents contains a 2 * 2 determinant that is known as the *minor* of the element in the first row. The value of the 3 * 3 determinant (Box 9.3 and Example 9.3) is then defined as:

$$\mathbf{A}| = a(ei - fh) - b(di - fg) + c(dh - eg)$$

Note the alternating signs, plus *a*, minus *b*, plus *c*, and so forth.

BOX 9.3 THE DETERMINANT OF A MATRIX

If **A** is a 3 * 3 square matrix such that

$$\mathbf{A} = \begin{vmatrix} a & b & c \\ d & e & f \\ g & h & i \end{vmatrix} \text{ then } |\mathbf{A}| = a * \begin{vmatrix} e & f \\ h & i \end{vmatrix} - b * \begin{vmatrix} d & f \\ g & i \end{vmatrix} + c * \begin{vmatrix} d & e \\ g & h \end{vmatrix}$$

Thus, the determinant of **A** = |**A**| = $a(ei - fh) - b(di - fg) + c(dh - eg)$.

If |**A**| = 0 the determinant is singular, for a 4 * 4 square matrix the determinant becomes

$$\begin{vmatrix} a & b & c & p \\ d & e & f & q \\ g & h & i & r \\ j & k & l & s \end{vmatrix} = a * \begin{vmatrix} e & f & q \\ h & i & r \\ k & l & s \end{vmatrix} - b * \begin{vmatrix} d & f & q \\ g & i & r \\ j & l & s \end{vmatrix}$$

$$+ c * \begin{vmatrix} d & e & q \\ g & h & r \\ j & k & s \end{vmatrix} - p * \begin{vmatrix} d & e & f \\ g & h & i \\ j & k & l \end{vmatrix}$$

where each of the 3 * 3 minor determinants has to be evaluated as above.

Note the overall symmetry of the operation and the alternating signs + – + – + –, and so forth.

If for instance two (or more) rows (or columns) differ by a common factor, then |**A**| will be zero. For instance, if in the example in Box 9.3, row 3 is a multiple of row 2 so that instead of [*g h i*] we have [*k* * *d k* * *e k* * *f*] where *k* is a constant, then we have

$$|\mathbf{A}| = a(e * k * f - f * k * e) - b(d * k * i - f * k * d) + c(d * k * e - e * k * d) = 0$$

The related matrix will therefore be singular.

If we change the rows in the matrix **A** into columns to form a new matrix, we call this the *transpose* of **A** and label it $\mathbf{A}^T$. In the case of a square matrix, since the calculations of the value of a determinant are symmetrical between rows and columns, the determinant of the transpose has the same numerical value as that of the original so $|\mathbf{A}| = |\mathbf{A}^T|$.

Also, if **A** has *n* rows and *n* columns, then if we multiply all the terms of **A** by a scalar quantity *k*, then $|k\mathbf{A}| = k^n|\mathbf{A}|$. Thus, if the elements of **A** are doubled in a 3 * 3 square matrix to give a new matrix **B** = 2**A**, then the value of $|\mathbf{B}| = 2^3|\mathbf{A}| = 8|\mathbf{A}|$—see Example 9.4.

Note that if *k* = −1 then $|-\mathbf{A}| = (-1)^n|\mathbf{A}|$ and not as one might expect −|**A**| which will only occur if *n* is an odd number.

Example 9.3 A 3 * 3 Determinant

If the matrix **A** is $\begin{pmatrix} 1 & 4 & 3 \\ 2 & 5 & 4 \\ 7 & 6 & 7 \end{pmatrix}$ as in Example 9.2 then

$$|\mathbf{A}| = 1*(5*7 - 4*6) - 4*(2*7 - 4*7) + 3*(2*6 - 5*7) = 11 + 56 - 69 = -2$$

If $\mathbf{B} = \begin{pmatrix} -5.5 & 5 & -0.5 \\ -7 & 7 & -1 \\ 11.5 & -11 & 1.5 \end{pmatrix}$ again as in Example 9.2 then

$$\begin{aligned}|\mathbf{B}| &= -5.5*\{7*1.5 - 1*11\} - 5*\{-7*1.5 - (-1)*11.5\} + (-0.5)* \\ &\quad \{(-7)*(-11) - (7*11.5)\} \\ &= 2.75 - 5 + 1.75 = -0.5\end{aligned}$$

Note: If the determinant of **A** * **B** = |**I**| = 1 then |**A**| * |**B**| = |**B**| * |**A**| = |**A** * **B**| = |**B** * **A**|. This applies generally for determinants so long as **A** and **B** are both conformable, even though in general **A** * **B** ≠ **B** * **A**.

If instead of multiplying all of the elements of **A** by *k*, we multiply only one row or one column by *k′* then the determinant of this matrix will only be *k′* times as big and hence have a value = *k′* |**A**|. If we interchange two rows (or two columns) then the determinant will be the same size but the opposite sign. For instance, if in |**B**| in Example 9.4, we interchange columns 2 and 3 to give a new matrix **C**, then

$$\begin{aligned}|\mathbf{C}| = \begin{vmatrix} 2 & 6 & 4 \\ 8 & 10 & 12 \\ 18 & 16 & 14 \end{vmatrix} &= 2 * \begin{vmatrix} 10 & 12 \\ 16 & 14 \end{vmatrix} - 6 * \begin{vmatrix} 8 & 12 \\ 18 & 14 \end{vmatrix} + 4 * \begin{vmatrix} 8 & 10 \\ 18 & 16 \end{vmatrix} \\ &= 2 * (140 - 192) - 6 * (112 - 216) + 4 * (128 - 180) \\ &= 104 + 624 - 208 = +312 = 8 * 39\end{aligned}$$

This is the same magnitude as |**B**| in Example 9.4.

Furthermore, consider two square matrices

$$\mathbf{A} = \begin{pmatrix} a_{11} & a_{12} & a_{13} \\ a_{21} & a_{22} & a_{23} \\ a_{31} & a_{32} & a_{33} \end{pmatrix} \text{ and } \mathbf{B} = \begin{pmatrix} b_{11} & b_{12} & b_{13} \\ b_{21} & b_{22} & b_{23} \\ b_{31} & b_{32} & b_{33} \end{pmatrix}$$

Example 9.4 Increasing the Elements in a Determinant by a Factor

Let

$$|\mathbf{A}| = \begin{vmatrix} 1 & 2 & 3 \\ 4 & 6 & 5 \\ 9 & 7 & 8 \end{vmatrix} = 1 * \begin{vmatrix} 6 & 5 \\ 7 & 8 \end{vmatrix} - 2 * \begin{vmatrix} 4 & 5 \\ 9 & 8 \end{vmatrix} + 3 * \begin{vmatrix} 4 & 6 \\ 9 & 7 \end{vmatrix}$$

$$|\mathbf{A}| = 1 * (48-35) - 2 * (32-45) + 3 * (28-54) = 13 + 26 - 78 = -39 = -1 * 39$$

Let the elements in **A** be doubled to give

$$|\mathbf{B}| = \begin{vmatrix} 2 & 4 & 6 \\ 8 & 12 & 10 \\ 18 & 14 & 16 \end{vmatrix} = 2 * \begin{vmatrix} 12 & 10 \\ 14 & 16 \end{vmatrix} - 4 * \begin{vmatrix} 8 & 10 \\ 18 & 16 \end{vmatrix} + 6 * \begin{vmatrix} 8 & 12 \\ 18 & 14 \end{vmatrix}$$

$$|\mathbf{B}| = 2 * (192-140) - 4 * (128-180) + 6 * (112-160) = 104 + 208 - 624 = -312$$
$$= -8 * 39$$

A and **B** are of order 3 * 3, hence, by doubling each of the elements in **A** to create **B**, the determinant of **B** = 2^3 = 8 times the determinant of **A**.

The product $\boldsymbol{A} * \boldsymbol{B} =$

$$\begin{pmatrix} a_{11}b_{11} + a_{12}b_{21} + a_{13}b_{31} & a_{11}b_{12} + a_{12}b_{22} + a_{13}b_{32} & a_{11}b_{13} + a_{12}b_{23} + a_{13}b_{33} \\ a_{21}b_{11} + a_{22}b_{21} + a_{23}b_{31} & a_{21}b_{12} + a_{22}b_{22} + a_{23}b_{32} & a_{21}b_{13} + a_{22}b_{23} + a_{23}b_{33} \\ a_{31}b_{11} + a_{32}b_{21} + a_{33}b_{31} & a_{31}b_{12} + a_{32}b_{22} + a_{33}b_{32} & a_{31}b_{13} + a_{32}b_{23} + a_{33}b_{33} \end{pmatrix}$$

The product $\boldsymbol{B} * \boldsymbol{A} =$

$$\begin{pmatrix} b_{11}a_{11} + b_{12}a_{21} + b_{13}a_{31} & b_{11}a_{12} + b_{12}a_{22} + b_{13}a_{32} & b_{11}a_{13} + b_{12}a_{23} + b_{13}a_{33} \\ b_{21}a_{11} + b_{22}a_{21} + b_{23}a_{31} & b_{21}a_{12} + b_{22}a_{22} + b_{23}a_{32} & b_{21}a_{13} + b_{22}a_{23} + b_{23}a_{33} \\ b_{31}a_{11} + b_{32}a_{21} + b_{33}a_{31} & b_{31}a_{12} + b_{32}a_{22} + b_{33}a_{32} & b_{31}a_{13} + b_{32}a_{23} + b_{33}a_{33} \end{pmatrix}$$

Or, on rearranging terms,

$$= \begin{pmatrix} a_{11}b_{11} + a_{21}b_{12} + a_{31}b_{13} & a_{12}b_{11} + a_{22}b_{12} + a_{32}b_{13} & a_{13}b_{11} + a_{12}b_{23} + a_{33}b_{13} \\ a_{11}b_{21} + a_{21}b_{22} + a_{31}b_{23} & a_{12}b_{21} + a_{22}b_{22} + a_{32}b_{23} & a_{13}b_{21} + a_{23}b_{22} + a_{33}b_{23} \\ a_{11}b_{31} + a_{21}b_{32} + a_{31}b_{33} & a_{12}b_{31} + a_{22}b_{32} + a_{32}b_{33} & a_{13}b_{31} + a_{23}b_{32} + a_{33}b_{33} \end{pmatrix}$$

It can be seen that the matrix product of **A** * **B** is not necessarily the same as **B** * **A**. However, since the determinant

$$|\mathbf{A}| = a_{11}(a_{22}a_{33} - a_{23}a_{32}) - a_{12}(a_{21}a_{33} - a_{23}a_{31}) + a_{13}(a_{21}a_{32} - a_{22}a_{31})$$

and

$$|\mathbf{B}| = b_{11}(b_{22}b_{33} - b_{23}b_{32}) - b_{12}(b_{21}b_{33} - b_{23}b_{31}) + b_{13}(b_{21}b_{32} - b_{22}b_{31})$$

then with a little manipulation, it can be shown that

$$|\mathbf{A} * \mathbf{B}| = |\mathbf{B} * \mathbf{A}| = |\mathbf{A}| * |\mathbf{B}| = |\mathbf{B}| * |\mathbf{A}|$$

9.3 RELATED MATRICES

The idea of changing rows into columns in a matrix **A** to form its transpose $\mathbf{A}^T$ is not restricted to square matrices. Thus, if

$$\mathbf{A} = \begin{pmatrix} a & b & c & d \\ e & f & g & h \end{pmatrix}, \text{ then } \mathbf{A}^T = \begin{pmatrix} a & e \\ b & f \\ c & g \\ d & h \end{pmatrix}$$

If **A** has *m* rows and *n* columns, then $\mathbf{A}^T$ will have *n* rows and *m* columns and we can always multiply **A** * $\mathbf{A}^T$ and also $\mathbf{A}^T$ * **A** although the results may differ (Box 9.4).

In the particular case where **A** = (*x y z*), then $\mathbf{A}^T = \begin{pmatrix} x \\ y \\ z \end{pmatrix}$ and

$$\mathbf{A} * \mathbf{A}^T = (x^2 + y^2 + z^2) \text{ which is } a\ 1 * 1 \text{ matrix.}$$

On the other hand, if we multiply $A^T * A = \begin{pmatrix} x \\ y \\ z \end{pmatrix}(x\ y\ z)$, we obtain

$$\mathbf{A}^T * \mathbf{A} = \begin{pmatrix} xx & xy & xz \\ yx & yy & yz \\ zx & zy & zz \end{pmatrix} \text{ which is a } 3 * 3 \text{ matrix.}$$

BOX 9.4 MATRIX MULTIPLICATION (2)

If $\mathbf{A} = \begin{pmatrix} a & b \\ c & d \end{pmatrix}$ and $\mathbf{B} = \begin{pmatrix} e & f \\ g & h \end{pmatrix}$ then

$$\mathbf{A} * \mathbf{B} = \begin{pmatrix} ae+bg & af+bh \\ ce+dg & cf+dh \end{pmatrix} \text{ and } \mathbf{B} * \mathbf{A} = \begin{pmatrix} ae+cf & be+df \\ ag+ch & bg+dh \end{pmatrix}$$

$$\mathbf{A}^{\mathrm{T}}\begin{pmatrix} a & c \\ b & d \end{pmatrix} \text{ hence } \mathbf{A} * \mathbf{A}^{\mathrm{T}} = \begin{pmatrix} a^2 + b^2 ac + bd \\ ac + bdc^2 + d^2 \end{pmatrix} \text{ while } \mathbf{A}^{\mathrm{T}} * \mathbf{A} = \begin{pmatrix} a^2 + c^2 ab + cd \\ ab + cdb^2 + d^2 \end{pmatrix}$$

In general (though not in all cases),

$$\mathbf{A} * \mathbf{B} \neq \mathbf{B} * \mathbf{A} \text{ and } \mathbf{A} * \mathbf{A}^{\mathrm{T}} \neq \mathbf{A}^{\mathrm{T}} * \mathbf{A}$$

The order in which matrices are multiplied together is of crucial importance.

Earlier, we introduced the idea of labeling every element according to its row and column. Rather than saying $\mathbf{A} = (a\ b\ c\ d)$, we can say that $\mathbf{A} = (a_{11}\ a_{12}\ a_{13}\ a_{14})$.

For a 4 * 4 matrix,

$$\mathbf{A} = \begin{pmatrix} a_{11} & a_{12} & a_{13} & a_{14} \\ a_{21} & a_{22} & a_{23} & a_{24} \\ a_{31} & a_{32} & a_{33} & a_{34} \\ a_{41} & a_{42} & a_{43} & a_{44} \end{pmatrix}$$

In general, a_{ij} is the element in the ith row and jth column. In the transpose $\mathbf{A}^{\mathrm{T}}$ it becomes the jth row of the ith column or a_{ji}.

We can think of a square matrix **A** as having elements occupying the squares as in a chessboard (shown in Figure 9.1 as 8 * 8 but we will assume that it is $n * n$ where n is a positive integer). The board would contain an element such as a_{ij} for the number in the ith row and jth column. If we strip out the ith row and jth column as in Figure 9.1b, then we are left with $(n - 1) * (n - 1)$ matrix that is called the *submatrix* of a_{ij}. The determinant of this submatrix of a_{ij} is known as the *minor* of a_{ij}.

If we give the minor of a_{ij} the name $\mathbf{M}_{ij}$ then, just as we saw in Box 9.3, when we consider the determinants of **A**, we need to alternate the signs (plus then minus then plus then minus then plus, etc.). We can think of black squares in the original as plus and white squares as minus or we can say that the sign is $(-1)^{(i-j)}$. We now define the *cofactor* of the element a_{ij} in **A** as the determinant of the matrix $(-1)^{(i-j)}\ \mathbf{M}_{ij}$ and write it as $\mathbf{A}_{ij}$. A cofactor is a determinant and reduces to a single number; it is the signed (i.e., plus or minus) minor of a_{ij}.

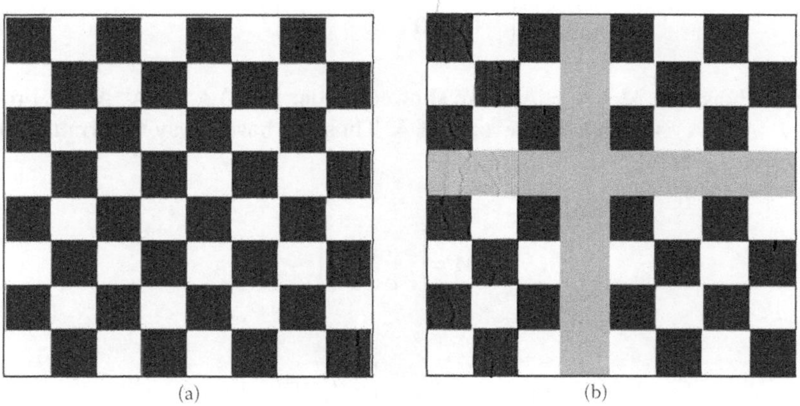

FIGURE 9.1 **(a, b)** Submatrices and minors.

The overall value of the determinant of the matrix **A** is obtained from any row and its related cofactors in the form

$$|\mathbf{A}| = a_{11}A_{11} + a_{12}A_{12} + a_{13}A_{13} + \cdots + a_{1n}A_{1n}$$

Alternatively, we can use row *i* so that

$$|\mathbf{A}| = a_{i1}A_{i1} + a_{i2}A_{i2} + a_{i3}A_{i3} + \cdots + a_{in}A_{in}$$

This is based on the calculation using the elements in row *i*.

Thus, an *n* * *n* matrix has a determinant that is the sum of *n* products between the row elements and their respective cofactors. Each cofactor can be evaluated as the sum of its minors, each minor being a (*n* − 1) * (*n* − 1) matrix, and so on down to the elements of a determinant that is 2 * 2. This is an extension of what we showed in Box 9.3.

We now define the *adjugate* or *adjoint* matrix **A** * of a square matrix **A** as a new matrix formed by cofactors of the elements of **A** in the transposed position. Thus, for a 4 * 4 matrix the terms are

$$\mathbf{A}^* = \begin{pmatrix} A_{11} & A_{21} & A_{31} & A_{41} \\ A_{12} & A_{22} & A_{32} & A_{42} \\ A_{13} & A_{23} & A_{33} & A_{43} \\ A_{14} & A_{24} & A_{34} & A_{44} \end{pmatrix}$$

If

$$\mathbf{A} = \begin{pmatrix} a_{11} & a_{12} & a_{13} & a_{14} \\ a_{21} & a_{22} & a_{23} & a_{24} \\ a_{31} & a_{32} & a_{33} & a_{34} \\ a_{41} & a_{42} & a_{43} & a_{44} \end{pmatrix}$$

then we can show that

$$\mathbf{AA}^* = \begin{pmatrix} |\mathbf{A}| & 0 & 0 & 0 \\ 0 & |\mathbf{A}| & 0 & 0 \\ 0 & 0 & |\mathbf{A}| & 0 \\ 0 & 0 & 0 & |\mathbf{A}| \end{pmatrix} = |\mathbf{A}|\,\mathbf{I}$$

We can also show that $\mathbf{A}^* * \mathbf{A} = |\mathbf{A}|\mathbf{I}$. We showed earlier that $\mathbf{AA}^{-1} = \mathbf{A}^{-1}\mathbf{A} = \mathbf{I}$. From this it follows that $\mathbf{A}^* \,/\, |\mathbf{A}| = \mathbf{A}^{-1}$ which is the inverse of **A**. Thus, we have a way to form the inverse of any matrix by computing its adjoint.

With a 2 * 2 matrix,

$$\mathbf{M} = \begin{bmatrix} a & b \\ c & d \end{bmatrix}$$

then

$$\mathbf{M}^{-1} = \frac{1}{(ad - bc)} \begin{bmatrix} d & -b \\ -c & a \end{bmatrix}$$

where the determinant $|\mathbf{M}| = (ad - bc)$.

If we have a 3 * 3 matrix

$$\mathbf{M} = \begin{pmatrix} abc \\ def \\ ghk \end{pmatrix}$$

then

$$|\mathbf{M}| = a(ek - fh) - b(dk - fg) + c(dh - eg)$$

$$\mathbf{M}^{-1} = \begin{pmatrix} a & b & c \\ d & e & f \\ g & h & k \end{pmatrix}^{-1} = \frac{1}{|\mathbf{M}|}\begin{pmatrix} A & B & C \\ D & E & F \\ G & H & K \end{pmatrix}^{T} = \frac{1}{|\mathbf{M}|}\begin{pmatrix} A & D & G \\ B & E & H \\ C & F & K \end{pmatrix}$$

where A is the cofactor of a in **M**. This means that:

$$\text{Determinant } |\mathbf{M}| = a(ek - fh) - b(dk - fg) + c(dh - eg)$$

$$\begin{aligned} &A = ek - fh; && B = -(dk - fg); && C = (dh - eg); \\ &D = -(bk - ch); && E = (ak - cg); && F = -(ah - bg); \\ &G = (bf - ec); && H = -(af - cd); && K = (ae - bd) \end{aligned}$$

Thus,

$$\mathbf{M}^{-1} = \frac{1}{a(ek - fh) - b(dk - fg) + c(dh - eg)}\begin{bmatrix} (ek - fh)(ch - bk)(bf - ec) \\ (fg - dk)(ak - cg)(cd - af) \\ (dh - eg)(bg - ah)(ae - bd) \end{bmatrix}$$

If we have a 4 * 4 matrix $\mathbf{M} = \begin{bmatrix} a & b & c & d \\ e & f & g & h \\ j & k & l & m \\ n & p & q & r \end{bmatrix}$, then its determinant |**M**| =

$$= a\begin{vmatrix} f & g & h \\ k & l & m \\ p & q & r \end{vmatrix} - b\begin{vmatrix} e & g & h \\ j & l & m \\ n & q & r \end{vmatrix} + c\begin{vmatrix} e & f & h \\ j & k & m \\ n & p & r \end{vmatrix} - d\begin{vmatrix} e & f & g \\ j & k & l \\ n & p & q \end{vmatrix}$$

or

$$= -e\begin{vmatrix} b & c & d \\ k & l & m \\ p & q & r \end{vmatrix} + f\begin{vmatrix} a & c & d \\ j & l & m \\ n & q & r \end{vmatrix} - g\begin{vmatrix} a & b & d \\ j & k & m \\ n & p & r \end{vmatrix} + h\begin{vmatrix} a & b & c \\ j & k & l \\ n & p & q \end{vmatrix}$$

or

$$= j\begin{vmatrix} b & c & d \\ f & g & h \\ p & q & r \end{vmatrix} - k\begin{vmatrix} a & c & d \\ e & g & h \\ n & q & r \end{vmatrix} + l\begin{vmatrix} a & b & d \\ e & f & h \\ n & p & r \end{vmatrix} - m\begin{vmatrix} a & b & c \\ e & f & g \\ n & p & q \end{vmatrix}$$

or

$$= -n\begin{vmatrix} b & c & d \\ f & g & h \\ k & l & m \end{vmatrix} + p\begin{vmatrix} a & c & d \\ e & g & h \\ j & l & m \end{vmatrix} - q\begin{vmatrix} a & b & d \\ e & f & h \\ j & k & m \end{vmatrix} + r\begin{vmatrix} a & b & c \\ e & f & g \\ j & k & l \end{vmatrix}$$

On expanding, this gives the determinant

$$|M| = a\{f(lr - mq) - g(kr - mp) + h(kq - lp)\} - b\{e(lr - mq) - g(jr - mn) + h(jp - kn)\}$$
$$+ c\{e(kr - pm) - f(jr - mn) + h(jp - kn)\} - d\{e(kq - lp) - f(jq - ln) + g(jp - kn)\}$$

or

$$= -e\{b(lr - mq) - c(kr - mp) + d(kq - lp)\} + f\{a(lr - mq) - c(jr - mn) + d(jq - ln)\}$$
$$- g\{a(kr - mp) - b(jr - mn) + d(jp - kn)\} + h\{a(kq - lp) - b(jq - ln) + c(jp - kn)\}$$

or

$$= j\{b(gr - hq) - c(fr - hp) + d(fq - gp)\} - k\{a(gr - hq) - c(er - hn) + d(eq - gn)\}$$
$$+ l\{a(fr - hp) - b(er - hn) + d(ep - fn)\} - m\{a(fq - gp) - b(eq - gn) + c(ep - fn)\}$$

or

$$= -n\{b(gm - hl) - c(fm - hk) + d(fl - gk)\} + p\{a(gm - hl) - c(em - hj) + d(el - gj)\}$$
$$- q\{a(fm - hk) - b(em - hj) + d(ek - fj)\} + r\{a(fl - gk) - b(el - gk) + c(ek - fj)\}$$

All of these have the same value.

If the cofactors of **M** are $\begin{bmatrix} A & B & C & D \\ E & F & G & H \\ J & K & L & M \\ N & P & Q & R \end{bmatrix}$, then the adjugate of $\mathbf{M} = \mathbf{M}^* = \begin{bmatrix} A & E & J & N \\ B & F & K & P \\ C & G & L & Q \\ D & H & M & R \end{bmatrix}$

or

$$\begin{bmatrix} \{f(lr - mq) - g(kr - mp) + h(kq - lp)\} & \{-b(lr - mq) + c(kr - mp) - d(kq - lp)\} \\ \{-e(lr - mq) + g(jr - mn) - h(jp - kn)\} & \{a(lr - mq) - c(jr - mn) + d(jq - ln)\} \\ \{e(kr - pm) - f(jr - mn) + h(jp - kn)\} & \{-a(kr - mp) + b(jr - mn) - d(jp - kn)\} \\ \{-e(kq - lp) + f(jq - ln) - g(jp - kn)\} & \{a(kq - lp) - b(jq - ln) + c(jp - kn)\} \end{bmatrix}$$

$$\begin{bmatrix} \{b(gr - hq) - c(fr - hp) + d(fq - gp)\} & \{-b(gm - hl) + c(fm - hk) - d(fl - gk)\} \\ \{-a(gr - hq) + c(er - hn) - d(eq - gn)\} & \{a(gm - hl) - c(em - hj) + d(el - gj)\} \\ \{a(fr - hp) - b(er - hn) + d(ep - fn)\} & \{-a(fm - hk) + b(em - hj) - d(ek - fj)\} \\ \{-a(fq - gp) + b(je - gn) - c(ep - fn)\} & \{a(fl - gk) - b(el - gk) + c(ek - fj)\} \end{bmatrix}$$

The inverse of $\mathbf{M} = \mathbf{M}^{-1} = (1 / |\mathbf{M}|) * \mathbf{M}^*$. An illustration is given in Example 9.5 and the results are summarized in Box 9.5.

With a matrix larger than 4 * 4 it becomes extremely messy when we express this in separate computational terms. The process can, however, be treated in blocks, for instance, **A**, **B**, **C**, and **D** within a given matrix. Provided that the matrices **A** and **D** are square and that both **A** and $(\mathbf{B} - \mathbf{CA}^{-1}\mathbf{B})$ are not singular, it can then be shown that

$$\begin{bmatrix} A & B \\ C & D \end{bmatrix}^{-1} = \begin{bmatrix} A^{-1} + A^{-1}B\left(D - CA^{-1}B\right)^{-1}CA^{-1} & -A^{-1}B\left(D - CA^{-1}B\right)^{-1} \\ -\left(D - CA^{-1}B\right)^{-1}CA^{-1} & \left(D - CA^{-1}B\right)^{-1} \end{bmatrix}$$

Example 9.5 Inverting A 4 * 4 Matrix

$$\text{If } \mathbf{A} = \begin{bmatrix} 3 & 1 & 4 & 2 \\ 6 & 2 & 1 & 3 \\ -1 & 2 & 4 & 2 \\ 4 & 1 & 3 & 7 \end{bmatrix} \text{ then its adjoint } \mathbf{A}^* = \begin{bmatrix} 27 & 27 & -36 & -9 \\ -60 & 75 & 98 & -43 \\ 75 & -33 & -1 & -7 \\ -39 & -12 & 7 & 49 \end{bmatrix}$$

$$\mathbf{AA}^* = \begin{bmatrix} 243 & 0 & 0 & 0 \\ 0 & 243 & 0 & 0 \\ 0 & 0 & 243 & 0 \\ 0 & 0 & 0 & 243 \end{bmatrix} \text{ showing the determinant of } A = |A| = 243.$$

This reduces the operation to a more manageable form. We will, however, not prove this here, although we will discuss partitioning matrices in Section 9.6.

9.4 APPLYING MATRICES

Let us now consider some simple applications of matrices, recalling that matrices are a mathematical form of shorthand and only really come into their own when dealing with large sets of numbers. Matrices are made up of numbers such as a_{ij} and a_{kl} that will normally be subject to the four simple operations of addition, subtraction, multiplication, and division. The answers to their manipulation are then placed in the appropriate location in the new resulting matrix.

BOX 9.5 INVERSES AND TRANSPOSES OF SQUARE MATRICES

$$\text{If } \mathbf{C} = \mathbf{AB}, \text{ then } \mathbf{C}^T = \mathbf{B}^T\mathbf{A}^T \text{ and } \mathbf{C}^{-1} = \mathbf{B}^{-1}\mathbf{A}^{-1}$$

$$\text{Similarly, if } \mathbf{D} = \mathbf{ABC}, \text{ then } \mathbf{D}^T = \mathbf{C}^T\mathbf{B}^T\mathbf{A}^T \text{ and } \mathbf{D}^{-1} = \mathbf{C}^{-1}\mathbf{B}^{-1}\mathbf{A}^{-1}$$

where $\mathbf{A}^{-1}$, and so on, is the inverse of the square matrix **A**, having the property that $\mathbf{AA}^{-1} = \mathbf{A}^{-1}\mathbf{A} = \mathbf{I}$.

More generally $(\mathbf{A}^{-1})^{-1} = \mathbf{A}$; $(\mathbf{A}^T)^{-1} = (\mathbf{A}^{-1})^T$; $(k\mathbf{A})^{-1} = k^{-1}\mathbf{A}^{-1}$ where k is a scalar, $\mathbf{A}^*$ is the adjugate or adjoint of the square matrix **A**, made up of the cofactors of **A** in the transposed positions. $\mathbf{A}^*\mathbf{A} = |\mathbf{A}|\mathbf{I}$ or $\mathbf{A}^* / |\mathbf{A}| = \mathbf{A}^{-1}$

The processes of manipulating matrices are ideally suited to handling by computer as the operations are sequential and routine. Consider a very simple example that involves the intersection of two straight lines illustrated in Box 9.6. The calculation involves the conversion of a square matrix into its inverse (**A** into $\mathbf{A}^{-1}$) and the multiplication of two matrices ($\mathbf{A}^{-1} * \mathbf{B}$).

BOX 9.6 THE INTERSECTION OF TWO LINES

In Chapter 4, we calculated the intersection of two straight lines.

Let us express these in the form

$$a_{11}x + a_{12}y = b_{11}$$
$$a_{21}x + a_{22}y = b_{21}$$

or

$$\begin{pmatrix} a_{11} & a_{12} \\ a_{21} & a_{22} \end{pmatrix}\begin{pmatrix} x \\ y \end{pmatrix} = \begin{pmatrix} b_{11} \\ b_{21} \end{pmatrix}$$

or

$$\mathbf{AX} = \mathbf{B}$$

Multiply both sides of the expression by $\mathbf{A}^{-1}$ to give $\mathbf{A}^{-1}\mathbf{AX} = \mathbf{A}^{-1}\mathbf{B}$.

Since $\mathbf{A}^{-1}\mathbf{A} = \mathbf{I}$, we have

$$\mathbf{X} = \mathbf{A}^{-1}\mathbf{B}$$

From previously at the end of Section 9.1,

$$\mathbf{A}^{-1} = \{1 / a_{11}a_{22} - a_{12}a_{21}\} * \begin{pmatrix} a_{22} & -a_{12} \\ -a_{21} & a_{11} \end{pmatrix}$$

Hence,

$$\mathbf{X} = \begin{pmatrix} x \\ y \end{pmatrix} = \{1 / (a_{11}a_{22} - a_{12}a_{21})\} * \begin{pmatrix} a_{22} & -a_{12} \\ -a_{21} & a_{11} \end{pmatrix}\begin{pmatrix} b_{11} \\ b_{21} \end{pmatrix}$$

or

$$\begin{pmatrix} x \\ y \end{pmatrix} = \{1 / (a_{11}a_{22} - a_{12}a_{21})\} * \begin{pmatrix} a_{22}b_{11} - a_{12}b_{21} \\ -a_{21}b_{11} + a_{11}b_{21} \end{pmatrix}$$

or

$$x = (a_{22}b_{11} - a_{12}b_{21}) / (a_{11}a_{22} - a_{12}a_{21})$$
$$y = (a_{11}b_{21} - a_{21}b_{11}) / (a_{11}a_{22} - a_{12}a_{21})$$

This is the point where the two lines intersect. For a worked example see Example 9.6.

Example 9.6 Intersecting Lines

Consider the lines $3x + 4y = 10$ and $5x - 7y = 3$. We can express these as

$$\mathbf{AX} = \mathbf{B}$$

where

$$\mathbf{A} = \begin{pmatrix} 3 & 4 \\ 5 & -7 \end{pmatrix}, \mathbf{X} = \begin{pmatrix} x \\ y \end{pmatrix}, \text{ and } \mathbf{B} = \begin{pmatrix} 10 \\ 3 \end{pmatrix}$$

Using the equations in Box 9.6,

$$\mathbf{A}^{-1} = \{1/(-21-20)\}\begin{pmatrix} -7 & -4 \\ -5 & 3 \end{pmatrix} \text{ hence,}$$

$$\mathbf{A}^{-1}\mathbf{B} = \{-1/41\}\begin{pmatrix} -7 & -4 \\ -5 & 3 \end{pmatrix}\begin{pmatrix} 10 \\ 3 \end{pmatrix} = \{-1/41\}\begin{pmatrix} -82 \\ -41 \end{pmatrix} = \begin{pmatrix} 2 \\ 1 \end{pmatrix} = \begin{pmatrix} x \\ y \end{pmatrix}$$

Thus, $\begin{pmatrix} x \\ y \end{pmatrix} = \begin{pmatrix} 2 \\ 1 \end{pmatrix}$. The lines intersect where $x = 2$ and $y = 1$.

In the calculations in Box 9.6 and Example 9.6, if the determinant $|\mathbf{A}|$ = zero (which happens when $a_{11}a_{22} - a_{12}a_{21} = 0$), there is no solution. This will happen when the two lines are parallel and therefore do not intersect in finite space.

9.5 ROTATIONS AND TRANSLATIONS

Matrix algebra is very useful when dealing with coordinate systems and changes in the position and orientation of the basic axes of reference. In two dimensions, let the origin be changed by an amount (a, b) from origin 1 to origin 2 (see Figure 9.2).

If the coordinates of any point P are $(x_1\ y_1)$ when referred to the old origin 1, then relative to the new origin 2, they are $(x_1 - a, y_1 - b)$. For a series of points:

$$\mathbf{X_n} = \mathbf{X_o} - \mathbf{A}$$

where n means new, o means old, and the matrix

$$\mathbf{X} = \begin{pmatrix} x_i \\ y_i \end{pmatrix} \text{ and } \mathbf{A} = \begin{pmatrix} a \\ b \end{pmatrix}.$$

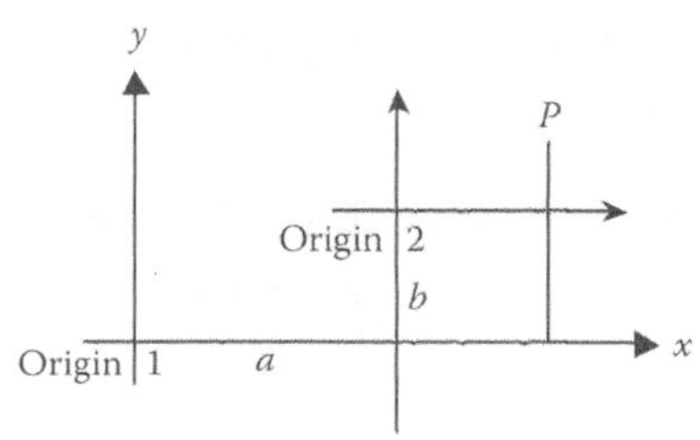

FIGURE 9.2 Shift or translation of origin.

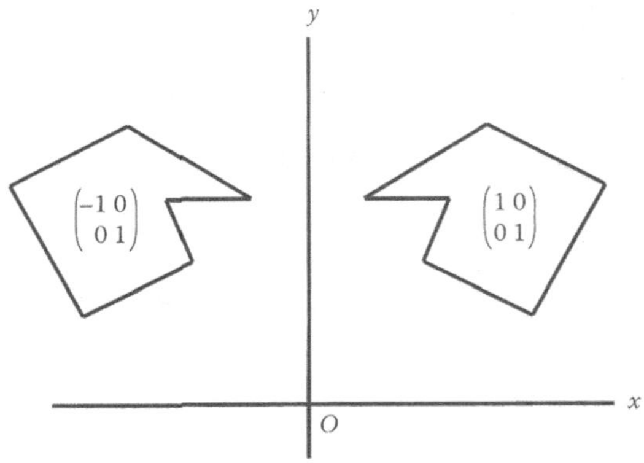

FIGURE 9.3 Reflection in the y-axis.

This relationship $\mathbf{X}_n = \mathbf{X}_o - \mathbf{A}$ will apply in any number of dimensions, especially three. It is often referred to as *translation*, meaning moving the origin from one point to another.

If we wish to obtain the reflection of an object in the y-axis (Figure 9.3), that is obtain the mirror image, then by applying the matrix

$$\begin{pmatrix} -1 & 0 \\ 0 & 1 \end{pmatrix} \text{to} \begin{pmatrix} x \\ y \end{pmatrix}$$

we obtain a new x value $x_n = -x$ while y_n remains $= y$.

Similarly, by applying $\begin{pmatrix} 1 & 0 \\ 0 & -1 \end{pmatrix}$ we obtain a reflection in the x-axis.

If we want to change the scale of our projection by a factor s then $\mathbf{X}_n = s\,\mathbf{X}$, that is, each of the values of x and y will be multiplied by s. If we want to change the scale in the x-direction by a different amount than in the y-direction, then we need to premultiply by $\mathbf{S} = \begin{pmatrix} s_1 & 0 \\ 0 & s_2 \end{pmatrix}$ so that $\mathbf{X}_n = \mathbf{SX}$.

Thus, if we wish to stretch the distances by a factor 2 in the x-direction and 3 in the y-direction we would apply the matrix $\begin{pmatrix} 2 & 0 \\ 0 & 3 \end{pmatrix}$. This will result in $x_n = 2x_o$ and $y_n = 3y_o$.

If we then change the origin, we obtain $\mathbf{X}_n = \mathbf{SX} - \mathbf{A}$. If we want to change the origin before we change the scale then $\mathbf{X}_n = \mathbf{S}(\mathbf{X} - \mathbf{A})$, which is not the same. Once again, the order in which we carry out operations makes a critical difference.

If we want to rotate the axes about their origin through an angle θ, then matrix algebra provides a convenient way to describe this (Figure 9.4).

Consider a point P and a set of axes OX_{old}, OY_{old}. The coordinates of P are given by $x_o = OA$ and $y_o = AP$. If the axes are rotated clockwise through an angle θ to OX_{new}, OY_{new}, then the new coordinates will be $x_n = OB$ and $y_n = BP$. Let angle $POA = \phi$ so that $PA = y_o = OP \sin\phi$ and $OA = x_o = OP\cos\phi$.

$$\begin{aligned} BP = y_n &= OP\sin(\theta + \varnothing) = OP(\sin\theta\cos\varnothing + \cos\theta\sin\varnothing) \\ &= OP\sin\theta\cos\varnothing + OP\cos\theta\sin\varnothing \end{aligned}$$

Hence,

$$y_n = y_o\cos\theta + x_o\sin\theta$$

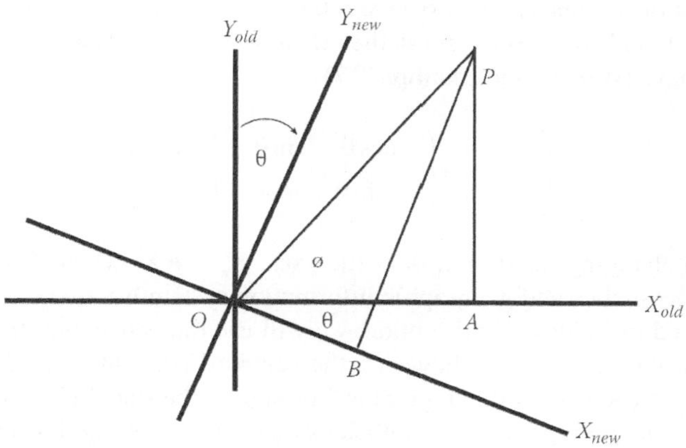

FIGURE 9.4 Rotation of axes.

Also,

$$OB = x_n = OP\ \cos(\theta + \varnothing) = OP\ (\cos\theta\ \cos\ \varnothing - \sin\theta \sin\varnothing)$$

Hence,

$$x_n = x_o \cos\theta - y_o \sin\theta$$

Put in matrix form

$$\begin{pmatrix} x_n \\ y_n \end{pmatrix} = \begin{pmatrix} \cos\theta & -\sin\theta \\ \sin\theta & \cos\theta \end{pmatrix} \begin{pmatrix} x_o \\ y_o \end{pmatrix} \text{ or } \mathbf{X_n} = \mathbf{RX_o}$$

where

$$\mathbf{R} = \begin{pmatrix} \cos\theta & -\sin\theta \\ \sin\theta & \cos\theta \end{pmatrix}$$

Note that the determinant of $\mathbf{R} = |\mathbf{R}| = \cos^2\theta + \sin^2\theta = 1$. The inverse of $\mathbf{R}$ is

$$\mathbf{R}^{-1} = \begin{pmatrix} \cos\theta & \sin\theta \\ -\sin\theta & \cos\theta \end{pmatrix}$$

Hence,

$$\mathbf{RR}^{-1} = \begin{pmatrix} \cos^2\theta + \sin^2\theta & \sin\theta\cos\theta - \sin\theta\cos\theta \\ \sin\theta\cos\theta - \sin\theta\cos\theta & \sin^2\theta + \cos^2\theta \end{pmatrix} = \begin{pmatrix} 1 & 0 \\ 0 & 1 \end{pmatrix}$$

This is the identity matrix **I**. Any matrix that when multiplied by its inverse gives the identity matrix is called an *orthogonal matrix*.

The processes of rotating axes but retaining their rectangularity, moving the origin, and perhaps applying a uniform scale change is called a *similarity transformation*. It is often used in photogrammetry and in computer-generated models. We will consider an example in Chapter 11.

The above derivation assumes that a positive rotation is clockwise as we look down on the plan. Since cos (−θ) = cos θ and sin (−θ) = −sin θ, then if we were to regard an anticlockwise rotation as positive, we would have (Box 9.7 and Example 9.7)

$$\begin{pmatrix} x_n \\ y_n \end{pmatrix} = \begin{pmatrix} \cos\theta & \sin\theta \\ -\sin\theta & \cos\theta \end{pmatrix} \begin{pmatrix} x_o \\ y_o \end{pmatrix}$$

In general, when changing the direction of the axes, $\mathbf{X}_n = \mathbf{R}\mathbf{X}_o$ where **R** is called the *rotation matrix*. Because in geomatics and geographic information system we are often talking about bearings that are measured in a clockwise direction, we will use the convention that rotations are positive if clockwise. In photogrammetry, however, the standard convention uses the opposite, which inevitably leads to confusion. The crucial point is "the sign of the sine." The problem arises because what is clockwise when looking upward is anticlockwise when looking downward. An example of a photogrammetric calculation is given in Chapter 11.

BOX 9.7 ROTATION OF AXES

To transform old coordinates to new coordinates when rotating the x and y orthogonal axes through an angle θ measured as positive in a clockwise direction:

$$\begin{pmatrix} x_n \\ y_n \end{pmatrix} = \begin{pmatrix} \cos\theta & -\sin\theta \\ \sin\theta & \cos\theta \end{pmatrix} \begin{pmatrix} x_o \\ y_o \end{pmatrix}$$

(The same result will occur if we keep the axes fixed but rotate the object about the axes in an anticlockwise direction.)

R as defined above is an orthogonal matrix; but there are occasions when a rectangular system needs to be skewed. For example, in Figure 9.5, the old grid that is rectangular has been skewed to form a new grid for ease of representation on a flat piece of paper. The axes are no longer at right angles although the distances parallel to the corresponding x-axis and y-axis remain the same ($AP = AP'$).

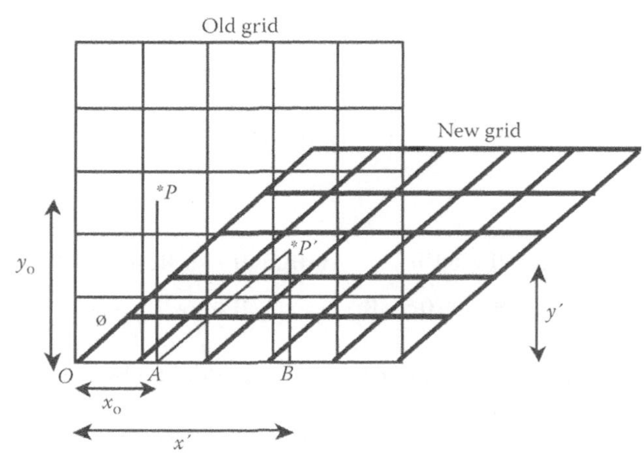

FIGURE 9.5 A skewed grid.

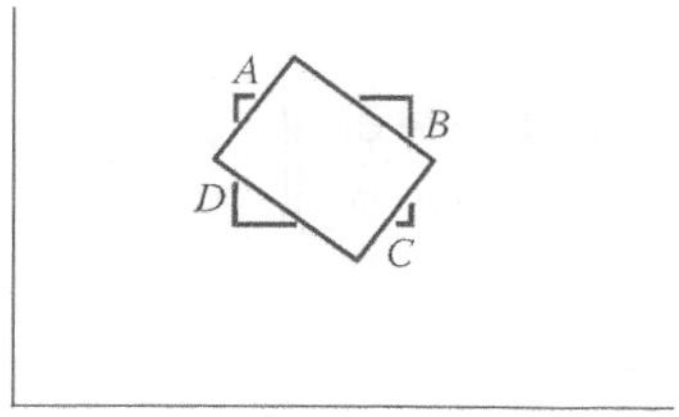

FIGURE 9.6 Rotating a rectangle.

Thus, in the old grid, the coordinates of P are (x_o, y_o) or (OA, AP), while the distances for point P' on the new grid are (OA, AP') unless there is any change in scale. However, relative to the old grid the coordinates of P' have become (x', y') or (OB, BP').

$$x' = OB = OA + AP'\sin\phi = x + y\sin\phi$$

Example 9.7 Rotating an Object in 2D

Consider the rectangle $ABCD$ in Figure 9.6 with coordinates A (4, 6), B (8, 6), C (8, 3), and D (4, 3). Its center is (6, 4.5).

If the coordinates (x_o, y_o) based on the origin are (x_i, y_i) with i = 1 to 4, and its center is $\mathbf{O} = \begin{pmatrix} 6 \\ 4.5 \end{pmatrix}$ then, based on the center of the rectangle, the new values (x_n, y_n) are

$$\mathbf{X_n} = \mathbf{X_o} - \mathbf{O} \text{ or } \begin{pmatrix} x_n \\ y_n \end{pmatrix} = \begin{pmatrix} x_o - 6 \\ y_o - 4.5 \end{pmatrix}$$

Let us rotate the rectangle by 40° clockwise about its center, where sin 40 = 0.6428 and cos 40 = 0.7660. Applying the rotation $\begin{pmatrix} 0.7660 & 0.6428 \\ -0.6428 & 0.7660 \end{pmatrix}$ we obtain:

$$\begin{pmatrix} 0.7660 & 0.6428 \\ -0.6428 & 0.7660 \end{pmatrix}\begin{pmatrix} x_n \\ y_n \end{pmatrix} = \begin{pmatrix} 0.7660 & 0.6428 \\ -0.6428 & 0.7660 \end{pmatrix}\begin{pmatrix} x_o - 6.0 \\ y_o - 4.5 \end{pmatrix}$$

$$= \begin{pmatrix} 0.7660x_o - 4.5960 + 0.6428y_o - 2.8926 \\ -0.6428x_o + 3.8568 + 0.7660y_o - 3.4470 \end{pmatrix}$$

$$= \begin{pmatrix} 0.7660x_o + 0.6428y_o - 7.4886 \\ -0.6428x_o + 0.7660y_o + 0.4098 \end{pmatrix}$$

Note that we are rotating the object, not the axes.

Next we must return to the original origin by adding 6 to the x value and 4.5 to the y. This gives

$$\begin{pmatrix} 0.7660x_o + 0.6428y_o - 1.4886 \\ -0.6428x_o + 0.766y_o + 4.9098 \end{pmatrix}$$

for the new coordinates after rotation. A (4, 6) becomes (5.4322, 6.9346); B (8, 6) becomes (8.4962, 4.3634); C (8, 3) becomes (6.5678, 2.0654); and D (4, 3) becomes (3.5038, 4.6366).

where ø is the clockwise angle by which the y-axis has been rotated.

$$y' = BP' = AP'\cos\phi = y\cos\phi$$

Thus,

$$\begin{pmatrix} x' \\ y' \end{pmatrix} = \begin{pmatrix} 1 & \sin\phi \\ 0 & \cos\phi \end{pmatrix} \begin{pmatrix} x \\ y \end{pmatrix} \text{ or } \mathbf{X}' = \mathbf{MX}$$

where

$$\mathbf{M} = \begin{pmatrix} 1 & \sin\phi \\ 0 & \cos\phi \end{pmatrix}$$

Here **M** is *not* an orthogonal transformation.

If we want to reduce the scale in the y-direction by a factor s_y so that y would become $s_y y$ while in the x-direction we applied a scale factor of s_x so that x becomes $s_x x$, then

$$\begin{pmatrix} x' \\ y' \end{pmatrix} = \begin{pmatrix} 1 & \sin\phi \\ 0 & \cos\phi \end{pmatrix} \begin{pmatrix} s_x & 0 \\ 0 & s_y \end{pmatrix} \begin{pmatrix} x \\ y \end{pmatrix} = \begin{pmatrix} s_x & s_y \sin\phi \\ 0 & s_y \cos\phi \end{pmatrix} \begin{pmatrix} x \\ y \end{pmatrix}$$

Note that

$$\begin{pmatrix} 1 & \sin\phi \\ 0 & \cos\phi \end{pmatrix} \begin{pmatrix} s_x & 0 \\ 0 & s_y \end{pmatrix} = \begin{pmatrix} s_x & s_y \sin\phi \\ 0 & s_y \cos\phi \end{pmatrix}$$

while

$$\begin{pmatrix} s_x & 0 \\ 0 & s_y \end{pmatrix} \begin{pmatrix} 1 & \sin\phi \\ 0 & \cos\phi \end{pmatrix} = \begin{pmatrix} s_x & s_x \sin\phi \\ 0 & s_y \cos\phi \end{pmatrix}$$

This is different. In the first case, we scale before we transform the coordinates while in the second, we apply the scale factor after the transformation.

This sequencing is especially important when considering the rotation of axes in three dimensions. The measurement of angles in a clockwise direction is sometimes referred to as the *left-hand rule* because, if your left thumb is pointing upward (in the Z-direction) and the fingers move from straight to bent, their pointing moves clockwise as you look down on your thumb. If the index finger remains straight it represents the *Y*-direction. See Figure 9.7. With the right-hand rule, *X* and *Y* are reversed. In each case, the fingers or thumb point in the positive direction.

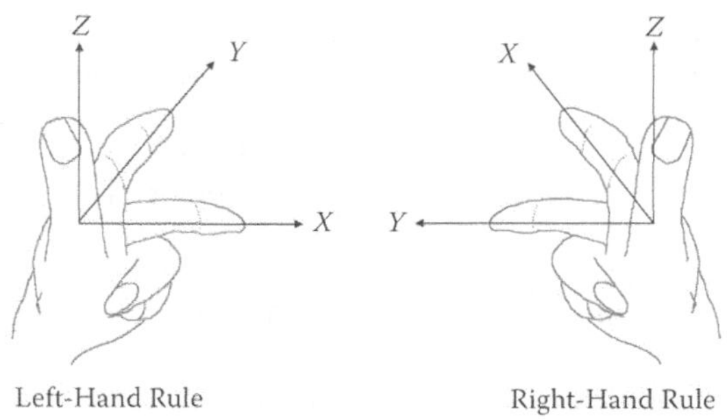

FIGURE 9.7 Left- and right-hand rules.

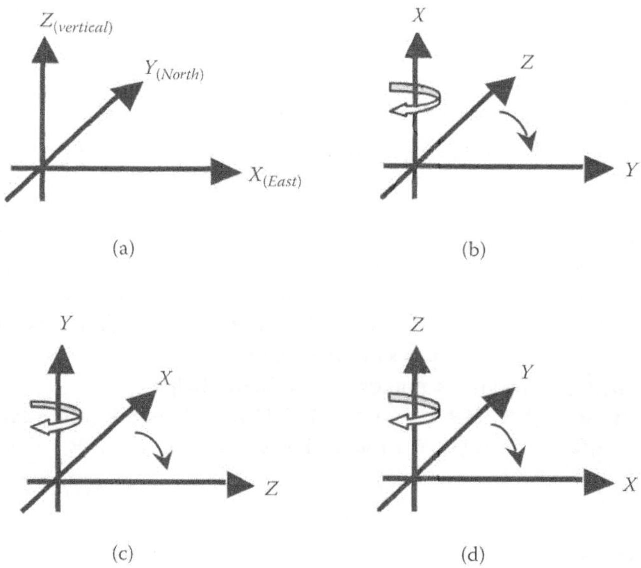

FIGURE 9.8 **(a–d)** Positive rotations around each axis—left-hand rule.

In Figure 9.8a, we have the axes *X* (east), *Y* (north) in the plane of the paper and *Z* perpendicular to the surface. Figure 9.8b, c, and d shows the effect of a clockwise turn about the *X*-axis, then the *Y*-axis then the *Z*-axis, as if looking along each axis in a positive direction.

It should however be stressed that, especially in photogrammetry, the opposite (right-hand) rule is used and positive rotations have the opposite sense. The confusion is compounded by whether an object is rotated about the axes or the axes are rotated about the object. For now, we will consider the rotation of axes; the key point is, as we have already observed, the sign of the sine.

If we are dealing in three dimensions (*x*, *y*, *z*), then the rotation matrix **R** about the *Z*-axis as shown in the two-dimensional case is

$$\mathbf{R}_z = \begin{pmatrix} \cos\phi_z & -\sin\phi_z & 0 \\ \sin\phi_z & \cos\phi_z & 0 \\ 0 & 0 & 1 \end{pmatrix}$$

where $\varnothing_z$ is the angle of clockwise rotation looking along the *Z*-axis. Similarly,

$$\mathbf{R}_x = \begin{pmatrix} 1 & 0 & 0 \\ 0 & \cos\phi_x & -\sin\phi_x \\ 0 & \sin\phi_x & \cos\phi_x \end{pmatrix} \text{ and } \mathbf{R}_y = \begin{pmatrix} \cos\phi_y & 0 & -\sin\phi_y \\ 0 & 1 & 0 \\ \sin\phi_y & 0 & \cos\phi_y \end{pmatrix}$$

Thus, if the axes are rotated in the positive direction as defined in Figure 9.8b, c, and d, then

$$\mathbf{X}_n = \begin{pmatrix} x_n \\ y_n \\ z_n \end{pmatrix} = \mathbf{R}_x \mathbf{R}_y \mathbf{R}_z \mathbf{X}_o$$

where $\mathbf{X}_o$ = the original coordinates

$$\begin{pmatrix} x_o \\ y_o \\ z_o \end{pmatrix}$$

It is important to keep in mind that the sequence in which the rotations takes place is significant since in matrix algebra **A** * **B** does not normally equal **B** * **A**. The rotations $\mathbf{R}_y$ $\mathbf{R}_x$ $\mathbf{R}_z$ in that order (applying $\mathbf{R}_x$ to $\mathbf{R}_z$ and then $\mathbf{R}_y$ to the result; or $\mathbf{R}_y$ to $\mathbf{R}_x$ and then the resulting product of $\mathbf{R}_y\mathbf{R}_x$ to $\mathbf{R}_z$) will give a different answer to that of $\mathbf{R}_x$ $\mathbf{R}_y$ $\mathbf{R}_z$. In the latter case, we rotate first about the z-axis, then about the y-axis, and then about the x (working from right to left). As we saw with dice in Figure 2.1 in Chapter 2, the sequence in rotations makes a significant difference.

Fortunately, **A** * **B** * **C** = **A** * (**B** * **C**) = (**A** * **B**) * **C** so it does not matter whether we calculate **A** * **B** first and then apply the result to **C** or whether we first calculate **B** * **C** and then premultiply by **A**.

9.6 SIMPLIFYING MATRICES

Matrices can be *partitioned,* that is, treated in parts. Consider the two matrices **A** with 4 rows and 5 columns and **B** with 5 rows and 2 columns.

$$A * B = \begin{pmatrix} a_{11} & a_{12} & a_{13} & a_{14} & a_{15} \\ a_{21} & a_{22} & a_{23} & a_{24} & a_{25} \\ a_{31} & a_{32} & a_{33} & a_{34} & a_{35} \\ a_{41} & a_{42} & a_{43} & a_{44} & a_{45} \end{pmatrix} * \begin{pmatrix} b_{11} & b_{12} \\ b_{21} & b_{22} \\ b_{31} & b_{32} \\ b_{41} & b_{42} \\ b_{51} & b_{52} \end{pmatrix}$$

We can write these in the form

$$A * B = \left(\begin{array}{ccc||cc} a_{11} & a_{12} & a_{13} & a_{14} & a_{15} \\ a_{21} & a_{22} & a_{23} & a_{24} & a_{25} \\ a_{31} & a_{32} & a_{33} & a_{34} & a_{35} \\ a_{31} & a_{32} & a_{43} & a_{44} & a_{45} \end{array}\right) * \begin{pmatrix} b_{11} & b_{12} \\ b_{21} & b_{22} \\ b_{31} & b_{32} \\ b_{41} & b_{42} \\ b_{51} & b_{52} \end{pmatrix}$$

or

$$\mathbf{A} * \mathbf{B} = \begin{pmatrix} \mathbf{A}_1 & \mathbf{A}_2 \end{pmatrix} * \begin{pmatrix} \mathbf{B}_1 \\ \mathbf{B}_2 \end{pmatrix}$$

where

$$\mathbf{A}_1 = \begin{pmatrix} a_{11} & a_{12} & a_{13} \\ a_{21} & a_{22} & a_{23} \\ a_{31} & a_{32} & a_{33} \\ a_{41} & a_{42} & a_{43} \end{pmatrix}, \mathbf{A}_2 = \begin{pmatrix} a_{14} & a_{15} \\ a_{24} & a_{25} \\ a_{34} & a_{35} \\ a_{44} & a_{45} \end{pmatrix}$$

and

$$\mathbf{B}_1 = \begin{pmatrix} b_{11} & b_{12} \\ b_{21} & b_{22} \\ b_{31} & b_{32} \end{pmatrix}, \mathbf{B}_1 = \begin{pmatrix} b_{41} & b_{42} \\ b_{51} & b_{52} \end{pmatrix}$$

$\mathbf{A}_1$, $\mathbf{A}_2$, $\mathbf{B}_1$, and $\mathbf{B}_2$ are partitions of the matrix. The original multiplication $\mathbf{A} * \mathbf{B}$ is the same and can be expressed as

$$\mathbf{A} * \mathbf{B} = \begin{pmatrix} \mathbf{A}_1\mathbf{B}_1 & \mathbf{A}_1\mathbf{B}_2 \\ \mathbf{A}_2\mathbf{B}_1 & \mathbf{A}_2\mathbf{B}_2 \end{pmatrix}$$

We can also simplify matrices by making them triangular, where $\mathbf{A}_U$ is called an *upper triangular matrix*, while $\mathbf{A}_L$ is a *lower triangular matrix*:

$$\mathbf{A}_U = \begin{pmatrix} a_{11} & a_{12} & a_{13} & a_{14} & a_{15} \\ 0 & a_{22} & a_{23} & a_{24} & a_{25} \\ 0 & 0 & a_{33} & a_{34} & a_{35} \\ 0 & 0 & 0 & a_{44} & a_{45} \\ 0 & 0 & 0 & 0 & a_{55} \end{pmatrix}, \mathbf{A}_L = \begin{pmatrix} a_{11} & 0 & 0 & 0 & 0 \\ a_{21} & a_{22} & 0 & 0 & 0 \\ a_{31} & a_{32} & a_{33} & 0 & 0 \\ a_{41} & a_{42} & a_{43} & a_{44} & 0 \\ a_{51} & a_{52} & a_{53} & a_{54} & a_{55} \end{pmatrix}$$

To illustrate how this works, consider five equations for five unknowns (x_1, x_2, x_3, x_4, x_5)

$$a_{11}x_1 + a_{12}x_2 + a_{13}x_3 + a_{14}x_4 + a_{15}x_5 = b_1$$
$$a_{21}x_2 + a_{22}x_2 + a_{23}x_3 + a_{24}x_4 + a_{25}x_5 = b_2$$
$$a_{31}x_3 + a_{32}x_2 + a_{33}x_3 + a_{34}x_4 + a_{35}x_5 = b_3$$
$$a_{41}x_4 + a_{42}x_2 + a_{43}x_3 + a_{44}x_4 + a_{45}x_5 = b_4$$
$$a_{51}x_5 + a_{52}x_2 + a_{53}x_3 + a_{54}x_4 + a_{55}x_5 = b_5$$

where a_{11}, a_{12}, and so on, and b_1, b_2, and so on, are known quantities. We can express this in the form

$$\begin{pmatrix} a_{11} & a_{12} & a_{13} & a_{14} & a_{15} \\ a_{21} & a_{22} & a_{23} & a_{24} & a_{25} \\ a_{31} & a_{32} & a_{33} & a_{34} & a_{35} \\ a_{41} & a_{42} & a_{43} & a_{44} & a_{45} \\ a_{51} & a_{52} & a_{53} & a_{54} & a_{55} \end{pmatrix} * \begin{pmatrix} x_1 \\ x_2 \\ x_3 \\ x_4 \\ x_5 \end{pmatrix} = \begin{pmatrix} b_1 \\ b_2 \\ b_3 \\ b_4 \\ b_5 \end{pmatrix}$$

or as

$$\mathbf{AX} = \mathbf{B}$$

Now premultiply both sides of $\mathbf{AX} = \mathbf{B}$ by the diagonal matrix $\mathbf{C}_1$ whose leading diagonal is made from the reciprocals of the numbers in column 1. Thus, the diagonal elements of $\mathbf{C}_1$ are $\{1/a_{11}, 1/a_{21}, 1/a_{31}, 1/a_{41}, 1/a_{51}\}$.

This gives us $\mathbf{C_1AX} = \mathbf{C_1B}$, or

$$\begin{pmatrix} a_{11}/a_{11} & a_{12}/a_{11} & a_{13}/a_{11} & a_{14}/a_{11} & a_{15}/a_{11} \\ a_{21}/a_{21} & a_{22}/a_{21} & a_{23}/a_{21} & a_{24}/a_{21} & a_{25}/a_{21} \\ a_{31}/a_{31} & a_{32}/a_{31} & a_{33}/a_{31} & a_{34}/a_{31} & a_{35}/a_{31} \\ a_{41}/a_{41} & a_{42}/a_{41} & a_{43}/a_{41} & a_{44}/a_{41} & a_{45}/a_{41} \\ a_{51}/a_{51} & a_{52}/a_{51} & a_{53}/a_{51} & a_{54}/a_{51} & a_{55}/a_{51} \end{pmatrix} * \begin{pmatrix} x_1 \\ x_2 \\ x_3 \\ x_4 \\ x_5 \end{pmatrix} = \begin{pmatrix} b_1/a_{11} \\ b_2/a_{21} \\ b_3/a_{31} \\ b_4/a_{41} \\ b_5/a_{51} \end{pmatrix}$$

or

$$\begin{pmatrix} 1 & a_{12} / a_{11} & a_{13} / a_{11} & a_{14} / a_{11} & a_{15} / a_{11} \\ 1 & a_{22} / a_{21} & a_{23} / a_{21} & a_{24} / a_{21} & a_{25} / a_{21} \\ 1 & a_{32} / \mathrm{a}_{31} & a_{33} / a_{31} & a_{34} / a_{31} & a_{35} / a_{31} \\ 1 & a_{42} / a_{41} & a_{43} / a_{41} & a_{44} / a_{41} & a_{45} / a_{41} \\ 1 & a_{52} / a_{51} & a_{53} / a_{51} & a_{54} / a_{51} & a_{55} / a_{51} \end{pmatrix} * \begin{pmatrix} x_1 \\ x_2 \\ x_3 \\ x_4 \\ x_5 \end{pmatrix} = \begin{pmatrix} b_1 / a_{11} \\ b_2 / a_{21} \\ b_3 / a_{31} \\ b_4 / a_{41} \\ b_5 / a_{51} \end{pmatrix}$$

Now subtract row 1 from the remaining rows to give

$$\begin{pmatrix} 1 & a_{12} / a_{11} & a_{13} / a_{11} & a_{14} / a_{11} & a_{15} / a_{11} \\ 0 & a_{22} / a_{21} - a_{12} / a_{11} & a_{23} / \mathrm{a}_{21} - a_{13} / a_{11} & a_{24} / a_{21} - a_{14} / a_{11} & a_{25} / a_{21} - a_{15} / a_{11} \\ 0 & a_{32} / a_{31} - a_{12} / a_{11} & a_{33} / a_{31} - a_{13} / a_{11} & a_{34} / a_{31} - a_{14} / a_{11} & a_{35} / a_{31} - a_{15} / a_{11} \\ 0 & a_{42} / a_{41} - a_{12} / a_{11} & a_{43} / a_{41} - a_{13} / a_{11} & a_{44} / a_{41} - a_{14} / a_{11} & a_{45} / a_{41} - a_{15} / a_{11} \\ 0 & a_{52} / a_{51} - a_{12} / a_{11} & a_{53} / a_{51} - a_{13} / a_{11} & a_{54} / a_{51} - a_{14} / a_{11} & a_{55} / a_{51} - a_{15} / a_{11} \end{pmatrix}$$

$$* \begin{pmatrix} x_1 \\ x_2 \\ x_3 \\ x_4 \\ x_5 \end{pmatrix} = \begin{pmatrix} b_1 / a_{11} \\ b_2 / a_{21} - b_1 / a_{11} \\ b_3 / a_{31} - b_1 / a_{11} \\ b_4 / a_{41} - b_1 / a_{11} \\ b_5 / a_{51} - b_1 / a_{11} \end{pmatrix}$$

We can then repeat this process ignoring the first row and first column and considering only the (4 * 4) matrix of rows 2 to 5 and columns 2 to 5 (that is, we consider the new submatrix of the original row one column one). We then repeat this again on the (3 * 3) and again on the (2 * 2) finishing up with an upper triangular matrix

$$\mathbf{B}' = \begin{pmatrix} 1 & a_{12} / a_{11} & a_{13} / a_{11} & a_{14} / a_{11} & a_{15} / a_{11} \\ 0 & 1 & a'_{23} & a'_{24} & a'_{25} \\ 0 & 0 & 1 & a'_{34} & a'_{35} \\ 0 & 0 & 0 & 1 & a'_{45} \\ 0 & 0 & 0 & 0 & 1 \end{pmatrix} * \begin{pmatrix} x_1 \\ x_2 \\ x_3 \\ x_4 \\ x_5 \end{pmatrix} = \begin{pmatrix} b_1 / a_{11} \\ b'_2 \\ b'_3 \\ b'_4 \\ b'_5 \end{pmatrix}$$

The process is boring and repetitive, ideally suited to automate in a computer program that can routinely evaluate all the a' and b'. A worked example is given in Example 9.8.

The numbers **X** are found by back substitution. Thus, $x_5 = b_5'$, the number that one finishes up with in the last row of the modified matrix **B′**. Moving up a row in matrix **B′** $x_4 + a_{45}'x_5 = b_4'$ or $x_4 = b_4' - a_{45}'x_5$, and so on.

If we return for a moment to the original expression of **AX** = **B**, namely,

$$\begin{pmatrix} a_{11} & a_{12} & a_{13} & a_{14} & a_{15} \\ a_{21} & a_{22} & a_{23} & a_{24} & a_{25} \\ a_{31} & a_{32} & a_{33} & a_{34} & a_{35} \\ a_{41} & a_{42} & a_{43} & a_{44} & a_{45} \\ a_{51} & a_{52} & a_{53} & a_{54} & a_{55} \end{pmatrix} * \begin{pmatrix} x_1 \\ x_2 \\ x_3 \\ x_4 \\ x_5 \end{pmatrix} = \begin{pmatrix} b_1 \\ b_2 \\ b_3 \\ b_4 \\ b_5 \end{pmatrix}$$

The manipulation was all in terms of *a* and *b*. We could have written the equation in one solid block of numbers as

$$\begin{pmatrix} a_{11} & a_{12} & a_{13} & a_{14} & a_{15} & b_1 \\ a_{21} & a_{22} & a_{23} & a_{24} & a_{25} & b_2 \\ a_{31} & a_{32} & a_{33} & a_{34} & a_{35} & b_3 \\ a_{41} & a_{42} & a_{43} & a_{44} & a_{45} & b_4 \\ a_{51} & a_{52} & a_{53} & a_{54} & a_{55} & b_5 \end{pmatrix}$$

where the column **B** is a column *vector of constants*, while the *a* terms form the *matrix of coefficients*. The combined expression is known as an *augmented matrix*.

Example 9.8 The Solution of Simultaneous Equations

Consider the four equations

$$x + y + z + w = 11; 2x - 6y + 3z + 7w = 24; 7x + 2y + 5z - 3w = 10; 9x - 3y + 2z + 2w = 21$$

In matrix form

$$\begin{pmatrix} 1 & 1 & 1 & 1 \\ 2 & -6 & 3 & 7 \\ 7 & 2 & 5 & -3 \\ 9 & -3 & 2 & 2 \end{pmatrix} \begin{pmatrix} x \\ y \\ z \\ w \end{pmatrix} = \begin{pmatrix} 11 \\ 24 \\ 10 \\ 21 \end{pmatrix}$$

On reducing column 1 to unity (e.g., dividing all of row 2 by 2), we obtain

$$\begin{pmatrix} 1 & 1 & 1 & 1 \\ 1 & -3 & 1.5 & 3.5 \\ 1 & 0.286 & 0.714 & -0.429 \\ 1 & -0.333 & 0.222 & 0.222 \end{pmatrix} * \begin{pmatrix} x \\ y \\ z \\ w \end{pmatrix} = \begin{pmatrix} 11 \\ 12 \\ 1.429 \\ 2.333 \end{pmatrix}$$

Then subtracting row 1 from each of the remaining rows

$$\begin{pmatrix} 1 & 1 & 1 & 1 \\ 0 & -4 & 0.5 & 2.5 \\ 0 & -0.714 & -0.286 & -1.429 \\ 0 & -1.333 & -0.778 & -0.778 \end{pmatrix} \begin{pmatrix} x \\ y \\ z \\ w \end{pmatrix} = \begin{pmatrix} 11 \\ 1 \\ -9.571 \\ -8.667 \end{pmatrix}$$

And repeating

$$\begin{pmatrix} 1 & 1 & 1 & 1 \\ 0 & 1 & -0.125 & -0.625 \\ 0 & 1 & 0.4 & 2 \\ 0 & 1 & 0.584 & 0.584 \end{pmatrix}\begin{pmatrix} x \\ y \\ z \\ w \end{pmatrix} = \begin{pmatrix} 11 \\ -0.25 \\ 13.4 \\ 6.5 \end{pmatrix}$$

Then by subtracting row 2 from rows 3 and 4

$$\begin{pmatrix} 1 & 1 & 1 & 1 \\ 0 & 1 & -0.125 & -0.625 \\ 0 & 0 & 0.525 & 2.625 \\ 0 & 0 & 0.71 & 1.21 \end{pmatrix}\begin{pmatrix} x \\ y \\ z \\ w \end{pmatrix} = \begin{pmatrix} 11 \\ -0.25 \\ 13.65 \\ 6.75 \end{pmatrix}$$

$$\begin{pmatrix} 1 & 1 & 1 & 1 \\ 0 & 1 & -0.125 & -0.625 \\ 0 & 0 & 1 & 5 \\ 0 & 0 & 1 & 1.7 \end{pmatrix} * \begin{pmatrix} x \\ y \\ z \\ w \end{pmatrix} = \begin{pmatrix} 11 \\ -0.25 \\ 26 \\ 9.5 \end{pmatrix}$$

And repeating

$$\begin{pmatrix} 1 & 1 & 1 & 1 \\ 0 & 1 & -0.125 & -0.625 \\ 0 & 0 & 1 & 5 \\ 0 & 0 & 0 & -3.3 \end{pmatrix} * \begin{pmatrix} x \\ y \\ z \\ w \end{pmatrix} = \begin{pmatrix} 11 \\ -0.25 \\ 26 \\ -16.5 \end{pmatrix}$$

And again,

$$\begin{pmatrix} 1 & 1 & 1 & 1 \\ 0 & 1 & -0.125 & -0.625 \\ 0 & 0 & 1 & 5 \\ 0 & 0 & 0 & 1 \end{pmatrix}\begin{pmatrix} x \\ y \\ z \\ w \end{pmatrix} = \begin{pmatrix} 11 \\ -0.25 \\ 26 \\ 5 \end{pmatrix}$$

And finally,

$$\begin{pmatrix} 1 & 1 & 1 & 1 \\ 0 & 1 & -0.125 & -0.625 \\ 0 & 0 & 1 & 5 \\ 0 & 0 & 0 & 1 \end{pmatrix}\begin{pmatrix} x \\ y \\ z \\ w \end{pmatrix} = \begin{pmatrix} 11 \\ -0.25 \\ 26 \\ 5 \end{pmatrix}$$

Thus, from the fourth row of this matrix,

$$w = 5$$

From the third row, $z + 5w = 26,$ so $z = 1$

From the second row, $y - 0.125z - 0.625w = -0.25,$ so $y = 3$

From the first row, $x + y + z + w = 11,$ so $x = 2$

When reducing a matrix to diagonal form by the method outlined above, a problem can arise where one of the coefficients used for division is zero. This will happen when the five equations are not independent in which case there is no solution—there must at least be the same number of independent equations as there are unknowns in order to find a unique solution. If there are extra equations, as can happen with land survey measurements, where there are redundant measurements in order to improve the accuracy of the result, then a most probable solution can be found using statistical techniques such as the least square solution that is discussed in Chapter 15.

GLOSSARY

Adjoint matrix Also known as an *adjugate matrix*, it is the matrix $\mathbf{A}^*$ formed from a *square matrix* $\mathbf{A}$ with the *cofactors* of $\mathbf{A}$ placed in their *transposed* positions. $\mathbf{A}^* * \mathbf{A} = |\mathbf{A}|\mathbf{I}$ where $|\mathbf{A}|$ is the *determinant* of $\mathbf{A}$. From this it follows that $\mathbf{A}^* / |\mathbf{A}| = \mathbf{A}^{-1}$ which is the *inverse* of $\mathbf{A}$.

Array A two-dimensional rectangular arrangement of numbers or symbols in rows and columns.

Augmented matrix A matrix that combines a set of coefficients with a column of constant values, as, for example, when solving a set of simultaneous equations.

Cofactor The signed value of the *minor.* If the minor is $\mathbf{M}_{ij}$, then the cofactor = $(-1)^{(i-j)} * \mathbf{M}_{ij}$. Being based on a determinant, a cofactor is a single number.

Conformable The number of elements in each row of a matrix equals the number of elements in each column of the second.

Determinant A *scalar quantity* that is calculated from the alternating sum (±) of elements of a *square matrix.* Normally written with straight-line brackets, for example, $|\mathbf{A}|$.

Diagonal matrix A *square matrix* that has zero for every *element* except where the row number equals the column number.

Element A number or symbol at a particular row and column within an array. The element in the ith row and jth column of the matrix $\mathbf{A}$ is written as a_{ij}.

Identity matrix A *diagonal matrix* in which the nonzero elements are all equal to 1. Often written as $\mathbf{I}$.

$$\mathbf{I} * \mathbf{A} = \mathbf{A} * \mathbf{I} = \mathbf{A}$$

Inverse matrix If a square matrix $\mathbf{A}$ and a square matrix $\mathbf{B}$ are such that $\mathbf{A} * \mathbf{B} = \mathbf{I}$ (the *identity matrix*) then $\mathbf{B}$ is said to be the inverse of $\mathbf{A}$ and is written as $\mathbf{A}^{-1}$.

$$\mathbf{A} * \mathbf{A}^{-1} = \mathbf{A}^{-1} * \mathbf{A} = \mathbf{I}$$

Leading diagonal The line of elements in a square matrix or determinant that runs from the top left corner to the bottom right.

Matrix A set of numbers aligned in rows and columns forming a rectangular *array.* Often written in bold as $\mathbf{M}$. The number of rows is not necessarily the same as the number of columns but each row must be the same length, as must each column.

Matrix addition or subtraction Two matrices can be added or subtracted only if they have the same number of rows and the same number of columns. If matrix $\mathbf{A}$ has elements a_{ij} and matrix $\mathbf{B}$ has elements b_{ij} then $\mathbf{A} \pm \mathbf{B}$ has elements $a_{ij} \pm b_{ij}$.

Matrix multiplication Can only occur when the number of columns in the first matrix $\mathbf{A}$ is the same as the number of rows in the second matrix $\mathbf{B}$. The ith element of a row in $\mathbf{A}$ is used to multiply the ith element in the appropriate column in matrix $\mathbf{B}$. If A has i rows and j columns and $\mathbf{B}$ has j rows and k columns then $\mathbf{A} * \mathbf{B}$ has i rows and k columns, each element being made up from the sum of i multiplications.

In general, $\mathbf{A} * \mathbf{B} \neq \mathbf{B} * \mathbf{A}$ unless $\mathbf{A} * \mathbf{B} = \mathbf{I}$.

Minor The *determinant* of a square *submatrix* of a given matrix.

Modulus The absolute value of a quantity regardless of its sign.

Orthogonal matrix A *square matrix* that when multiplied by its *inverse* gives the *identity matrix*. Thus, if $\mathbf{AA}^{-1} = \mathbf{I}$ then **A** is orthogonal.

Partitioning The division of a matrix into *conformable* submatrices.

Rotation matrix A *matrix* used to calculate the result of rotating the axes of a reference coordinate system.

Scalar quantity A quantity that has magnitude but no direction, such as distance.

Similarity transformation A transformation that involves translation, rotation, and uniform scale change.

Singular matrix A square matrix that has no *inverse* since its *determinant* is zero.

Square matrix A *matrix* that has the same number of rows as columns.

Submatrix A matrix derived from a given matrix by removing some of its rows and columns. In particular, if a square matrix has the element a_{ij} for its ith row and jth column, then the submatrix for element a_{ij} is the matrix remaining after excluding all elements in row i and column j.

Translation The movement of the origin of a set of coordinates.

Transpose matrix The matrix formed by changing rows into columns. The transpose of **A** is $\mathbf{A}^T$.

Triangular matrix A matrix having zeros for all elements of the array below the *leading diagonal* in the case of an upper triangular matrix and above the leading diagonal for a lower triangular matrix.

10 Differential and Integral Calculus

10.1 DIFFERENTIATION

Calculus is a branch of mathematics the origins of which were the basis of an acrimonious dispute between two great mathematicians—Sir Isaac Newton and Gottfried Leibniz—both of whom claimed to be the discoverer. *Differential calculus* is concerned with the rate at which a function changes, while *integral calculus* extends the idea that the sum of a finite number of separate values can be used to form a continuous value as, for instance, in determining the area under a curve.

First, consider a function such as the parabola plotted as *ABCD*, and so forth, in Figure 10.1. It can take the form $y = ax^2$ where a is some constant. If we travel from a point B with coordinates (x, y) on the curve to a point C, the direction in which we are traveling along the curve changes from BB' to CC'. These directions are called the *tangents* at B and C. They are related to the idea of the tangent of an angle, as discussed in Chapter 6, since the slope of the straight line from B to C is the ratio shown as $\delta y / \delta x$ in Figure 10.2 (δ being the Greek letter "delta" and is shown here in lowercase). $\delta y / \delta x$ is the tangent of the angle θ (the Greek letter "theta") of the slope of the straight line BC.

If B is (x, y) and the distance marked as "δx" is small (δ standing for "a small amount of") and if, on moving from B to C along the curve, the y value increases by an equivalent small amount δy, then the coordinates of C must be: $(x + \delta x, y + \delta y)$.

These must also satisfy the basic equation $y = ax^2$. Hence:

$$y + \delta y = a\left(x + \delta x\right)^2 = ax^2 + 2ax\delta x + a\delta x^2$$

Since for point B, $y = ax^2$, it follows that

$$\delta y = 2ax \cdot \delta x + a\delta x^2 = \left(2ax + a\delta x\right)\delta x$$

Dividing both sides by δx,

$$\delta y / \delta x = 2ax + a\delta x$$

As δx gets smaller and smaller, so does δy and the ratio between the two, namely, $\delta y / \delta x$ tends to the value $= 2ax$. In the limit, we use the ordinary alphabet and say that $\mathrm{d}y / \mathrm{d}x = 2ax$. This process is known as *differentiation*. Thus, if we differentiate y with respect to x where $y = ax^2$, we obtain $\mathrm{d}y / \mathrm{d}x = 2ax$.

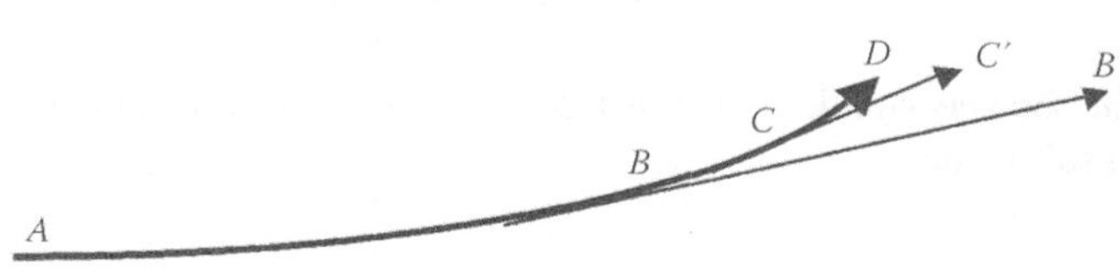

FIGURE 10.1 Tangents to a curve.

DOI: 10.1201/9781003464402-13

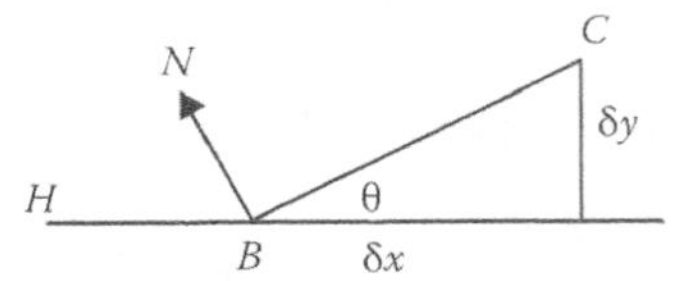

FIGURE 10.2 The slope and the normal to a curve.

Note that δx means "a little bit of x" but dy / dx means "in the limit the value of $\delta y / \delta x$ when both δy and δx have become very, very small."

If we have a function of the form $y = ax^2 + bx + c$, then

$$y + \delta y = a(x + \delta x)^2 + b(x + \delta x) + c = ax^2 + 2ax\delta x + a(\delta x)^2 + bx + b\delta x + c$$

or

$$\delta y = 2ax\delta x + a(\delta x)^2 + b\delta x$$

On dividing both sides by δx,

$$\delta y / \delta x = 2ax + b + a\delta x$$

In the limit as δx tends to zero,

$$dy / dx = 2ax + b$$

We can extend the argument to a cubic and find that if

$$y = ax^3 + bx^2 + cx + e$$

then

$$dy / dx = 3ax^2 + 2bx + c$$

And in general, if $y = ax^n$, then $dy / dx = nax^{(n-1)}$.

If we have two functions u and v that are both functions of x and if $y = uv$, then

$$y + \delta y = (u + \delta u)(v + \delta v) = uv + u\delta v + v\delta u + \delta u \cdot \delta v$$

Thus,

$$\delta y = u\delta v + v\delta u + \delta u \cdot \delta v$$

hence

$$\delta y / \delta x = u\delta v / \delta x + v\delta u / \delta x + \delta u.\delta v / \delta x$$

The expression $\delta u \cdot \delta v$ gets smaller twice as fast as either δu or δv separately; hence $+\delta u \cdot \delta v / \delta x$ tends to zero. The result is that

If $y = uv$, then

$$dy / dx = u \; dv / dx + v \; du / dx$$

(Note: If $y = x^2 = x * x$, then $u = v = x$ and $dy / dx = x(dx / dx) + x(dx / dx) = 2x$, as we would expect.) This is summarized in Box 10.1.

"dy / dx" is the rate of change of the function y with respect to x or, looked at another way as in Figure 10.1, it is the slope of the curve at point (x, y). The line that is perpendicular to the tangent (*BN* in Figure 10.2) is known as the *normal* to the curve.

The slope of $BN = \theta + 90°$. Also,

$$\begin{aligned}\tan(\theta+90°) &= \sin(\theta+90°)/\cos(\theta+90°)\\ &= -\cos\theta/\sin\theta = -\cot\theta\\ &= -(\delta x/\delta y) = -1/(\delta y/\delta x)\end{aligned}$$

Thus, the tangent of the slope of the normal = −1 divided by the tangent of the slope of the curve.

In general, if y is a function of x, we can write this as $y = f(x)$ where we use the expression "$f(x)$" to mean "the function (x)," which in turn means "the value of the function given the value x." When we differentiate the function $y = f(x)$, we obtain dy / dx which we can write as $f'(x)$. This can in turn be differentiated to give the *second derivative*.

BOX 10.1 BASIC DIFFERENTIALS

If $y = ax^n + bx^{(n-1)} + cx^{(n-2)} + \cdots \cdots px^2 + qx + r$, then

$$\frac{dy}{dx} = nax^{(n-1)} + (n-1)bx^{(n-2)} + (n-2)cx^{(n-3)} + \cdots \cdots 2px + q$$

If $y = uv$ where u and v are both functions of x, then

$$dy/dx = u\,dv/dx + v\,du/dx$$

This can be written either as d^2y / dx^2 or as $f''(x)$. Thus, if $y = f(x) = a\,x^n$, then

$$dy/dx = f'(x) = n\,a\,x^{(n-1)}$$

and

$$d^2y/dx^2 = f''(x) = n(n-1)a\,x^{(n-2)}$$

Thus, if

$$y = ax^3 + bx^2 + cx + e$$

then

$$\begin{aligned}dy/dx &= f'(x) = 3ax^2 + 2bx + c\\ d^2y/dx^2 &= f''(x) = 6ax + 2b\\ d^3y/dx^3 &= f'''(x) = 6a\\ d^4y/dx^4 &= f^{IV}(x) = 0\end{aligned}$$

TABLE 10.1
Data for $y = 1 + 9x - 6x^2 + x^3$

x	−3	−2	−1	−0.5	0	0.5	1	1.5	2	2.5	3	3.5	4	4.5	5	5.5	6
y	−107	−49	−15	−5.1	1	4.1	5	4.4	3	1.6	1	1.9	5	11.1	21	35.4	55

($f'''(x)$ is the *third derivative* while f^{IV} is the *fourth derivative*, which for a cubic is zero.)

If dy/dx represents the rate of change of a curve then d^2y/dx^2 represents the rate of change; put in another context, if y is a distance, dy/dx is the rate of change of the rate of change distance or the velocity or speed, and d^2y/dx^2 is the acceleration. d^3y/dx^3 would be the rate of change of the acceleration.

Consider the function

$$y = 1 + 9x - 6x^2 + x^3$$
$$dy/dx = 9 - 12x + 3x^2 = 3(1-x)(3-x)$$
$$d^2y/dx^2 = -12 + 6x$$
$$d^2y/dx^2 = 0 \text{ when } x = 2 \text{ and } y = 3$$
$$dy/dx = 0 \text{ when } x = 1 \text{ and } y = 5$$
$$\text{and when } x = 3 \text{ and } y = 1$$

In Table 10.1, we list some values of x and y for this function and have interpolated a curve between the points in Figure 10.3a. It will be seen from Figure 10.3a that there are two points where the slope is horizontal, that is, $dy/dx = 0$ {here (1,5) and (3,1)}. These points are known as the *maxima* or *minima* for the curve. A cubic has one maximum and one minimum point—although note that this does not mean a maximum value for all points on the curve, which can come from and go off to ± infinity.

Maxima and minima are turning points where the curve having gone upward now turns downward or vice versa.

A point where $d^2y/dx^2 = 0$ is known as a *point of inflection* and a cubic curve has one of them. At a point of inflection the curve stops bending in one direction and starts bending in the other; in effect, the curve crosses its tangent at that point. In the cubic in Figure 10.3a, the curve starting from the bottom left turns clockwise until it reaches the point of inflection where $x = 2$ and $y = 3$ and then starts bending in an anticlockwise direction. An example based on a quartic equation is given in Example 10.1 and illustrated in Figure 10.3b.

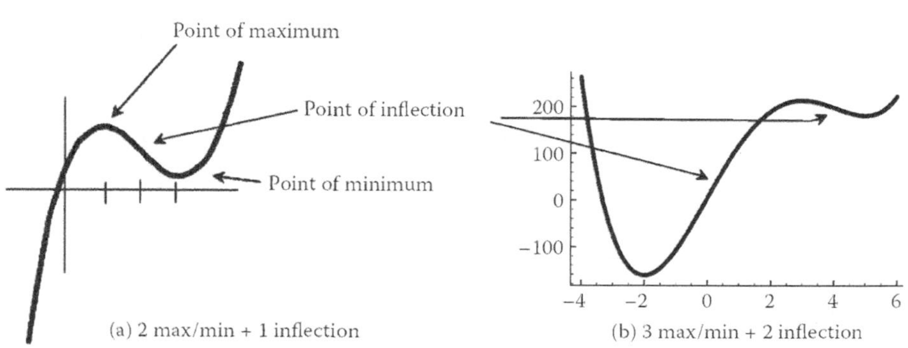

FIGURE 10.3 (a) A cubic curve and (b) a quartic.

Example 10.1 Points of Inflection

The quartic equation $y = x^4 - 8x^3 - 2x^2 + 120x + 6$ has three maxima/minima and two points of inflection.

$$dy / dx = 4x^3 - 24x^2 - 4x + 120 = 4(x+2)(x-3)(x-5)$$

This is zero when $x = -2$, $+3$, and $+5$. Hence, there are three places where the curve reaches a peak or a trough, namely, (−2, −162), (3, 213), (5, 181)

$$d^2y / dx^2 = 12x^2 - 48x - 4 = 4(3x^2 - 12x - 1)$$

Using the formula from Chapter 4 where the solution to the quadratic equation

$ax^2 + bx + c = 0$ was given as $x = \left\{-b \pm \sqrt{(b^2 - 4ac)}\right\} / 2a$, then

$(3x^2 - 12x - 1)$ is zero when $x = \left\{12 \pm \sqrt{(144 + 23)}\right\} / 6$ or $x \approx -0.08$ or 4.08.

In Chapter 13, we discuss the idea of a radius of curvature at a point on a curve; at a point of inflection the radius of curvature is infinitely large.

10.2 DIFFERENTIATING TRIGONOMETRIC FUNCTIONS

Before illustrating the use of differentiation, we need to consider the trigonometric functions and how they can be differentiated. We showed in Chapter 6 that

$$\sin(A + B) = \sin A \cos B + \cos A \sin B$$

Hence,

$$\sin(\theta + \delta\theta) = \sin\theta \cos\delta\theta + \cos\theta\ \sin\delta\theta$$

In Chapter 6, we defined "sine" as side opposite divided by the hypotenuse = BC / AC in Figure 10.4 with "cosine" as side adjacent over side opposite = AB / AC. As BC gets smaller and smaller, that is, the angle $\delta\theta$ approaches zero, then ultimately $AB = AC$. Thus, cosine $\delta\theta = AB / AC = 1$ as $\delta\theta$ tends to zero. Now if $AC = R$ and $\delta\theta$ is measured in radians then $BC = R\delta\theta$.

As $\delta\theta$ gets smaller, BC / AC or sin $\delta\theta$ becomes $R\delta\theta / R = \delta\theta$. Hence, for small values of $\delta\theta$, $\sin \delta\theta = \delta\theta$. If $y = \sin\theta$, then

$$\begin{aligned} y + \delta y &= \sin(\theta + \delta\theta) = \sin\theta\cos\delta\theta + \cos\theta\sin\delta\theta \\ &= \sin\theta + \delta\theta\ \cos\theta \end{aligned}$$

Since $y = \sin\theta$, it means that $\delta y = \delta\theta \cos\theta$, or $\delta y / \delta\theta = \cos\theta$.
If we revert to the notation that $y = \sin x$, then $dy / dx = \cos x$.

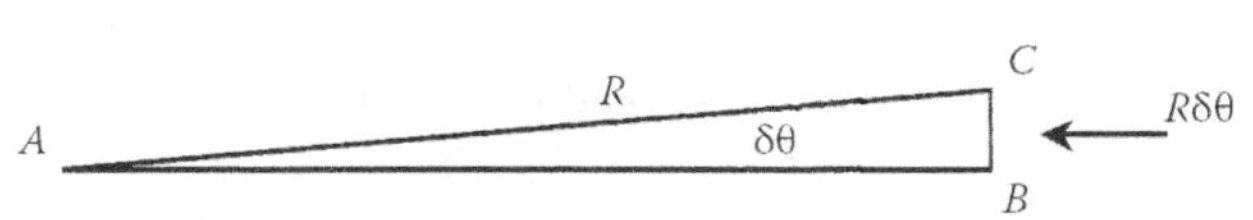

FIGURE 10.4 Small angles.

We have also seen in Chapter 6 that cos $(A + B) = \cos A \cos B - \sin A \sin B$.

Hence, if $y = \cos\theta$ then

$$y + \delta y = \cos(\theta + \delta\theta) = \cos\theta\cos\delta\theta - \sin\theta\sin\delta\theta$$
$$= \cos\theta - \delta\theta\sin\theta$$

Therefore, $\delta y / \delta\vartheta = -\sin\vartheta$. Thus, if $y = \cos x$, then $dy / dx = -\sin x$.

The results are summarized in Box 10.2. See also Example 10.2.

Before considering the other trigonometric functions, we need to consider how to differentiate products. Consider $y = \sin^2 x = \sin x * \sin x$. Let $u = \sin x$, then $du / dx = \cos x$. If $y = \sin^2 x$, then $y = u^2$, and $dy / du = 2u$. Remembering that dy / du means the limit of $\delta y / \delta u$ as δy and δu approach zero, then

Example 10.2 The Rate of Change of Sin x

We have shown that if $y = \sin x$, then $dy / dx = \cos x$. To illustrate this, let $x = 30° = \pi / 6$ radians $= 0.5235987756^r$

$$\sin(0.5235988^r) = 0.50000002113249$$

$$\sin(0.5235989^r) = 0.50000010773503$$

The difference in sines = 0.00000008660254 = rate of change for sin (30°) for a difference in angle of 0.0000001 radians

$$\cos 30° = 0.8660254$$

Hence, the rate of change of sin (30°) = cos (30°).

$$\frac{\delta y}{\delta x} = \frac{\delta y * \delta u}{\delta x * \delta u} = \frac{\delta y * \delta u}{\delta u * \delta x} = \frac{\delta y}{\delta u} * \frac{\delta u}{\delta x}$$

In the limit,

$$(dy / dx) = (dy / du) \ * \ (du / dx) = 2u \ * \ \cos x = 2\sin\ x\cos\ x$$

Thus, if

$$y = \sin^2 x$$

then

$$dy / dx = 2\ \sin x\cos x$$

Similarly, if

$$y = \sin^3 x$$

then

$$dy / dx = 3\ \sin^2\ x\ \cos\ x, \text{ and so on.}$$

Hence, if

$$y = \sin^n x$$

then

$$dy / dx = n\ \sin^{(n-1)}\ x\ \cos\ x$$

We showed above that if $y = u * v$ where u and v are both functions of x, then

$$dy / dx = u \, dv / dx + v \, du / dx$$

We can use this to differentiate tan x.
If

$$y = \tan x = (\sin x)/(\cos x)$$

then let

$$u = \sin x$$

for which

$$du / dx = \cos x$$

If

$$v = \sec x = 1/(\cos x) = (\cos x)^{-1}$$

then

$$dv / dx = -1 * (\cos x)^{-2} * (d(\cos x)/dx)$$

Hence,

$$dv / dx = -1 * (\cos x)^{-2} * (-\sin x) = \sin x / (\cos^2 x) = \tan x \sec x$$

(We can write this as $y = \sec x$ and $dy / dx = \sec x \tan x$.)
If

$$y = \tan x = (\sin x)/(\cos x) = u * v$$

then

$$du / dx = \cos x \text{ and } dv / dx = \sec x \tan x$$

By substituting these values in "{$dy / dx = u \, dv / dx + v \, du / dx$}," we obtain for

$$y = \tan x$$

then

$$\begin{aligned} dy / dx &= \sin x * \{\sin x / (\cos^2 x)\} + (1/\cos x) * \cos x \\ &= \sin^2 x / \cos^2 x + 1 = (\sin^2 x + \cos^2 x)/\cos^2 x \\ &= 1/\cos^2 x = \sec^2 x \end{aligned}$$

(since $\sin^2 + \cos^2 = 1$). Hence, if

$$y = \tan x, \text{ then } dy / dx = \sec^2 x$$

Similar relationships exist for cosec x and cot x. These are summarized in Box 10.2.

BOX 10.2 DIFFERENTIATING TRIGONOMETRICAL FUNCTIONS

If y = sin x	then	dy / dx = cos x
If y = cos x		dy / dx = −sin x
If y = tan x		dy / dx = sec² x
If y = cosec x		dy / dx = −cosec x cot x
If y = sec x		dy / dx = sec x tan x
If y = cot x		dy / dx = −cosec² x

10.3 POLYNOMIAL FUNCTIONS

Let us return to the general polynomial of the form:

$$y = f(x) = a_0 + a_1x + a_2x^2 + a_3x^3 + a_4x^4 + \cdots + a_nx^n$$

(Note as an aside that $a_0 = a_0x^0$ while $a_1x = a_1x^1$ and that $f(0)$, the value of $f(x)$ when $x = 0$ is such that $f(0) = a_0$.)

$$\mathrm{d}y / \mathrm{d}x = f'(x) = a_1 + 2a_2x + 3a_3x^2 + 4a_4x^3 + \cdots + na_nx^{(n-1)}$$

When x becomes zero then $f'(0) = a_1$ or $a_1 = f'(0) / 1$. If we differentiate $f'(x)$, we obtain $f''(x)$ and so:

$$\mathrm{d}^2y / \mathrm{d}x^2 = f''(x) = 2a_2 + 2 * 3a_3x + 3 * 4a_4x^2 + \cdots + (n-1)na_nx^{(n-2)}$$

Thus,

$$f''(0) = 2a_2$$

or

$$a_2 = f''(0) / (1 * 2)$$

Again,

$$\mathrm{d}^3y / \mathrm{d}x^3 = f'''(x) = 1 * 2 * 3a_3 + 2 * 3 * 4a_4x + \cdots + (n-2)(n-1)nx^{(n-3)}$$

Thus,

$$f'''(0) = 1 * 2 * 3 * a_3$$

or

$$a_3 = f'''(0) / (1 * 2 * 3)$$

If we keep on repeating this process, we will find that

$$a_n = f^n(0) / (1 * 2 * 3 * \cdots * n) = f^n(0) / n!$$

where $f^n(0)$ is the value of the nth derivative when $x = 0$. Hence, we can write our original equation as

$$y = f(0) + x^1f'(0)/1! + x^2f''(0)/2! + x^3f'''(0)/3! + \cdots + x^nf^n(0)/n!$$

This is known as Maclaurin's theorem after the 18th-century Scottish mathematician Colin Maclaurin. Its particular relevance relates to functions such as $f(x) = \sin x$ because then we have $f'(x) = \cos x$, $f''(x) = -\sin x$, $f'''(x) = -\cos x$, and $f^{iv}(x) = \sin x$.

Since sin (0) = 0 and cos (0) = 1, then by measuring x in radians and using Maclaurin's formula, we obtain:

$$\sin x = x - x^3/3! + x^5/5! - x^7/7! + \cdots$$

Likewise,

$$\cos x = 1 - x^2/2! + x^4/4! - x^6/6! + \cdots$$

Thus, we have a way of calculating sin x and cos x.

If we define the exponential function $f(x) = e^x$ as the function for which its differential is itself, then $f'(x) = e^x = f''(x)$, and so on. Since $e^0 = 1$ (anything to the power of zero = 1), then

$$\begin{aligned} e^x &= 1 + x^1/1! + x^2/2! + x^3/3! + x^4/4! + \cdots \\ e^1 &= 1 + 1/1! + 1/2! + 1/3! + 1/4! + \cdots \\ &= 2.7182818\ldots. \text{ As quoted earlier.} \end{aligned}$$

These series are infinite in that they go on forever, but they converge fairly rapidly allowing for the values of functions such as sine, cosine, and e^x to be computed relatively simply (Box 10.3 and Example 10.3). It is not possible to express all functions in the form of a polynomial but many can be expressed in this way.

Note that in the expansion of e^x given above, we can replace e^x by e^{nx} where n is some constant. The general term in the expansion of e^{nx} will be $(nx)^r / r! = n^r x^r / r!$ Differentiating,

$$d\left(n^r x^r / r!\right) / dx = r\, n^r x^{r-1} / r! = n\left\{n^{r-1} x^{r-1} / (r-1)!\right\}$$

This applies to every term in the expansion of e^{nx}. Hence,

$$\text{the differential of } e^{nx} = n * e^{nx}$$

or

$$d\left(e^{nx}\right) / dx = n * e^{nx}$$

BOX 10.3 MACLAURIN'S THEOREM

If $y = f(x)$, then

$$y = f(0) + xf'(0)/1! + x^2 f''(0)/2! + x^3 f'''(0)/3! + \cdots + x^n f^n(0)/n!$$

In particular,

$$e^x = 1 + x^1/1! + x^2/2! + x^3/3! + x^4/4! + \cdots.$$

And, where x is in radians,

$$\sin x = x - x^3/3! + x^5/5! - x^7/7! + \cdots \quad \cos x = 1 - x^2/2! + x^4/4! - x^6/6! + \cdots$$

Also (see Section 10.4), $\log_e(1 + x) = \ln(1 + x) = x - x^2/2 + x^3/3 - x^4/4 + \cdots$

In particular, the differential of $e^{-x} = d(e^{-x}) / dx = -e^{-x}$.

Also, if $y = e^x$, then $\log_e y = \log_e (e^x) = x \log_e e = x$ (as was shown in Chapter 2). Differentiating with respect to y rather than x we obtain:

$$\frac{d(\log_e y)}{dy} = dx / dy$$

But we defined e^x so that $dy / dx = e^x = y$. Hence,

$$dx / dy = 1 / y$$

Example 10.3 Sine and Cosine Values Using Maclaurin's Theorem

First, let us evaluate sin 30 where 30° = π / 6 radians = 0.5235987756r. Using the formula from Box 10.3, namely, $\sin x = x - x^3 / 3! + x^5 / 5! - x^7 / 7! + \cdots$ then,

$$\begin{aligned}\sin 30 &= 0.5235988 - 0.0239246 + 0.0003280 - 0.0000021 + \cdots \\ &= 0.5\end{aligned}$$

Using the formula for $\cos x = 1 - x^2 / 2! + x^4 / 4! - x^6 / 6! + \cdots$ then,

$$\begin{aligned}\cos 30 &= 1 - 0.1370778 + 0.0031317 - 0.0000286 + 0.0000001 - \cdots \\ &= 0.8660254\end{aligned}$$

These values agree with the tables for sine and cosine.

Thus,

$$\frac{d(\log_e y)}{dy} = 1 / y$$

In general,

$$\frac{d(\log_e x)}{dx} = 1 / x$$

Also,

$$\frac{d(\log_e y)}{dx} = \frac{d(\log_e y)}{dy} * dy / dx = (1 / y) * dy / dx$$

In the case of $y = e^{x^2}$, then $\log_e y = x^2$ and $(1 / y)dy / dx = 2x$. Hence,

$$\text{if } y = e^{x^2}$$

then

$$dy / dx = 2yx = 2xe^{x^2}$$

If we extend all these ideas to three or more dimensions, we will have functions such as

$$z = f(x, y) = 3x^3 + 2x^2y + y^3 + 4x + 5y$$

If we assume that y is treated as a constant, then we can differentiate z with respect to x. Similarly, we can keep x constant and differentiate with respect to y. These are known as the *partial derivatives* and are shown with a curly delta ∂. Thus, in this example,

$$\partial z / \partial x = 9x^2 + 4xy + 4$$

and

$$\partial z / \partial y = 2x^2 + 3y^2 + 5$$

We will use partial derivatives in later chapters.

10.4 LINEARIZATION

The process of converting complex relationships into linear combinations of variables is sometimes referred to as *linearization*. Sir Isaac Newton, for example, developed a way to solve quite complex nonlinear equations in which he reduced the problem to one dimension and repeated the process iteratively until he had a satisfactory solution. In Example 2.4 in Chapter 2, we showed how to obtain a square root by iteration, although we did not explain why it worked.

Consider the general case of $f(x) = k$ where k is a constant and $f(x)$ is a function of x such as the polynomial $ax^3 + bx^2 + cx$.

Example 10.4 Newton's Method for Solving Polynomials

In Table 10.1, we considered the function $y = f(x) = 1 + 9x - 6x^2 + x^3$ for which $dy / dx = f'(x) = 9 - 12x + 3x^2$. What is the value of x for which $y = 0$? From Table 10.1 an answer must lie between $x = -0.5$ and 0. Let us try

$$x_0 = -0.4 \text{ and use } x_{new} = x_{old} - \left\{f\left(x_{old}\right) - k\right\} / f'\left(x_{old}\right)$$

where

$$k = 1 + 9x - 6x^2 + x^3 \approx 0$$

Our next tries should be:

$$\begin{aligned} x_1 &= -0.4 - \{1 - 3.6 - 0.96 - 0.064\} / \{9 + 4.8 + 0.48\} \\ &= -0.4 + 3.624 / 14.28 = -0.146 \end{aligned}$$

So now try $x = -0.146$

$$x_2 = -0.146 - \{1 - 1.314 - 0.128 - 0.003\} / \{9 + 1.752 + 0.064\} = -0.146 + 0.445 / 10.816 = -0.105$$

So now try $x = -0.105$

$$\begin{aligned} x_3 &= -0.105 - \{1 - 945 - 0.066 - 0.001\} / \{9 + 1.260 + 0.033\} \\ &= -0.105 + 0.012 / 10.293 = -0.104 \end{aligned}$$

To three decimal places, the solution = −0.104

Solving the equations $ax^3 + bx^2 + cx = k$ can be complicated. In fact, *Newton's method* for solving equations (see Example 10.4) is particularly useful where the polynomial is of fifth or higher degree since there are no general algebraic methods for solving higher-order equations.

However, if we can take a guess at an approximate solution (in the case of a cubic there will of course be three possible answers, though some may involve imaginary numbers), then we can calculate

$$x_{\text{new}} = x_{\text{old}} - \left\{ \frac{f(x_{\text{old}} - k)}{f'(x_{\text{old}})} \right\} = x_{\text{old}} - \{f(x_{\text{old}} - k)\} / f'(x_{\text{old}})$$

where $f(x_{\text{old}})$ means the value of the function using the currently assumed value for x and f' is the first derivative evaluated using the same value.

We then repeat this process over and over until we have a sufficient number of significant figures for the answer to be acceptable. In fact, this is what happened in Example 2.4 where $f(x) = x^2 = k$ (with k being the number 27392834) and $f'(x) = 2x$.

We then set $x_{\text{new}} = [x_{\text{old}} - \{(x_{\text{old}})^2 - k\} / 2x_{\text{old}}] = [(1/2)\{x_{\text{old}} + k / x_{\text{old}}\}]$. On the third iteration we found the value we were seeking. Example 10.4 illustrates the procedure for a cubic.

In Section 10.3, we introduced Maclaurin's theorem in which

$$y = f(0) + x^1 f'(0)/1! + x^2 f''(0)/2! + x^3 f'''(0)/3! + \cdots + x^n f^n(0)/n!$$

Another 18th-century mathematician, Brook Taylor, has given his name to an extension of this theorem, which states that if $f(x)$ is a polynomial of the form $a_0 + a_1 x + a_2 x^2 + \cdots + a_n x^n$, then if δ is any number

$$f(x+\delta) = f(\delta) + xf'(\delta)/1! + x^2 f''(\delta)/2! + x^3 f'''(\delta)/3! + \cdots + x^n f^n(\delta)/n!$$

where f^n is the nth derivative of $f(x)$. The proof of this is given in Box 10.4. By putting $\delta = 0$, we have the same expression as in the Maclaurin theorem.

BOX 10.4 TAYLOR'S THEOREM

Given that

$$f(x) = a_0 + a_1 x + a_2 x^2 + \cdots + a_n x^n$$

we have

$$f'(x) = a_1 + 2a_2 x + \cdots + na_n x^{n-1}$$
$$f''(x) = 2a_2 + 3 * 2a_3 x + \cdots + n * (n-1) a_n x^{n-2}, \text{ and so on.}$$

Also, for

$$f(x+\delta) = a_0 + a_1(x+\delta) + a_2(x+\delta)^2 + \cdots + a_n(x+\delta)^n$$

then

$$f'(x+\delta) = a_1 + 2a_2(x+\delta) + \cdots + na_n(x+\delta)^{n-1}, \text{ and so on.}$$

$f''(x+\delta) = 2a_2 + 6a_3(x+\delta) + \cdots + n * (n-1)a_n(x+\delta)^{n-2}$, and so on, for all values of x. Hence, putting $x = 0$,

$$f(\delta) = a_0 + a_1(\delta) + a_2(\delta)^2 + \cdots + a_n(\delta)^n$$
$$f'(\delta) = a_1 + 2a_2(\delta) + \cdots + na_n(\delta)^{n-1}$$
$$f''(\delta) = 2a_2 + 6a_3(\delta) + \cdots + n * (n-1)a_n(\delta)^{n-2}, \text{ and so on.}$$

Combining all these, we obtain Taylor's theorem, which states that

$$f(x+\delta) = f(\delta) + xf'(\delta)/1! + x^2 f''(\delta)/2! + x^3 f'''(\delta)/3! + \cdots + x^n f^n(\delta)/n!$$

We can also express this as

$$f(x+\delta) = f(x) + \delta f'(x)/1! + \delta^2 f''(x)/2! + \delta^3 f'''(x)/3! + \cdots + \delta^n f^n(x)/n!$$

If, for instance, $y = f(x) = \log_e(x)$, then

$$f'(x) = (x)^{-1}; f''(x) = -(x)^{-2}; f'''(x) = +2(x)^{-3}; \ldots; f^n(x) = (-1)^{(n-1)}(n-1)!(x)^{(-n)}$$

If $\delta = 1$, then

$$\log_e(x+1) = f(1) + xf'(1)/1! + x^2 f''(1)/2! + x^3 f'''(1)/3! + \cdots + x^n f^n(1)/n! + \cdots$$

Since $\log_e(1) = 0$;

$$f^n(1) = (-1)^{(n-1)}(n-1)!\ (1)^{(-n)} \text{ and } x^n f^n(1)/n! = (-1)^{(n-1)} x^n/n$$

Hence,

$$\log_e(1+x) = \ln(1+x) = x - x^2/2 + x^3/3 - x^4/4 + \cdots$$

Given as shown in Box 10.4 that

$$f(x+\delta) = f(x) + \delta f'(x)/1! + \delta^2 f''(x)/2! + \delta^3 f'''(x)/3! + \cdots + \delta^n f^n(x)/n!$$

then if δ is small, we can ignore terms in δ^2 and rearrange this as

$$\delta = \left\{\frac{f(x+\delta) - f(x)}{f'(x)}\right\} = \{f(x+\delta) - f(x)\}/f'(x)$$

leading to

$$x_{\text{new}} = (x+\delta) = x_{\text{old}} - \{f(x_{\text{old}}) - k\}/f'(x_{\text{old}})$$

where k is the value of $f(x)$ that we are seeking. This is what we call *Newton's method* for solving complex equations.

In fact, Taylor's expansion can be extended for functions other than polynomials, including functions of the form $z = f(x, y)$. This involves partial derivatives of the form $\frac{\partial^{(r+s)} f(x,y)}{\partial x^r \partial y^s}$, which means all the combinations of the partial derivatives up to $n = r + s$. As stated at the end of Section 10.3, a first-order partial derivative is the differential of an expression in terms of one variable, all other variables being treated as constants. We do not intend to explore higher-order partial derivatives because here, fortunately, we can use only the first derivatives if we have made a reasonable approximation; if not, we will just have to repeat the process several times. This gives us:

$$f\{x+\delta x, y+\delta y\} = f(x,y) + (\partial f / \partial x)\delta x + (\partial f / \partial y)\delta y + \text{ some small terms}$$

where $\partial f / \partial x$ means the differential of $f(x, y)$ with respect to x, with y being treated as a constant.

We will not demonstrate the proof of this here, but simply note that if we can ignore the terms of higher order, then we are left with a linear relationship of the form

$$f_{\text{new}} = f_{\text{old}} + (\partial f_{\text{old}} / \partial x)\delta x + (\partial f_{\text{old}} / \partial y)\delta y$$

where f_{old} means the value that we use when inserting the initial or iterated values in the function $f(x, y)$ and its partial derivative with respect to x and the partial derivative with respect to y. We will work through an example of how this applies when we discuss least square adjustment in Chapter 15.

10.5 BASIC INTEGRATION

The process of differentiation can be put into reverse in what is known as *integration*. We have shown that if $y = x^n$, then $\mathrm{d}y / \mathrm{d}x = nx^{(n-1)}$. Consider the function $f(x) = c + x^{(n+1)} / (n + 1)$ where c is any constant number. If we differentiate this, we obtain $f'(x) = x^n$. The constant c that had been introduced disappears when we differentiate since it is in effect cx^0 and becomes 0.

If $y = x^n$, then the function $\{c + x^{(n+1)} / (n + 1)\}$ is said to be the *indefinite integral* of y. It is called *indefinite* since the number c is at this stage unknown. c is said to be the *constant of integration.*

The integral of y is often shown with an elongated letter s of the form $\int$. If $y = bx$, then:

$$\text{the integral of } y \text{ with respect to } x = \int y\,\mathrm{d}x = \int (bx)\,\mathrm{d}x = c + bx^2 / 2$$

Rather than keep writing "with respect to x," we use the notation from before and write "$\int y\,\mathrm{d}x$" meaning "what happens to y when we integrate with respect to x."

If $y = \sin x$, then $\int y\,\mathrm{d}x = c - \cos x$ while $\int \cos x\,\mathrm{d}x = c + \sin x$. Examples of the results of integration are given in Box 10.5.

As with basic arithmetic

$$\int \{f(x) + g(x)\}\mathrm{d}x = \int f(x)\mathrm{d}x + \int g(x)\mathrm{d}x$$

Also,

$$\int kf(x)\mathrm{d}x = k\int f(x)\,\mathrm{d}x \text{ where } k \text{ is a constant.}$$

We have seen that if

$$y = f(x) \; * \; g(x)$$

then

$$\mathrm{d}y / \mathrm{d}x = g(x) \; * \; \mathrm{d}f / \mathrm{d}x + f(x) \; * \mathrm{d}g / \mathrm{d}x$$

Unfortunately, $\int \{f(x) * g(x)\}\,\mathrm{d}x$ cannot be treated in such a neat way. It does not obey this rule.

Given that integration is the reverse of differentiation and given that the integral of x^n equals $\int x^n dx = c + \{1 / (n+1)\}x^{n+1}$, we can apply the same principles if n is negative, with one exception. Thus, the integral of $x^{-2} = \int x^{-2}dx = c - x^{-1}$.

The one exception occurs when $n = -1$. We cannot integrate x^{-1} in this way since $\{1 / (n + 1)\}$ would become 1 / 0 and be infinite. We have, however, seen that

$$\frac{d(\log_e x)}{dx} = 1/x = x^{-1}$$

BOX 10.5 SUMMARY OF INTEGRALS

$$\int x^n\ dx = c + \{1/(n+1)\}x^{n+1} \text{ provi ded } n \neq -1$$

$$\int x^{-1}\ dx = c + \log_e x$$

$$\int \sin x\ dx = c - \cos x$$

$$\int \sin 2x\ dx = c - (1/2)\cos 2x$$

$$\int \cos x\ dx = \sin x + c$$

$$\int e^{nx} = (1/n)e^{nx} + c$$

Also,

$$\int \tan x = \log_e(\sec x) + c$$

$$\int \cot x = \log_e(\sin x) + c$$

$$\int \sec x = \log_e(\sec x + \tan x) + c$$

{This is shown in Box 10.6 to equal $\log_e \tan(\pi / 4 + x / 2) + c$}

$$\int \text{cosec } x = \log_e(\text{cosec } x - \cot x) + c$$

{This is shown in Box 10.6 to equal $\log_e(\tan x / 2) + c$}

Knowing this,

$$\int x^{-1}\ dx = c + \log_e x$$

Since the differential of $e^x = e^x$, it follows that $\int e^x dx = e^x$ + constant. Also,

$$\int e^{nx}\ dx = (1/n)e^{nx} + \text{constant.}$$

For trigonometric functions, the easiest way to show that the integral of $\tan x = \int \tan x = \log_e(\sec x)$ is by differentiating the answer. Thus,

$$\text{if } v = \sec x$$

then

$$dv / dx = \sec x \tan x$$

Also,

$$d(\log_e v) / dv = 1 / v$$

Hence,

$$d(\log_e (\sec x)) / dx = \{d(\log_e v) / dv\} * \{(dv / dx)\} = (1 / \sec x) * \sec x \tan x = \tan x$$

BOX 10.6 COMBINED FUNCTIONS

In Box 10.5, we quote the fact that ($\sec x + \tan x$) = $\tan (\pi / 4 + x / 2)$. To prove this, we use the formulae for $\sin (A + B)$ and $\cos (A + B)$ in Chapter 6, Box 6.3,

$$\tan (\pi/4 + x/2) = \sin (\pi/4 + x/2) / \cos (\pi/4 + x/2)$$

$$= \{\sin \pi/4 \cos x/2 + \cos \pi/4 \sin x/2\} / \{\cos \pi / 4 \cos x/2 - \sin \pi\ 4 \sin x/2\}$$

$$= \{\cos x/2 + \sin x/2\} / \{\cos x/2 - \sin x/2\}$$

(since $\sin \pi / 4 = \cos \pi / 4 = 1 / \sqrt{2}$).
Multiplying above and below by $\{\cos x / 2 + \sin x / 2\}$, then,

$$\tan (\pi / 4 + x / 2) = \{\cos^2 x / 2 + \sin^2 x / 2 + 2 \cos x / 2 \sin x / 2\} / \{\cos^2 x / 2 - \sin^2 x / 2\}$$

But $\cos 2A = \cos^2 A - \sin^2 A$, $\sin 2A = 2 \sin A \cos A$, and $\cos^2 A + \sin^2 A = 1$. Hence,

$$\tan (\pi / 4 + x / 2) = \{1 + \sin x\} / \cos x = \sec x + \tan x$$

Thus,

$$\int \sec x = \log_e (\sec x + \tan x) + c = \log_e \{\tan(\pi / 4 + x / 2)\} + c$$

Similarly,

$$\operatorname{cosec} x - \cot x = \{1 - \cos x\} / \sin x = \{2 \sin^2 x / 2\} / \{2 \sin x / 2 \cos x / 2\}$$

Hence,

$$\operatorname{cosec} x - \cot x = \tan x / 2$$

Thus,

$$\int \operatorname{cosec} x = \log_e (\operatorname{cosec} x - \cot x) + c = \log_e (\tan x / 2) + c$$

10.6 AREAS AND VOLUMES

Integration occurs in geomatics and geographic information system (GIS) in a number of operations, one of which is in determining area. Consider two points on a curve (Figure 10.5), for instance,

$$P(x, y) \text{ and } P'(x + \delta x, y + \delta y)$$

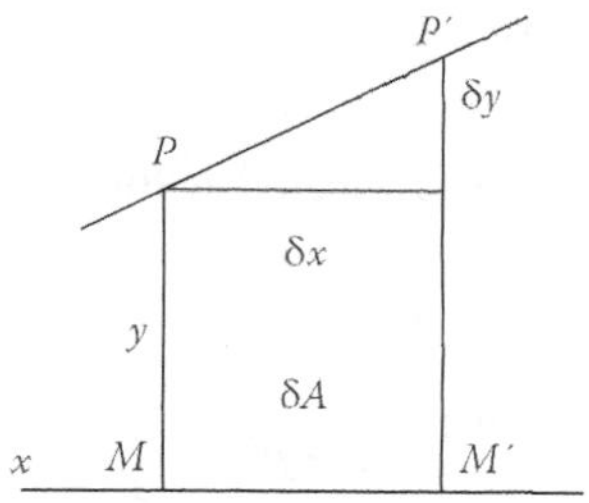

FIGURE 10.5 Area beneath a curve.

The area of the strip beneath the section of the curve $PP' = MPP'M'$ which we will call δA. Then, $\delta A = y\delta x + ½\,\delta y\delta x$ = a rectangle plus a triangle. As in previous discussions, as δy and δx get small, so

$$\delta A / \delta x = y + \frac{1}{2}\delta y$$

In the limit,

$$\mathrm{d}A / \mathrm{d}x = y$$

We can express this in another way by saying that if we add all the little bits of δA together as we move along the curve, we will obtain the whole area under the curve, namely, $\Sigma\delta A$ where Σ (sigma) means "the sum of."

More particularly, as δx tends to zero we can change to the differential notation and in so doing change the summation symbol from Σ to $\int$ yielding

$$A = \int \mathrm{d}A = \int y \;\mathrm{d}x$$

Thus, $\int y\,\mathrm{d}x$ is the area under a curve, between the curve and the x-axis.

If, for example, $y = x^2$ then the area under the curve $= c + x^3 / 3$. We still, of course, have the constant of integration that is unknown. If, however, we are interested in the area under the curve between the points, for instance, where $x = 1$ and $x = 2$, then it will be $\{(c + 2^3 / 3) - (c + 1^3 / 3)\}$ and the unknown c disappears, yielding 7 / 3.

By setting definite limits to the integration (say from $x = a$ to $x = b$), we obtain the *definite integral*. We write this as $\int_a^b$ which means "the value of the integral $\int y\,\mathrm{d}x$ when x equals b minus the value when x equals a."

If we set limits to the integration, then the result is sometimes written within squared brackets in the form $[I]_{x=a}^{x=b}$ where I is the indefinite integral that is evaluated at $x = b$ and $x = a$.

So,

$$\text{if } y = x^2$$

then

$$I = x^3 / 3 \text{ and } \left[\frac{x^3}{3}\right]_{x=a}^{x=b} = \left(b^3 - a^3\right) / 3$$

Returning to Figure 10.4, the area of the very narrow triangle ABC is

$$\frac{1}{2}\text{base} \;*\; \text{height} = 1/2R \;*\; R\delta\theta$$

If we integrate this between 0 and 2π,

$$\text{Area} = \int_0^{2\pi} 1/2R^2 d\theta = 1/2R^2\theta \text{ evaluated within the limits } \theta = 0 \text{ to } \theta = 2\pi$$

This is

$$\left[\frac{R^2\theta}{2}\right]_0^{2\pi} = \left[\pi R^2 - 0\right] = \pi R^2$$

This confirms the well-known formula for the area of a circle.

If $y = ax^2 + bx + c$ and is part of a parabola, then the area between y_1 at x_1 and y_2 at x_2 and the x-axis will be

$$\left[ax^3/3 + bx^2/2 + cx\right]_{x_1}^{x_2}$$

Area

$$= ax_2^{\,3}/3 + bx_2^{\,2}/2 + cx_2 - ax_1^{\,3}/3 - bx_1^{\,2}/2 - cx_1$$
$$= ax_2^{\,3}/3 - ax_1^{\,3}/3 + bx_2^{\,2}/2 - bx_1^{\,2}/2 + cx_2 - cx_1$$

Here, the lowest common denominator = 6. Also,

$$\left(x_2^{\,3} - x_1^{\,3}\right) = \left(x_2 - x_1\right)\left(x_2^{\,2} + x_2x_1 + x_1^{\,2}\right) \text{ and } \left(x_2^{\,2} - x_1^{\,2}\right) = \left(x_2 - x_1\right)\left(x_2 + x_1\right)$$

Hence,

$$\text{Area} = (1/6)\left(x_2 - x_1\right)\left\{2ax_2^{\,2} + 2ax_2x_1 + 2ax_1^{\,2} + 3bx_2 + 3bx_1 + 6c\right\}$$
$$= (1/6)\left(x_2 - x_1\right)\left\{ax_2^{\,2} + bx_2 + c + ax_1^{\,2} + bx_1 + c + ax_2^{\,2} + 2ax_2x_1 + ax_1^{\,2} + 2bx_2 + 2bx_1 + 4c\right\}$$
$$= (1/6)\left(x_2 - x_1\right)\left\{y_2 + y_1 + a\left(x_2 + x_1\right)^2 + 2b\left(x_2 + x_1\right) + 4c\right\}$$
$$= (1/6)\left(x_2 - x_1\right)\left\{y_2 + y_1 + 4a\left(\frac{x_2 + x_1}{2}\right)^2 + 4b\left(\frac{x_2 + x_1}{2}\right) + 4c\right\}$$
$$= (1/6)\left(x_2 - x_1\right)\left\{y_2 + y_1 + 4y_m\right\}$$

where y_m is the value of y at the point midway between x_1 and x_2 where x has the value $(x_1 + x_2)/2$.

This formula, known as *Simpson's rule* after the 18th-century mathematician Thomas Simpson, provides a way in which to determine areas of irregular shapes drawn on a map. Thus, in Figure 10.6 we can divide an area up into a series of strips of equal width and measure the

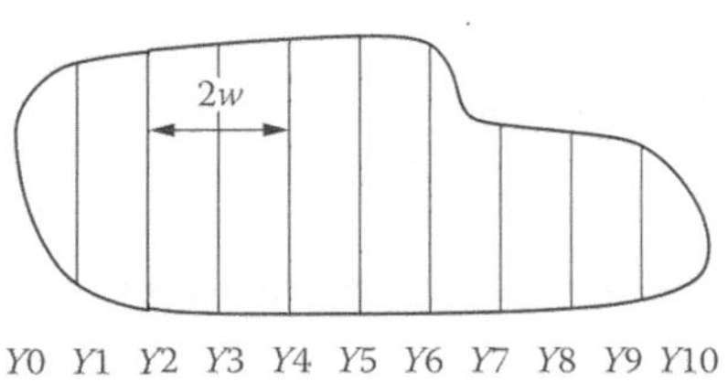

FIGURE 10.6 Area of an irregular shape.

length of each straight line. If we assume that each strip is of width w then, if we treat each strip as a parallelogram, its area is

$$w * (Y_r + Y_{r+1})/2$$

For Figure 10.6,

$$\text{Area} = (w/2) * (Y_0 + 2Y_1 + 2Y_2 + 2Y_3 + 2Y_4 + 2Y_5 + 2Y_6 + 2Y_7 + 2Y_8 + 2Y_9 + Y_{10})$$

This assumes that the boundaries are all straight lines rather than curves. If we assume that each end of a strip is part of a parabola, we can use Simpson's rule (shown in Box 10.7). To do so, we would first need to make sure that there is an even number of strips and then treat each pair of strips. The area of the first pair, for example, is:

$$\text{Area of double strip} = (1/6) * 2w * (Y_0 + 4Y_1 + Y_2)$$

The area for the whole shape would then be

$$\text{Area} = (w/3)\{Y_0 + 4Y_1 + 2Y_2 + 4Y_3 + 2Y_4 + 4Y_5 + 2Y_6 + 4Y_7 + 2Y_8 + 4Y_9 + Y_{10}\}$$

This will usually give a better approximation as it assumes that the boundaries are curves, not straight lines. In Example 10.5, we apply the formula to an ellipse to show how reasonable even with only four strips, the area calculation can be.

BOX 10.7 AREA OF IRREGULAR SHAPES

For strips of width w and lengths Y_i ($i = 0$ to n) then:

By Trapezium rule:

$$\text{Area} = (w/2) * (Y_0 + 2Y_1 + 2Y_2 + 2Y_3 + 2Y_4 + 2Y_5 + \cdots + 2Y_{n-2} + 2Y_{n-1} + Y_n)$$

By Simpson's rule:

$$\text{Area} = (w/3)\{Y_0 + 4Y_1 + 2Y_2 + 4Y_3 + 2Y_4 + 4Y_5 + \cdots + 2Y_{n-2} + 4Y_{n-1} + Y_n\}$$

where n is an even number.

Example 10.5 The Area of an Ellipse

In Chapter 5, Section 5.6, we quoted the area of an ellipse as "πab" where a is the semimajor axis and b the semiminor. The reason for this can be seen from Figure 5.16 in Chapter 5, which shows the auxiliary circle whose area is πa^2. Since in the ellipse the dimensions parallel to the major axis are unchanged but in the direction of the minor axis, the dimensions are all reduced by a factor b / a. Thus, the area of the ellipse is (b / a) times the area of its auxiliary circle, namely, $(b / a) * (\pi a^2)$ or πab.

Let us divide the ellipse into four strips of equal width parallel to the minor axis (Figure 10.7). If spaced at an interval along the major axis of $a / 2$, then $y_0 = 0 = y_4$; $y_2 = 2b$; and $y_1 = b\sqrt{3} = y_3$ since the ellipse is defined by $x^2 / a^2 + y^2 / b^2 = 1$ and the x values at y_1 and y_3 are $\pm a / 2$ giving half the lengths as $b(\sqrt{3}) / 2$.

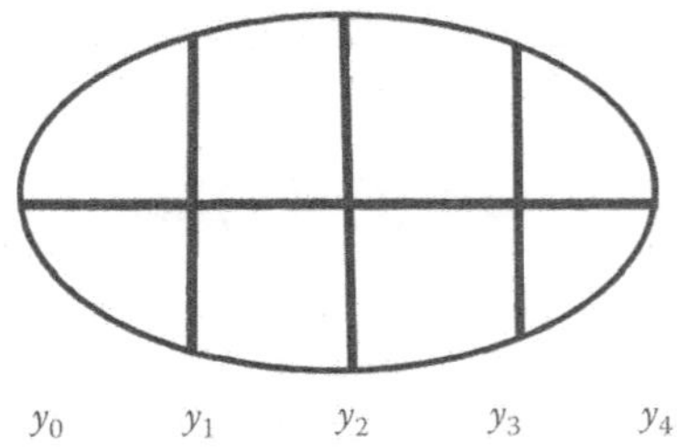

FIGURE 10.7 An ellipse cut into four strips.

Using Simpson's rule (Box 10.7), $w = a / 2$.

$$\text{The area of the ellipse} = (a/6)\left(0+4 * b\sqrt{3}+2 * 2b+4 * b\sqrt{3}+0\right)$$
$$= ab\left(4+8\sqrt{3}\right)/6 \approx 3ab$$

This is not far short of the true value (πab or 3.14 * ab), and we only divided the area into four strips.

The principles of integral calculus can also be used to determine volumes. The volume of a cylinder, whatever the shape of its base = base area * height provided the sides are parallel. Thus, the volume of the cylinder in Figure 10.8 = Ah.

For the cone (be it a pyramid or any other base shape with a base area A and height h) the area A_z of cross-sectional slice at height z above the base will be in proportion to the base A but reduced both in terms of length and breadth to an amount

$$\{h-z)/h\}$$

Its area will be

$$A_z = \{(h-z)/h\} * \{(h-z)/h\} * A = \{A/h^2\}\{h^2-2hz+z^2\}$$

If the cross-section is an amount δz thick, then its volume will be

$$\delta V = \left(\frac{A}{h^2}\right)\left(h^2-2hz+z^2\right)\delta z$$

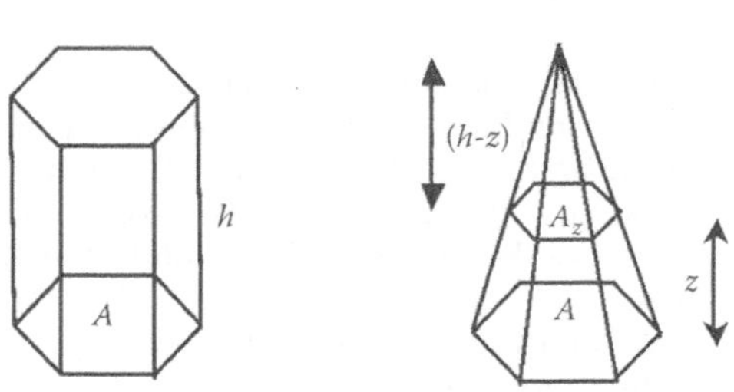

FIGURE 10.8 Volumes of a cylinder and cone.

If we sum all of this from the base to the apex

$$V = \left(A/h^2\right) = \left(A/h^2\right)\left[h^2 z - hz^2 + z^3/3\right]_{z=0}^{z=h}$$
$$= (1/3)Ah^3/h^2 = (1/3)Ah$$
$$= \text{one-third of base area} * \text{height}$$

10.7 NUMERICAL METHODS

The scope of problems encountered by GIS practitioners extends through a broad spectrum of applied sciences. Many of these disciplines benefit from creating mathematical models of processes and phenomena being studied. Many such models include differential or integral equations, and evaluating integrals often becomes a necessity. As described in earlier parts of the chapter, a definite integral is an infinite sum of small quantities which may or may not converge. In cases where this sum converges, it is possible to evaluate it either algebraically or numerically.

To proceed with calculating the integral we must first have a function, say $f(x)$, which is a formula that we can use to calculate each one of the small quantities that we will add together in the process of integration. We partition the domain of interest into equal intervals (Figure 10.9).

The rationale behind this is that due to the linearity of an integral, if we add up the results of integration of each interval, we will obtain the value of the integral over the entire domain we started with.

Our next challenge is to evaluate this integral in each interval. If we continue with the same idea of splitting the interval into even smaller parts, we will enter an infinite process. This is the moment when an approximation comes into play. If the size of one interval is small enough, the shape of the function over this interval might look rather simple (Figure 10.10a).

In the simplest case it will be approximately a straight horizontal line (Figure 10.10b). Although on closer examination it might still look a bit curved (Figure 10.10c). Depending on the type of function we might see that it still has a complicated shape even at this scale (Figure 10.10d). In such a case we attempt to reduce the size of each interval (and increase the number of intervals at the same time) until the function looks reasonably simple.

Method of rectangles: The simplest case is when we consider only one of the endpoints (Figure 10.11a), and assume that the entire interval has this height.

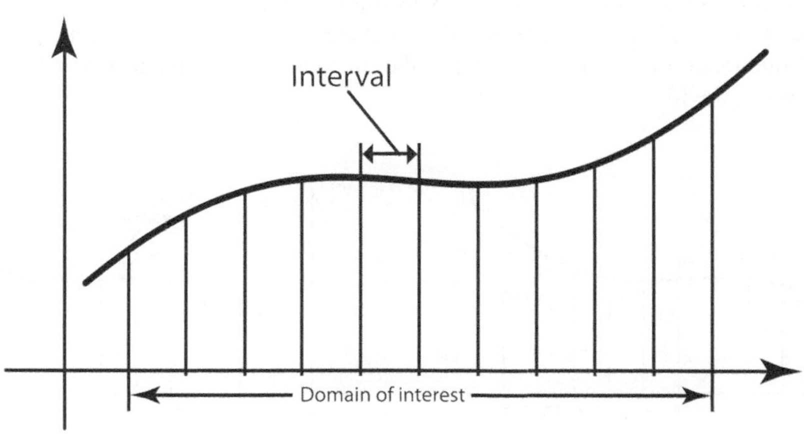

FIGURE 10.9 Subdividing an integral into intervals.

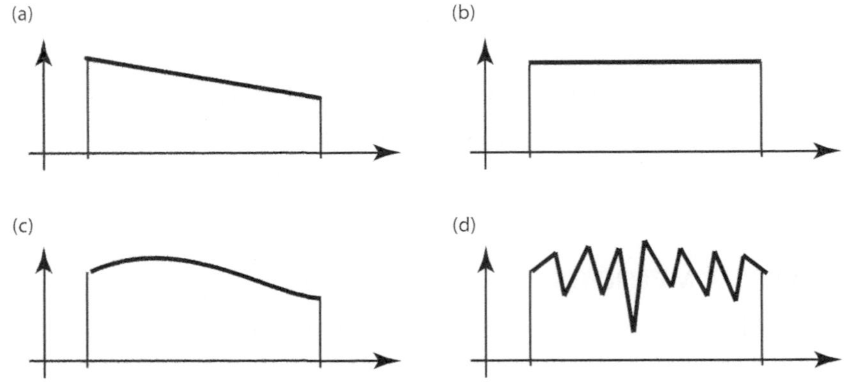

FIGURE 10.10 (a–d) Possible shapes of the function in an interval.

Method of trapezoids: The next simplest case is that we take the average of the two endpoints (Figure 10.11b) of the interval, multiply it by the length, d, of the interval and that is the value I_i of the integral for this interval.

$$I_i = \frac{f(x)+f(x+d)}{2}d$$

In this case, the value of the integral over the domain of interest [A,B] is:

$$I = \sum_{i=1}^{n} I_i = \sum_{i=1}^{n} \frac{f(x)+f(x+d)}{2}d = \left[\frac{f(A)+f(B)}{2} + \sum_{i=1}^{n-1} f(A+id)\right]d$$

where n is the number of intervals and, therefore:

$$d = \frac{B-A}{n}$$

If the function over a small interval is not a straight line this method produces an error, the size of which depends on the size of the interval and on the function. To reduce this error, one could increase the number of intervals. Theoretically by doing so the error can be reduced to any desired value. As the number of intervals increases the computer will have to perform more computations, and the total amount of time it takes to complete the calculation will depend on how time-consuming each evaluation of the function $f(x)$ is.

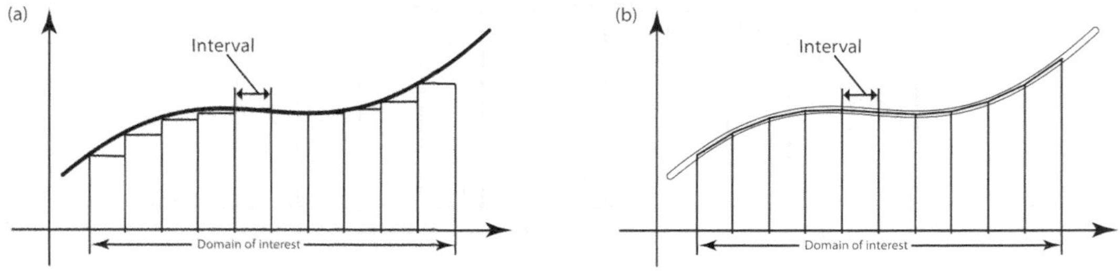

FIGURE 10.11 (a, b) Model each interval with a single value or a trapezoid.

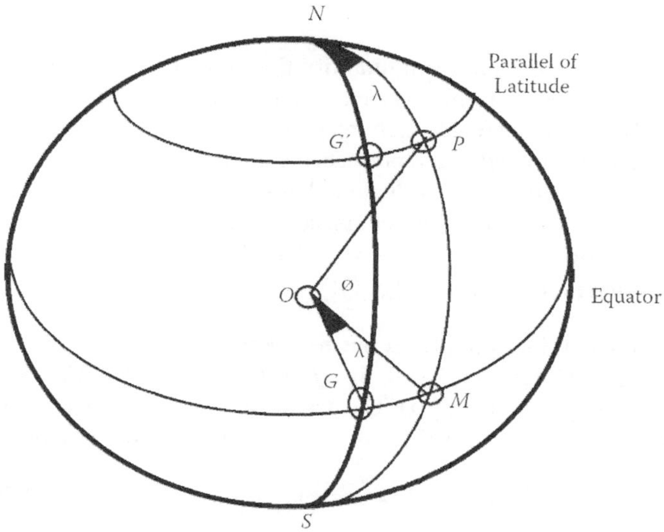

GG′ is the Greenwich Meridian of Origin (Zero Longitude)
GM is the Equator (Zero Latitude)
O is the center of the spherical Earth
The angle *GOM* is the *longitude* of $P = \lambda$
The angle *MOP* is the *latitude* of $P = \varnothing$

FIGURE 10.12 A rectangle bounded by constant latitude and longitude lines.

We can also apply such numerical methods to spherical geometry problems. A common problem is the area calculation of a larger rectangle bounded by constant latitude and longitude lines, or, constant meridian and parallel lines. Suppose we want to calculate the area of a rectangle that spans 10 degrees longitude and 60 degrees latitude, and whose southernmost bound is the equator, as in Figure 10.12. The simplest approximation is to calculate this area as a flat trapezoid. This approximation, incidentally, is how manufacturers of globes utilize in order to cover the surface of the sphere with patches of printed maps.

Here we proceed to calculate the trapezoid's area.

1. The length of the lower horizontal edge is 10 degrees longitude at equator, which converts to the radius of the earth times 10 degrees converted to radians, and equals to 1113.17 km.
2. The upper horizontal edge is 10 degrees longitude at 60 degrees latitude, whose length turns out to be the same number times the cosine of 60 degrees. This turns out to be 556.59 km.
3. The vertical distance at both left and right edges are calculated by the radius of the earth times 60 degrees converted to radians, which turns out to be 6679.02 km.
4. Using the formula to calculate the area of a trapezoid, the approximate area of this "rectangle" is 5,576,163.5, or roughly 5.6 million square kilometers.

The above approximation is the simplest and would have larger error than if one subdivides the large trapezoid we created above into many smaller trapezoid-shaped areas. The more subdivisions are made, the more accurate the result will be.

GLOSSARY

Constant of integration An unknown quantity that must be added after *integration*. Thus, $\int x \, dx = \frac{1}{2}x^2 + c$ where c is the constant of integration.

Definite integral The value obtained when integrating a function between defined limits. In the case of a polynomial $y = f(x)$, it represents the area between a curve and the x-axis and the x-limits between which it is evaluated. Thus, from $x = a$ to $x = b$, it is $\int f(x)$ evaluated at b minus $\int f(x)$ evaluated at a. This eliminates the need for a *constant of integration*. If $g(x) = \int f(x)$, then the value would be $g(b) - g(a)$. The limits are written at the top and bottom of the integral sign in the form $\int_a^b$ or $\left[g(x)\right]_{x=a}^{x=b}$.

Delta (δ) A small amount, for example, δx means a very small increase in x.

Derivative The limit of the function $\{f(x + \delta) - f(x)\} / \delta$ as δ tends to zero. Where $y = f(x)$, this is written either as dy / dx or as $f'(x)$. The derivative of a derivative (called the *second derivative*) is written as d^2y / dx^2 or as $f''(x)$. In the case of $y = x^n$, the derivative $dy / dx = nx^{(n-1)}$ then,

$$d^2y / dx^2 = n(n-1)x^{(n-2)} \; d\{\sin(x)\} / dx = \cos(x) \; d\{\cos(x)\} / dx = -\sin(x) d\{\tan(x)\} / dx = \sec^2(x) d\{\text{cosec}(x)\} / dx = -\text{cosec}(x)\cot(x) d\{\sec(x)\} / dx = \sec(x)\tan(x) d\{\cot(x)\} / dx = -\text{cosec}^2(x)$$

The derivative of $\log(x) = 1 / x$. For compound functions of the form $y = f(x) * g(x)$, then $dy / dx = f(x) * g'(x) + g(x) * f'(x)$ where $f'(x)$ and $g'(x)$ are the first derivatives of $f(x)$ and $g(x)$.

If $y = u * v$ where u and v are both functions of x then,

$$dy / dx = u \, dv / dx + v \, du / dx$$

Differential calculus The branch of mathematics that studies *derivatives*.

Differentiation The operation of determining a lower *derivative* of a function.

Indefinite integral The function that arises when a given function is integrated within unknown limits, hence the need to introduce an unknown quantity called the *constant of integration*.

Integral Written with the symbol $\int$ and sometimes known as an *antiderivative*. It is the function, which when differentiated, gives the function on which the *integration* is carried out. Thus, $\int x^n \, dx = \{x^{(n+1)}\} / (n + 1) + c$ where c is a *constant of integration*.

Integral calculus The branch of mathematics that studies *integration*.

Integration The process of determining the *integral* of a function.

Linearization The process of converting complex relationships into more simple linear combinations of variables.

Maclaurin's theorem If the function "f" is infinitely differentiable then,

$$f(x) = f(0) + x^1 f'(0) / 1! + x^2 f''(0) / 2! + \cdots\cdots x^n f^n(0) / n! + \cdots\cdots$$

where $f^n(0)$ is the value of the nth derivative when x is set to zero. From this

$$\sin(x) = x - x^3 / 3! + x^5 / 5! - x^7 / 7! + \cdots\cdots \; \cos(x) = 1 - x^2 / 2! + x^4 / 4! - x^6 / 6! + \cdots \; e^x = 1 + x^1 / 1! + x^2 / 2! + x^3 / 3! + x^4 / x! + \cdots\cdots$$

Maxima and minima Points on a curve where $dy / dx = 0$.

Newton's method A method for solving complex equations by iteration using

$$x_{\text{new}} = x_{\text{old}} - \left\{ \frac{f(x_{\text{old}} - k)}{f'(x_{\text{old}})} \right\}$$

Normal The line at a point on a curve perpendicular to the *tangent.*

Partial derivative Normally written as a curly delta ∂. Differentiation of a function with two or more variables with respect to one of the variables, all the others being assumed constant. If

$$z = ax^2 + by^2 + cxy + dx + ey + f$$

then,

$$\partial z / \partial x = 2ax + cy + d \quad \partial z / \partial y = 2by + cx + e$$

Point of inflection Point on a curve where $d^2y / dx^2 = 0$ and changes from positive to negative or vice versa. At this point the curve crosses from one side of its tangent to the other.

Points of maximum or minimum Points on a curve where $dy / dx = 0$.

Second derivative (third and fourth) The *derivative* of dy / dx. Shown as d^2y / dx^2. It is the rate of change of dy / dx. The third derivative is the derivative of the second, and so on.

Simpson's rule If a closed shape is divided into an even number of strips of width w by straight lines of length $y_0, y_1, \ldots, y_n$ where n is an even number then the area of the irregular shape

$$\approx (w/3) \ * \ \left(y_0 + 4y_1 + 2y_2 + 4y_3 + 2y_4 + \cdots\cdots 2y_{(n-2)} + 4y_{(n-1)} + y_n\right).$$

Tangent A line that touches a curve so that its slope = dy / dx.

Taylor's theorem If $f(x)$ is a polynomial of the form $a_0 + a_1x + a_2x^2 + \cdots + a_nx^n$ and if δ is any number, then

$$f(x+\delta) = f(\delta) + x\, f'(\delta)/1! + x^2 f''(\delta)/2! + x^3 f'''(\delta)/3! + \cdots x^n f^n(\delta)/n!$$

For instance,

$$\log_e(1+x) = \ln(1+x) = x - x^2/2 + x^3/3 - x^4/4 + \cdots$$

Trapezium rule If a closed shape is divided into a number of strips of width w by straight lines of length $y_0, y_1, \ldots, y_n$, then the area of the irregular shape is approximately

$$(w/2) \ * \ \left(y_0 + 2y_1 + 2y_2 + 2y_3 + 2y_4 + \cdots\cdots 2y_{(n-2)} + 2y_{(n-1)} + y_n\right).$$

Volume of pyramid or cone Volume = one-third of base area * height.

Unit IV

Modeling Space

Geographic information system (GIS) practitioners can't get away from maps. Before the invention of modern-day computers, maps are the instruments created to store and analyze almost all geographic data. As a human cognitive invention maps have evolved throughout history, from the ancient schematic diagrams to the modern-day maps that precisely displays the location of places and things, guided by grid tics and measurable by a ruler. However, maps are not authentic representations of the Earth's surface, not only due to possible inaccuracies, but due to the fundamental problem that maps are two-dimensional, while the Earth's surface is three-dimensional. Despite this fundamental problem, mathematicians starting from the Late Middle Ages have invented methods to ensure that correctly made maps are useful and can be relied upon. Unit IV in the book is about understanding the precise mathematical methods behind the modeling of the three-dimensional geometry of the Earth's surface via two-dimensional maps of various types, and related mathematical methods used in photogrammetry. As a branch of mapping science, photogrammetry in concerned with measuring the precise location of objects on earth by cameras on board airplanes and satellites.

Although GIS practitioners generally rely on standardized algorithms programmed in GIS software packages for map projection tasks, understanding the mathematical foundations of map projection enables practitioners to understand the pros and cons of specific projection types, including the distortions that must be properly considered. In Chapter 11, we explore the mathematical methods behind transformation of three-dimensional coordinates to two-dimensional ones, which is the basis behind the ability to create flat, two-dimensional maps via projecting the Earth's three-dimensional surface. In Chapter 12, we explore the geometric structures behind each type of commonly used map projection and their pros and cons. In Chapter 13, we explore mathematical methods behind modeling curves and surfaces, such as lines and areas on the Earth's surface.

DOI: 10.1201/9781003464402-14

11 2D/3D Transformations

11.1 HOMOGENEOUS COORDINATES

In this chapter, we will consider the two-dimensional (2D) representation of three-dimensional (3D) objects in what is sometimes called *visualization*. We often need to display the landscape in perspective on a screen or on paper. The problem is how best to transfer the 3D coordinates of a point in space onto a flat surface. In Chapter 12, we will look at the specific case of points on the surface of the Earth and how they can be displayed on a map, for instance, in an atlas. Here, we are looking at 3D Euclidean space in general and how that may be transformed into two dimensions.

First, we need to extend the idea of 3D coordinates (x, y, z) to homogeneous coordinates (x, y, z, w) where in effect the w is a scaling factor and not a fourth dimension. The actual 3D coordinates in normal Euclidean space are then $(x / w, y / w, z / w)$. In the 2D case, the homogenous coordinates would be (x, y, w) with the traditional values with which we normally deal being $(x / w, y / w)$. The reason for introducing this extra complexity is to allow us to describe points that are at infinity.

Consider the line AB and a point C in Figure 11.1. Every line through C cuts the line AB somewhere (e.g., CD, CE, or CF); but when the line through C is parallel to AB, it is an infinite distance away, although in a known direction. It is not just anywhere but specifically in the direction in which AB is pointing.

If in conventional terms AB is the line $y = mx + c$, then, if we use homogeneous coordinates, we can express the point at infinity as $(x, mx, 0)$ or even as $(1, m, 0)$ and still retain the relationship that y is basically m times x as shown in Example 11.1. We can use this relationship to draw perspective, as parallel lines are made to appear to converge on the horizon as illustrated in Figure 11.2 at what are called the *vanishing points*.

In Box 13.1 in Chapter 13, we show various equations in parametric form; for instance, the straight line was given as

$$x = p + t \cos\theta; y = q + t \sin\theta$$

where (p, q) is a point on the line of slope θ, and t is the distance along that line from (p, q).

If $l = \cos\theta$ and $n = \sin\theta$, then $x = p + lt$ and $y = q + nt$. If we let $w = 1 / t$, then in homogeneous coordinates, any point on the line can be represented as

$$\left(pw + l,\ qw + n,\ w\right)$$

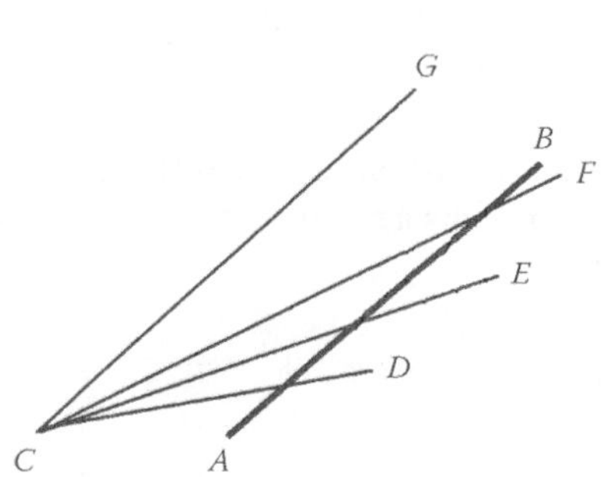

FIGURE 11.1 Points at infinity.

DOI: 10.1201/9781003464402-15

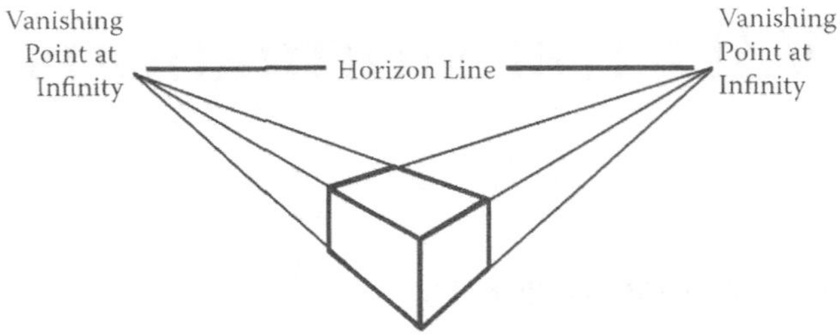

FIGURE 11.2 Vanishing points for a rectangular block.

As the number t goes off to infinity, w tends to zero and we move to point $(l, n, 0)$.

Note that if k is any constant, then the point $\{k(pw + l), k(qw + n), kw)\}$ represents the same Euclidean value as $(pw + l, qw + n, w)$ and hence $(kl, kn, 0)$ is the same point as $(l, n, 0)$. This is the point at infinity in the direction of the slope angle θ, namely, $\tan^{-1}(n / l)$.

Example 11.1 Homogeneous Coordinates

Consider the lines (a) $y = 2x + 3$ and (b) $y = 2x + 5$. Point (1, 5) lies on line (a) since $5 = 2 * 1 + 3$ but does not satisfy line (b) since $5 \neq 2 * 1 + 5$.

The two lines are parallel, and therefore in Euclidean space in terms of ordinary coordinates, they do not meet.

If we express any point on the lines as (x, y, w) in homogeneous coordinates, then we can express the first equation as $(y / w) = 2(x / w) + 3$ or equation (a) as $y = 2x + 3w$.

Similarly, equation (b) takes the form $y = 2x + 5w$.

In homogeneous coordinates, point (1, 2, 0) satisfies both equation (a) and equation (b) and hence that is a point at which the two parallel lines meet. It is known as the point at infinity on the line $y = 2x + c$.

(In the nonhomogeneous version, (1, 2) represented point (1, 2, 1) which does not lie on the line $y = 2x + 3$.)

11.2 ROTATING AN OBJECT

Consider a point whose 3D homogeneous coordinates are (x, y, z, w). In Chapter 9, we discussed the rotation of axes. In order to rotate an object or the axes that determines its position, we can apply a 4 * 4 matrix of the form:

$$\begin{pmatrix} \cos\theta & -\sin\theta & 0 & 0 \\ \sin\theta & \cos\theta & 0 & 0 \\ 0 & 0 & 1 & 0 \\ 0 & 0 & 0 & 1 \end{pmatrix} * \begin{pmatrix} x \\ y \\ z \\ w \end{pmatrix} = \begin{pmatrix} x\cos\theta - y\sin\theta \\ x\sin\theta + y\cos\theta \\ z \\ w \end{pmatrix}$$

This results in new homogeneous coordinates $\{(x\cos\theta - y\sin\theta), (x\sin\theta + y\cos\theta), z, w\}$.

As we saw in Chapter 9, this represents a rotation about the z-axis and we can call the matrix $\mathbf{R}_z$. Similarly, about the y-axis, we have

$$\mathbf{R}_y = \begin{pmatrix} \cos\phi & 0 & -\sin\phi & 0 \\ 0 & 1 & 0 & 0 \\ \sin\phi & 0 & \cos\phi & 0 \\ 0 & 0 & 0 & 1 \end{pmatrix}$$

For a rotation about the x-axis,

$$\mathbf{R}_x = \begin{pmatrix} 1 & 0 & 0 & 0 \\ 0 & \cos\omega & -\sin\omega & 0 \\ 0 & \sin\omega & \cos\omega & 0 \\ 0 & 0 & 0 & 1 \end{pmatrix}$$

In Chapter 9, we called these two matrices $\mathbf{R}_y$ and $\mathbf{R}_x$ and discussed how we can combine them into one 3 * 3 matrix, which we will show here as a 4 * 4 matrix

$$\mathbf{R} = \begin{pmatrix} a & b & c & 0 \\ d & e & f & 0 \\ g & h & i & 0 \\ 0 & 0 & 0 & 1 \end{pmatrix}$$

The values of a, b, c, and so forth, will depend on the sequence in which the rotations take place, but so long as the original rotations represent orthogonal transformations, the effect will be to keep right angles as right angles and straight lines as straight lines. If $\mathbf{R}$ is not an orthogonal matrix, the axes will become skewed.

Consider next the transformation

$$\begin{pmatrix} 1 & 0 & 0 & t \\ 0 & 1 & 0 & u \\ 0 & 0 & 1 & v \\ 0 & 0 & 0 & 1 \end{pmatrix} * \begin{pmatrix} x \\ y \\ z \\ w \end{pmatrix} = \begin{pmatrix} x+tw \\ y+uw \\ z+vw \\ w \end{pmatrix}$$

This results in new coordinates $x' = x/w + t$, $y' = y/w + u$, $z' = z/w + v$. This is a simple translation of the origin from its original point to a new origin at $(-t, -u, -v)$.

Now consider

$$\begin{pmatrix} 1 & 0 & 0 & 0 \\ 0 & 1 & 0 & 0 \\ 0 & 0 & 1 & 0 \\ 0 & 0 & 0 & s \end{pmatrix} * \begin{pmatrix} x \\ y \\ z \\ w \end{pmatrix} = \begin{pmatrix} x \\ y \\ z \\ sw \end{pmatrix}$$

This results in a new set of coordinates that in more conventional form are:

$$x' = x / sw, y' = y / sw, \text{ and } z' = z / sw$$

This means that all dimensions have been changed in scale by an amount = $1 / s$.

Finally, consider:

$$\begin{pmatrix} 1 & 0 & 0 & 0 \\ 0 & 1 & 0 & 0 \\ 0 & 0 & 1 & 0 \\ 0 & 0 & 0 & s \end{pmatrix} * \begin{pmatrix} x \\ y \\ z \\ w \end{pmatrix} = \begin{pmatrix} x \\ y \\ z \\ sw \end{pmatrix}$$

This results in

$$\begin{pmatrix} x \\ y \\ z \\ px + qy + rz + sw \end{pmatrix}$$

The expression $px + qy + rz + sw$ is the equation of a plane and all values will be reduced to a plane that is skew to our original axes, the amount of skew being dependent on the values p, q, and r. So taken overall, if we apply the matrix

$$\mathbf{M} = \begin{pmatrix} a & b & c & t \\ d & e & f & u \\ g & h & i & v \\ p & q & r & s \end{pmatrix}$$

to the set of coordinates $(x, y, z, w)^T$ then we will change the origin, rotate the axes, change the overall scale, and project the whole scene onto a plane surface. Thus, by suitable choice of the 16 elements, we can project any string of coordinates from three dimensions into two.

In Examples 11.2–11.5, we show the translation and rotation of a solid figure—in this case the barn shown in Figure 11.3. First, in Example 11.2 we move the origin to point A.

Next we will rotate the object about the z-axis and then about the x-axis, as shown in Example 11.3.

In Examples 11.2 and 11.3, we carried out a series of operations one at a time. These can be combined into a single operation as shown in Example 11.4.

It is important to note that the sequence of operations is critical and that $\mathbf{R}_x * \mathbf{R}_z * \mathbf{N}'$ as shown in Example 11.3 is not the same as $\mathbf{R}_z * \mathbf{R}_x * \mathbf{N}'$. This is demonstrated in Example 11.5 where the rotation about the x-axis is applied first, and then there is a rotation about the z-axis. Likewise, if the origin is moved after, rather than before the rotations, there will be a different solution.

The matrix N'_{xz} in Example 11.3 represents the coordinates in space of the 10 points after the rotation of the axes. There is no change in shape and no change in scale. The lengths of the sides are the same as before and right angles remain as right angles in 3D space. This is a *similarity transformation* in that it is a simple rotation and translation in which orthogonal lines remain orthogonal. A similarity transformation can also incorporate a uniform scale change. It is also *affine* in that it preserves collinearity and parallelism, straight lines remaining straight lines.

The image shown in Figure 11.4 treats the y values as zero and the image is displayed on the plane $y = 0$. From N'_{xz} in Examples 11.3 and 11.4, the nonhomogeneous coordinates are (x, z), namely,

$$A(0,0), B(8.7,-0.9), C(-1.3,-3.9), D(-10,-3), E(0,7.9), F(8.7,7), G(-1.3,4), H(-10,4.9),$$
$$L(4.3,11.4), M(-5.7,8.4)$$

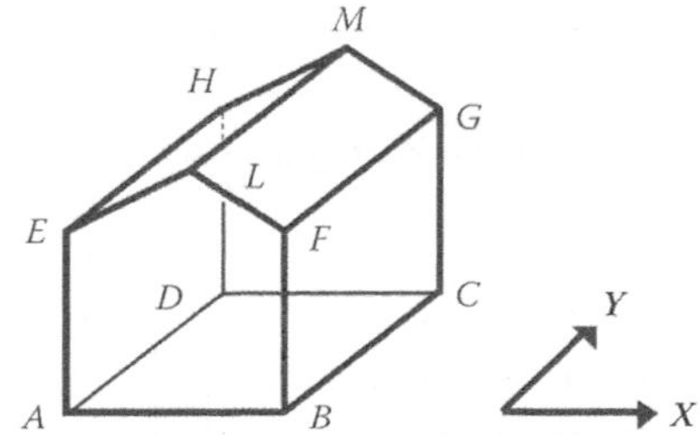

FIGURE 11.3 A barn.

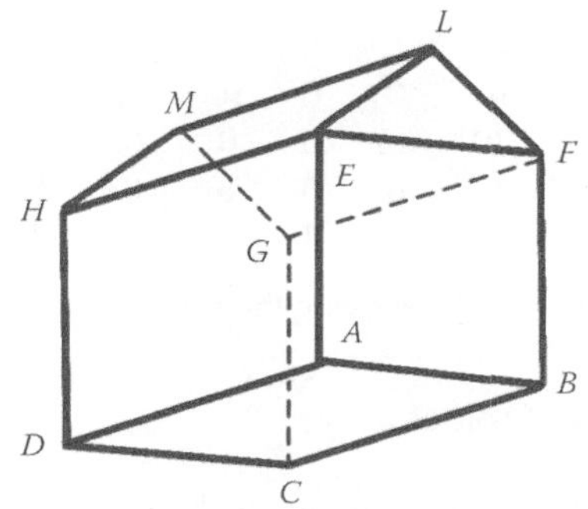

FIGURE 11.4 The barn after two rotations.

Example 11.2 Translation of Axes

Consider a barn (Figure 11.3) with a ridged roof for which the coordinates of the floor are

$$A(10,20,100), B(20,20,100), C(20,40,100), \text{ and } D(10,40,100).$$

Let the line of the eaves be

$$E(10,20,108), F(20,20,108), G(20,40,108), \text{ and } H(10,40,108)$$

and the ridge be

L(15,20,112), M(15,40,112)

The *AB* direction is x, the *AD* is y, and the *AE* is z.

Let us express all these 10 points as homogeneous coordinates in a matrix **N** where the columns represent the four coordinates of the corner points.

$$\begin{array}{cccccccccc} A & B & C & D & E & F & G & H & L & M \end{array}$$

$$N = \begin{pmatrix} 10 & 20 & 20 & 10 & 10 & 20 & 20 & 10 & 15 & 15 \\ 20 & 20 & 40 & 40 & 20 & 20 & 40 & 40 & 20 & 40 \\ 100 & 100 & 100 & 100 & 108 & 108 & 108 & 108 & 112 & 112 \\ 1 & 1 & 1 & 1 & 1 & 1 & 1 & 1 & 1 & 1 \end{pmatrix}$$

Now let us change the origin to point *A*(10, 20, 100, 1) by applying the matrix **T** where

$$\mathbf{T} = \begin{pmatrix} 1 & 0 & 0 & -10 \\ 0 & 1 & 0 & -20 \\ 0 & 0 & 1 & -100 \\ 0 & 0 & 0 & 1 \end{pmatrix}$$

$$T * N =$$

$$\begin{pmatrix} 1 & 0 & 0 & -10 \\ 0 & 1 & 0 & -20 \\ 0 & 0 & 1 & -100 \\ 0 & 0 & 0 & 1 \end{pmatrix} * \begin{pmatrix} 10 & 20 & 20 & 10 & 10 & 20 & 20 & 10 & 15 & 15 \\ 20 & 20 & 40 & 40 & 20 & 20 & 40 & 40 & 20 & 40 \\ 100 & 100 & 100 & 100 & 108 & 108 & 108 & 108 & 112 & 112 \\ 1 & 1 & 1 & 1 & 1 & 1 & 1 & 1 & 1 & 1 \end{pmatrix}$$

Call this

$$\mathbf{N}' = \begin{pmatrix} 0 & 10 & 10 & 0 & 0 & 10 & 10 & 0 & 5 & 5 \\ 0 & 0 & 20 & 20 & 0 & 0 & 20 & 20 & 0 & 20 \\ 0 & 0 & 0 & 0 & 8 & 8 & 8 & 8 & 12 & 12 \\ 1 & 1 & 1 & 1 & 1 & 1 & 1 & 1 & 1 & 1 \end{pmatrix}$$

This is a simple translation of axes by moving the location of the origin.

Example 11.3 Rotation of an Object

Using the data in Example 11.2, now let us rotate the building about the *z*-axis of the new origin, by an amount of 30° (with cos 30 ≈ 0.866 and sin 30 = 0.5).

$$\mathbf{R}^{*}\mathbf{N}' = \begin{pmatrix} 0.866 & -0.5 & 0 & 0 \\ 0.5 & 0.866 & 0 & 0 \\ 0 & 0 & 1 & 0 \\ 0 & 0 & 0 & 1 \end{pmatrix} * \mathbf{N}' = \mathbf{N}'_z$$

Our 10 points then, subject to rounding errors, become:

$$\begin{matrix} & A & B & C & D & E & F & G & H & L & M \end{matrix}$$

$$\mathbf{N}'_z = \begin{pmatrix} 0 & 8.66 & -1.34 & -10 & 0 & 8.66 & -1.34 & -10 & 4.33 & -5.67 \\ 0 & 5 & 22.32 & 17.32 & 0 & 5 & 22.32 & 17.32 & 2.5 & 19.82 \\ 0 & 0 & 0 & 0 & 8 & 8 & 8 & 8 & 12 & 12 \\ 1 & 1 & 1 & 1 & 1 & 1 & 1 & 1 & 1 & 1 \end{pmatrix}$$

Hence, *A* has become (0, 0, 0, 1), *B* (8.66, 5, 0, 1), and so forth. Now let us tip the image backward by rotating the shape by 10° about the *x*-axis (clockwise for right-handed systems; cos 10 = 0.985 and sin 10 = 0.174). The transformation matrix is then

$$\mathbf{R}_x * \mathbf{N}'_z = \begin{pmatrix} 1 & 0 & 0 & 0 \\ 0 & 0.985 & 0.174 & 0 \\ 0 & -0.174 & 0.985 & 0 \\ 0 & 0 & 0 & 1 \end{pmatrix} * \mathbf{N}'_z = \mathbf{N}'_{xz}$$

from where

$$\mathbf{N}'_{xz} = \begin{pmatrix} 0 & 8.66 & -1.34 & -10 & 0 & 8.66 & -1.34 & -10 & 4.33 & -5.67 \\ 0 & 4.92 & 21.99 & 17.06 & 1.39 & 6.31 & 23.38 & 18.45 & 4.55 & 21.61 \\ 0 & -0.87 & -3.88 & -3.01 & 7.88 & 7.01 & 4.00 & 4.87 & 11.38 & 8.37 \\ 1 & 1 & 1 & 1 & 1 & 1 & 1 & 1 & 1 & 1 \end{pmatrix}$$

Example 11.4 Combining Rotations

In Example 11.3, we could have combined all three operations by applying one matrix **R** where

$$\mathbf{R} = \begin{pmatrix} 1 & 0 & 0 & 0 \\ 0 & 0.985 & 0.174 & 0 \\ 0 & -0.174 & 0.985 & 0 \\ 0 & 0 & 0 & 1 \end{pmatrix} * \begin{pmatrix} 0.866 & -0.5 & 0 & 0 \\ 0.5 & 0.866 & 0 & 0 \\ 0 & 0 & 1 & 0 \\ 0 & 0 & 0 & 1 \end{pmatrix} * \begin{pmatrix} 1 & 0 & 0 & -10 \\ 0 & 1 & 0 & -20 \\ 0 & 0 & 1 & -100 \\ 0 & 0 & 0 & 1 \end{pmatrix}$$

$$= \begin{pmatrix} 0.866 & -0.5 & 0 & 1.34 \\ 0.492 & 0.853 & 0.174 & -39.38 \\ -0.087 & -0.151 & 0.985 & -94.62 \\ 0 & 0 & 0 & 1 \end{pmatrix}$$

What we would then have, subject to rounding errors, is: **R** * **N** =

$$\begin{pmatrix} 0.866 & -0.5 & 0 & 1.34 \\ 0.492 & 0.853 & 0.174 & -39.38 \\ -0.087 & -0.151 & 0.985 & -94.62 \\ 0 & 0 & 0 & 1 \end{pmatrix} * \begin{pmatrix} 10 & 20 & 20 & 10 & 10 & 20 & 20 & 10 & 15 & 15 \\ 10 & 20 & 40 & 40 & 20 & 20 & 40 & 40 & 20 & 40 \\ 100 & 100 & 100 & 100 & 108 & 108 & 108 & 108 & 112 & 112 \\ 1 & 1 & 1 & 1 & 1 & 1 & 1 & 1 & 1 & 1 \end{pmatrix}$$

$$= \begin{pmatrix} 0 & 8.66 & -1.34 & -10 & 0 & 8.66 & -1.34 & -10 & 4.33 & -5.67 \\ 0 & 4.92 & 21.99 & 17.06 & 1.39 & 6.31 & 23.38 & 18.45 & 4.55 & 21.61 \\ 0 & -0.87 & -3.88 & -3.01 & 7.88 & 7.01 & 4.00 & 4.87 & 11.38 & 8.37 \\ 1 & 1 & 1 & 1 & 1 & 1 & 1 & 1 & 1 & 1 \end{pmatrix}$$

which is the same value for N'_{xz} as the one derived earlier in Example 11.3.

Example 11.5 The Sequence of Rotations

The sequence in which rotations are carried out is important. If in Example 11.3 we had carried out the rotation about the x-axis before we carried out the rotation about the z, we would obtain a different answer **R′**.

$$\mathbf{R}' = \begin{pmatrix} 0.866 & -0.5 & 0 & 0 \\ 0.5 & 0.866 & 0 & 0 \\ 0 & 0 & 1 & 0 \\ 0 & 0 & 0 & 1 \end{pmatrix} * \begin{pmatrix} 1 & 0 & 0 & 0 \\ 0 & 0.985 & 0.174 & 0 \\ 0 & -0.174 & 0.985 & 0 \\ 0 & 0 & 0 & 1 \end{pmatrix} * \begin{pmatrix} 1 & 0 & 0 & -10 \\ 0 & 1 & 0 & -20 \\ 0 & 0 & 1 & -100 \\ 0 & 0 & 0 & 1 \end{pmatrix}$$

$$= \begin{pmatrix} 0.866 & -0.492 & -0.082 & 9.89 \\ 0.5 & 0.853 & 0.151 & -37.13 \\ 0 & -0.174 & 0.985 & -95.02 \\ 0 & 0 & 0 & 1 \end{pmatrix}$$

This is not the same as the original **R** shown in Example 11.4.

In Figure 11.4, the 3D coordinates show that the lengths DH, AE, and FB are all the same as are DA, HE, and ML, and so forth. EF and AB are parallel and meet at a point on the x-axis that is an infinite distance away. This point can be given the coordinates (1, 0, 0, 0). If we multiply this by the skew matrix:

$$\mathbf{S}_x = \begin{pmatrix} 1 & 0 & 0 & 0 \\ 0 & 1 & 0 & 0 \\ 0 & 0 & 1 & 0 \\ 0.025 & 0 & 0 & 1 \end{pmatrix} \text{ then } \mathbf{S}_x * \begin{pmatrix} 1 \\ 0 \\ 0 \\ 0 \end{pmatrix} = \begin{pmatrix} 1 \\ 0 \\ 0 \\ 0.025 \end{pmatrix}$$

The point at infinity along the x-axis becomes the point (1, 0, 0, 0.025) or in conventional coordinates (40, 0, 0) since $1 / (0.025) = 40$. Similarly, lines in the y-direction can be drawn to the point at infinity in the y-direction, namely, point (0, 1, 0, 0). Thus,

$$\mathbf{S}_{xy} = \begin{pmatrix} 1 & 0 & 0 & 0 \\ 0 & 1 & 0 & 0 \\ 0 & 0 & 1 & 0 \\ 0.025 & 0.025 & 0 & 1 \end{pmatrix}$$

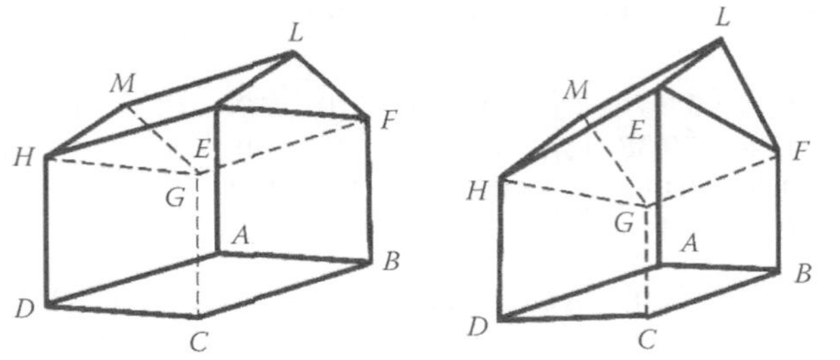

FIGURE 11.5 The affine and perspective projections of the barn.

will bring both the points at infinity in the x- and y-directions into a range that can be plotted on a sheet of paper. We then have a perspective view (see Figure 11.5) as shown in Example 11.6.

Figure 11.5 shows the barn after translation and rotation before and after applying perspective. The results from Example 11.6 provide the coordinates of the barn with a perspective view, again as in Figure 11.4, treating the y value as zero.

Example 11.6 Adding Perspective

Using the data from Examples 11.3 and 11.4,

$$\begin{pmatrix} 1 & 0 & 0 & 0 \\ 0 & 1 & 0 & 0 \\ 0 & 0 & 1 & 0 \\ 0.025 & 0.025 & 0 & 1 \end{pmatrix} * \mathbf{N}'_{xz}$$

$$= \begin{pmatrix} 1 & 0 & 0 & 0 \\ 0 & 1 & 0 & 0 \\ 0 & 0 & 1 & 0 \\ 0.025 & 0.025 & 0 & 1 \end{pmatrix} * \begin{bmatrix} 0 & 8.66 & -1.34 & -10 & 0 & 8.66 & -1.34 & -10 & 4.33 & -5.67 \\ 0 & 4.92 & 21.99 & 17.06 & 1.39 & 6.31 & 23.38 & 18.45 & 4.55 & 21.61 \\ 0 & -0.87 & -3.88 & -3.01 & 7.88 & 7.01 & 4.00 & 4.87 & 11.38 & 8.37 \\ 1 & 1 & 1 & 1 & 1 & 1 & 1 & 1 & 1 & 1 \end{bmatrix}$$

$$= \begin{pmatrix} 0 & 8.66 & -1.34 & -10 & 0 & 8.66 & -1.34 & -10 & 4.33 & -5.67 \\ 0 & 4.92 & 21.99 & 17.06 & 1.39 & 6.31 & 23.38 & 18.45 & 4.55 & 21.61 \\ 0 & -0.87 & -3.88 & -3.01 & 7.88 & 7.01 & 4.00 & 4.87 & 11.38 & 8.37 \\ 1 & 1.34 & 1.52 & 1.17 & 1.03 & 1.37 & 1.55 & 1.21 & 1.22 & 1.40 \end{pmatrix}$$

$$= \begin{pmatrix} 0 & 6.46 & -0.88 & -8.55 & 0 & 6.32 & -0.86 & -8.26 & 3.55 & -4.05 \\ 0 & 3.67 & 14.47 & 14.58 & 1.35 & 4.61 & 15.08 & 15.25 & 3.73 & 15.43 \\ 0 & -0.65 & -2.55 & -2.57 & 7.65 & 5.12 & 2.58 & 4.02 & 9.33 & 5.98 \\ 1 & 1 & 1 & 1 & 1 & 1 & 1 & 1 & 1 & 1 \end{pmatrix}$$

Prior to applying the perspective transformation, the nonhomogeneous coordinates were

$$A(0,0,0), B(8.7,4.9,-0.9), C(-1.3,22,-3.9), D(-10,17,-3), E(0,1.4,7.9),$$
$$F(8.7,6.3,7), G(-1.3,23.4,4), H(-10,18.4,4.9), L(4.3,4.6,11.4), M(-5.7,21.6,8.4)$$

After applying the perspective transformation they became:

$$A(0,0,0), B(6.5,3.7,-0.6), C(-0.9,14.5,-2.5), D(-8.5,14.6,-2.6), E(0,1.3,7.6)$$
$$F(6.3,4.6,5.1), G(-0.9,15.1,2.6), H(-8.3,15.2,4.0), L(3.5,2.7,9.3)$$
$$M(-4.0,15.4,6.0)$$

Note that the two points at infinity (1, 0, 0, 0) and (0, 1, 0, 0) under the original transformation became

$$\begin{pmatrix} 0.866 & -0.5 & 0 & 1.34 \\ 0.492 & 0.853 & 0.174 & -39.38 \\ -0.087 & -0.151 & 0.985 & -94.62 \\ 0 & 0 & 0 & 1 \end{pmatrix} * \begin{pmatrix} 1 & 0 \\ 0 & 1 \\ 0 & 0 \\ 0 & 0 \end{pmatrix} = \begin{pmatrix} 0.866 & -0.5 \\ 0.492 & 0.853 \\ -0.087 & -0.151 \\ 0 & 0 \end{pmatrix}$$

and were still at infinity. However, by applying the matrix $\mathbf{S}_{xy}$ we obtain

$$\begin{pmatrix} 1 & 0 & 0 & 0 \\ 0 & 1 & 0 & 0 \\ 0 & 0 & 1 & 0 \\ 0.025 & 0.025 & 0 & 1 \end{pmatrix} * \begin{pmatrix} 0.8666 & -0.5 \\ 0.492 & 0.853 \\ -0.087 & -0.151 \\ 0 & 0 \end{pmatrix} = \begin{pmatrix} 0.866 & -0.5 \\ 0.492 & 0.853 \\ -0.087 & -0.151 \\ 0.034 & 0.009 \end{pmatrix}$$
$$= \begin{pmatrix} 25.47 & -55.6 \\ 14.47 & 94.78 \\ -2.54 & -16.78 \\ 1 & 1 \end{pmatrix}$$

Thus, the lines *AB*, *EF*, *HG*, and *DC* converge at *P* (25.5, 14.5, –2.6) as shown in Figure 11.6, while *AD*, *EH*, *FG*, *BC*, and *LM* converge at (–55.6, 94.8, –16.8) (not shown).

Transformations of the kind illustrated above can be used, for example, to simulate a 3D landscape including the effects of changing viewing positions, for instance, for "fly-by" presentations. Each point or line can be transformed and then displayed at its new coordinates. Straight lines remain straight but parallel lines can be made to converge.

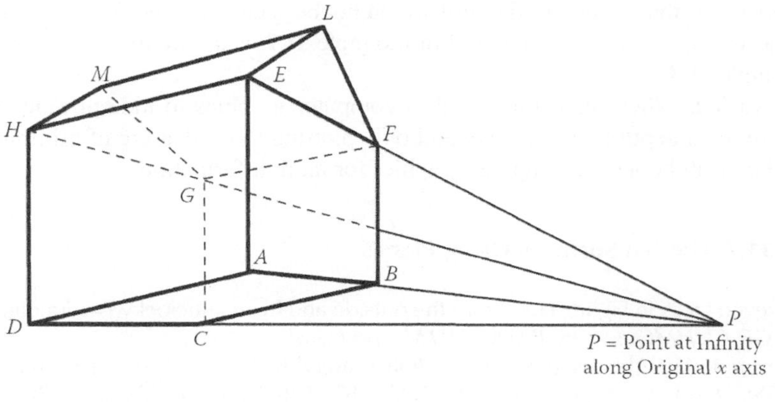

FIGURE 11.6 Transformation into a perspective view.

11.3 HIDDEN LINES AND SURFACES

Reverting to Euclidean space and nonhomogeneous coordinates, in Chapter 4 we noted that for two points on a line, A (x_A, y_A) and B (x_B, y_B), then

$$(y-y_A)=\frac{(y_B-y_A)}{(x_B-x_A)}*(x-x_A)$$

Given a point P (x_P, y_P), if we calculate

$$y=y_A+\frac{(y_B-y_A)}{(x_B-x_A)}*(x_P-x_A)$$

and if $(y_P > y)$, then P is above the line AB while if $y_P < y$, then P lies below the line.

If $y_P = y$, then P is on the line AB.

We can extend this idea to any plane through points ABC, which in Chapter 8 (Section 8.3) we showed to have the form

$$\begin{aligned}&(x-x_C)\{y_A(z_B-z_C)+y_B(z_C-z_A)+y_C(z_A-z_B)\}\\&+(y-y_C)\{z_A(x_B-x_C)+z_B(x_C-x_A)+z_C(x_A-x_B)\}\\&+(z-z_C)\{x_A(y_B-y_C)+x_B(y_C-y_A)+x_C(y_A-y_B)\}=0\end{aligned}$$

When we insert values for x, y, and z this quantity will be either positive or negative unless the point lies on the plane. Two points on the same side of the plane will share the same sign (i.e., both will give a positive value or both will give a negative).

We can use this fact to determine which lines are "hidden" when we view a solid object projected onto a flat surface. Every plane surface can be defined by three points so, for example, in Figure 11.6, for the surface defined by the four points A, D, H, and E, we only need the coordinates of three of these (see Example 11.7). If we take a coordinate value inside the building, say a point Q at (0, 5, 0) in the above example, and compare it with our viewing position, say somewhere along the y-axis at V (0, −1000, 0), then Q and V will be found to be on opposite sides of the plane ADE but on the same side of any of the planes that contain point G, which is therefore invisible. Each surface will need to be tested, but this sort of routine calculation is ideally suited to electronic data processing.

Note, however, that it is important to use a point that is well away from the object. Choosing a point close to the object but still outside the building—for instance (0, −100, 0) in Example 11.7—would give a negative amount for the plane *HELM*, showing it to be a hidden surface. This would, in fact, be correct, as that section of the roof would not be seen if viewed from so near the building.

An alternative approach to that adopted in Example 11.7 would be to use vectors. This is illustrated in Example 11.8.

There are various other algorithms used in computer graphics to hide unwanted surfaces, for instance, by giving a depth to every pixel and only plotting those that are of minimum depth. The reader should consult books on computer graphics for more information.

Example 11.7 Hidden Surfaces Using Planes

There are seven faces in Figure 11.6. From the outside and in an anticlockwise direction, these are *ABFLE, BCGF, CDHMG, DAEH, BADC, HELM,* and *FGML*.

From Example 11.6, the perspective coordinates are: *A* (0, 0, 0), *B* (6.46, 3.67, −0.65), *C* (−0.88, 14.47, −2.55), *D* (−8.55, 14.58, −2.57), *E* (0, 1.35, 7.65), *F* (6.32, 4.61, 5.12), *G* (−0.86, 15.08, 2.58), *H* (−8.26, 15.25, 4.02), *L* (3.55, 3.73, 9.33), and *M* (−4.05, 15.43, 5.98).

The plane of *ABF* is given by $ax + by + cz + d = 0$. Using the formulae for deriving *a, b, c, d,* then for *ABF* we have $21.79x - 36.88y + 6.31z = 0$ or dividing through by

$$\sqrt{(21.79^2 + 36.88^2 + 6.31^2)} \text{ to reduce } \sqrt{(a^2 + b^2 + c^2)} \text{ to } 1$$

Similarly,

ABF	$0.5x - 0.85y + 0.15z = 0$
BCG	$0.830x + 0.553y - 0.069z - 7.4 = 0$
CDH	$0.01x + 0.996y - 0.09z - 14.7 = 0$
DAE	$-0.866x - 0.492y + 0.087z = 0$
BAD	$-0.174y - 0.985z = 0$
HEL	$-0.421x - 0.013y + 0.907z - 6.9 = 0$
FGM	$0.631x + 0.562y + 0.534z - 9.3 = 0$

Now consider a point within the building (for instance (0, 5, 0) and a point well outside (0, −1000, 0). Inserting these values in the equations for the planes, we obtain

	Point Inside	**Point Outside**
ABF	−4.2	+ 850
BCG	−4.6	−560
CDH	−9.7	−1010
DAE	−0.2	+ 492
BAD	−0.1	+ 174
HEL	−7.0	+ 6
FGM	−6.6	−572

Here, those planes with opposite signs will be visible when viewed from a distance, whereas those with the same sign (that is, *BCG, CDH,* and *FGM*) will be hidden.

Example 11.8 Hidden Surfaces Using Vectors

Using the same data as in Example 11.7, the equation of the plane *ABF* is

$$0.5x - 0.85y + 0.15z = 0$$

The normal to the plane is therefore the unit vector $0.5\mathbf{i} - 0.85\mathbf{j} + 0.15\mathbf{k}$.

If our viewing point is (0, −50, 0), which has a unit vector (0, −1, 0) or $0 * \mathbf{i} - 1 * \mathbf{j} + 0 * \mathbf{k}$ then the cosine of the angle of incidence (cos ø) is given by the dot product with the plane *ABFLE*.

Thus,

$$\cos\varnothing = 0 * 0.5 + 1 * 0.85 + 0 * 0.15 = +0.85$$

For *BCGF*, cos ø = −0.553; for *CDHMG*, cos ø = −0.996; for *DAEH*, cos ø = +0.492; for *BADC*, cos ø = +0.174; for *HELM*, cos ø = +0.013; and for *FGML*, cos ø = −0.562.

The cosine of an angle that is greater than 90° is negative thus the normals to the planes *BCGF*, *CDHMG*, and *FGML* all point away from the viewer, and thus the surfaces are hidden.

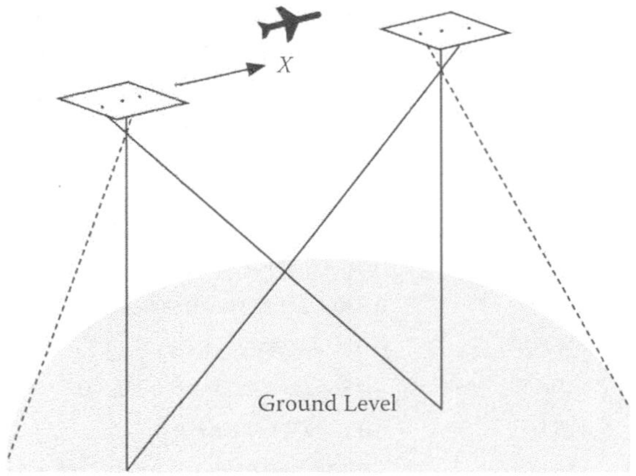

FIGURE 11.7 Stereo pair of aerial photographs.

11.4 PHOTOGRAMMETRIC MEASUREMENTS

Most topographic maps these days are produced using vertical aerial photography and a series of techniques known as *stereophotogrammetry*. In Figure 11.7, a camera takes a photograph of the ground from a point known as a base station and then flies forward to the next base station and takes a second photograph, the ground in between being covered by both photographs. The distance flown between the taking of each photograph is known as the *air base* and is approximately 40% of the width of each resulting aerial photograph (Figure 11.8). Rays of light from the ground pass through the camera lens and register as a negative image. If the focal length of the camera lens is f, the flying height is H, and the terrain height h (so that the height above the ground is $H - h$), then the scale of the photography will $\approx \frac{f}{H-h}$. Since h varies across the terrain, this means that the scale varies from one point on the photograph to the next.

The center of each photograph is known as the *principal point* and mapping takes place in the overlap area between successive principal points. Here a virtual 3D model of the terrain is formed and precise measurements within the model can be taken giving, with suitable photography and appropriate technology, ground value accuracies of the order of 1 centimeter or so. The coordinates of points visible on each photograph are measured and transformed, either by analogue or digital means, into points on a map.

The axis of the camera is said to be the z-axis (Figure 11.9); with the negative it points downward but with a positive print it points upward. The direction of flight for a series of aerial photographs is called the x-axis, while the horizontal line at right angles to the direction of flight is called the y-axis.

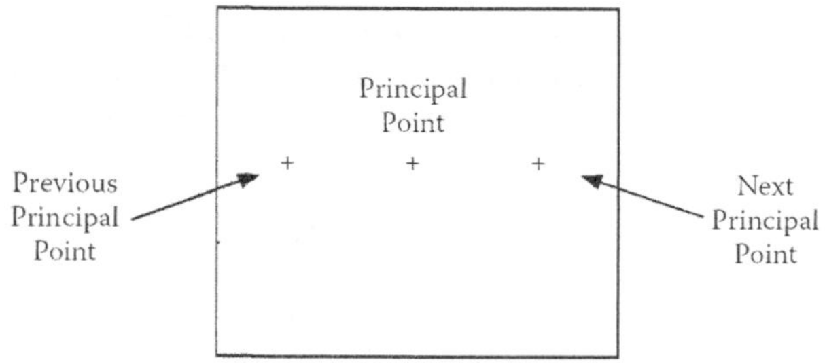

FIGURE 11.8 Principal point in the center of an aerial photograph.

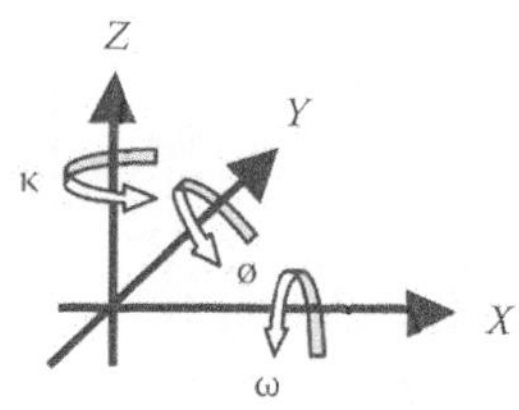

FIGURE 11.9 Photogrammetric rotations.

The photogrammetric process analyzes the rays of light by projecting them back out so that corresponding rays intersect, in effect turning the camera in Figure 11.7 into a projector. Using the two camera stations, a 3D model can be formed, the scale of which will depend on the distance between the two projectors, which in turn represents the length of the air base.

In order that the stereoscopic model is an accurate representation of the ground, three operations need to be carried out. The first, known as *inner* or *interior orientation*, is to ensure that the characteristics of the projectors match that of the cameras, either physically or by using appropriate mathematical transformations to ensure that the bundle of rays emerging from the projectors is exactly the same as that entering the camera lens and that any optical distortions are removed. This can be achieved either optically through the lens design and construction or, if the photographic image is held in digital form, by using an appropriate trend surface (see Chapter 15) to model the camera distortions.

The second operation is known as *relative orientation*. This means that the two projectors are adjusted so that their relative tips and tilts match those of the original camera positions.

Rotations of the camera about the x-axis or the roll of the aircraft are designated *omega*—the Greek letter "ω." A rotation about the z-axis, which for an aircraft is known as the *yaw*, is said to be *kappa*—the Greek letter "κ." Rotations about the y-axis are known as the *pitch* of the aircraft and are called *phi*—the Greek letter "ø." The rotations are right handed in that if your right-hand thumb points along the direction of the arrows, your fingers curl to point as shown in Figures 7.7 and 11.9.

If we rotate a camera / projector first about the x-axis (known as the *primary axis*) then about the y-axis, and finally the z-axis, the transformation will be of the form:

$$\mathbf{M} = \begin{pmatrix} \cos\kappa & \sin\kappa & 0 \\ -\sin\kappa & \cos\kappa & 0 \\ 0 & 0 & 1 \end{pmatrix} * \begin{pmatrix} \cos\varnothing & 0 & \sin\varnothing \\ 0 & 1 & 0 \\ -\sin\varnothing & 0 & \cos\varnothing \end{pmatrix} * \begin{pmatrix} 1 & 0 & 0 \\ 0 & \cos\omega & \sin\omega \\ 0 & -\sin\omega & \cos\omega \end{pmatrix}$$

$$= \begin{pmatrix} \cos\kappa\cos\varnothing & -\cos\kappa\sin\varnothing\sin\omega + \sin\kappa\cos\omega & \cos\kappa\sin\varnothing\cos\omega + \sin\kappa\sin\omega \\ -\sin\kappa\cos\varnothing & \sin\kappa\sin\varnothing\sin\omega + \cos\kappa\cos\omega & -\sin\kappa\sin\varnothing\cos\omega + \cos\kappa\sin\omega \\ -\sin\varnothing & -\cos\varnothing\sin\omega & \cos\varnothing\cos\omega \end{pmatrix}$$

In particular, if the angles through which we make the rotations are small and we call these δω, δø, and δκ, then we have shown in Chapter 6 that cos (δω), cos (δø), and cos (δκ) each equal 1, while the sine functions become δω, δø, and δκ in radians. Thus, for small rotations the matrix **M** reduces to

$$\begin{pmatrix} 1 & \delta\kappa & \delta\varnothing \\ -\delta\kappa & 1 & \delta\omega \\ -\delta\varnothing & -\delta\omega & 1 \end{pmatrix}$$

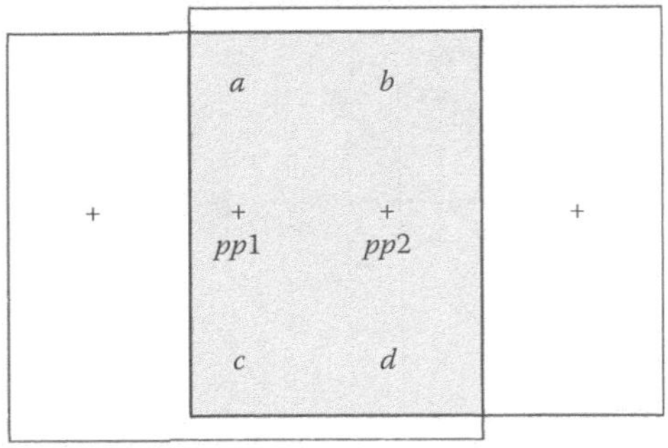

FIGURE 11.10 Model area for overlapping stereo pair of photographs.

Since there are two cameras or projectors (L for left and R for right), there are six unknowns, namely, $\delta\omega_L$, $\delta\phi_L$, $\delta\kappa_L$, and $\delta\omega_R$, $\delta\phi_R$, $\delta\kappa_R$. However, if we rotate the projectors by the same amount about the x-axis, the rays will still intersect and the values for $\delta\omega$ cannot at this stage be determined absolutely. Relative orientation allows us to determine the five values $\delta\phi_L$, $\delta\kappa_L$, $\delta\phi_R$, $\delta\kappa_R$, and the relative value $\delta\omega$.

This is achieved by choosing clearly identifiable points in the general area of the model (a, b, c, d, and $pp1$ and $pp2$), as shown in Figure 11.10 and measuring their locations on each flat-surface photograph and transforming their (x, y) values so that they match, using the rotation matrix **M** with the appropriate values for each projector. Finally, to bring the model into agreement with the ground coordinate values, we apply a 4 * 4 matrix transformation of the form discussed in Section 11.2.

The stereoscopic model must be leveled and adjusted for scale. This process, known as the *absolute orientation*, removes the uncertainties about tip and tilt. Three points are chosen that have known height and are suitably located to adjust the level of the model—such as a and b for roll and b and c for pitch in Figure 11.10. Likewise, the distance between two points (such as a to d) must be known in order that we can adjust for scale. These known points may have been established either by ground surveys or through what is known as *aerotriangulation*. Further details can be found in standard books on photogrammetry. Our interest here is in the mathematics behind the treatment of photogrammetric measurements.

GLOSSARY

Affine A transformation that preserves collinearity and parallelism so that points in a straight line remain in a straight line.

Hidden lines in a 3D object that cannot be seen by the viewer in 2D.

Homogeneous coordinates The introduction of an additional element in a set of coordinates that provides scale and can be used to describe points at infinity.

Orthogonality The preservation of right angles as right angles.

Principal point The optical center of an aerial photograph.

Rotation matrix In homogeneous coordinates, a 4 * 4 matrix of the form.

$$\begin{pmatrix} a & b & c & t & d & e & f & u & g & h & i & v & p & q & r & s \end{pmatrix}$$

in which the nine elements a, b … i represent the rotations, t, u, v the translation, and p, q, r the projection onto a plane; s is the scale change.

Similarity transformation A transformation by translation and rotation and uniform scale change that preserves *orthogonality*.

Stereophotogrammetry The measurement of position from overlapping pairs of photographs.

Vanishing point The point at infinity where parallel lines meet. It can be transformed into a finite point in space.

12 Map Projections

12.1 MAP PROJECTIONS

In this chapter, we will consider ways in which the location of points on the surface of the Earth can be represented in what are commonly referred to as *map projections*. The assumption behind the transformations considered in Chapter 11 has been that we have been dealing with straight lines or flat surfaces, in effect, a flat Earth. When we wish to represent the curved surface of the Earth, we need to take a different approach.

Consider a point A on the Earth's surface with latitude ϕ_A and longitude λ (Figure 12.1). The length of the arc from the equator (Q) to $A = R\,\phi_A$ where R is the radius of the Earth, assumed here to be a sphere, and ϕ is measured in radians. Let C be a point with latitude ϕ_C and longitude λ_C. Let the difference in longitude between A and $C = \Delta\lambda = \lambda_C - \lambda_A$ and the difference in latitude $= \Delta\phi = \phi_C - \phi_A$. Let B be on the same parallel of latitude as C and the same meridian of longitude as A. In triangle ABC then in terms of physical length: $AB = R\,\Delta\phi$. Since the radius of the circle for the parallel of latitude $\phi = PB$ and $PB = PC = R\cos\phi$, then

$$BC = PB\Delta\lambda = R\cos\phi_C\Delta\lambda$$

These quantities form the basis for plotting on a flat surface with a rectangular grid (see Figure 12.2). AB on the sphere becomes $A'B' = \Delta N$ and BC becomes $B'C' = \Delta E$, the differences in northings and eastings between A' and C'.

In the simplest map projection, known as the *simple cylindrical* or *plate carrée* projection, the distance north is plotted as $N = R\phi$ and the eastings as $R\lambda$ so that there is a rectangular grid. This means, in effect, that the distances north–south are treated as correct but distances east–west must be stretched from $(R\cos\phi)\,\lambda$ on the globe to $R\lambda$ on the grid. East–west distances on

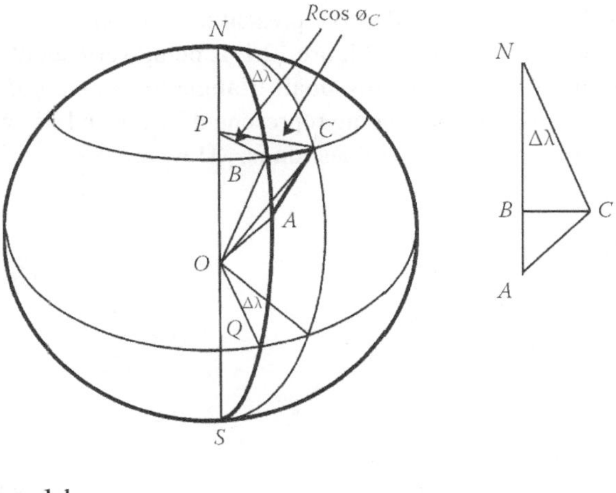

FIGURE 12.1 Points on a globe.

DOI: 10.1201/9781003464402-16

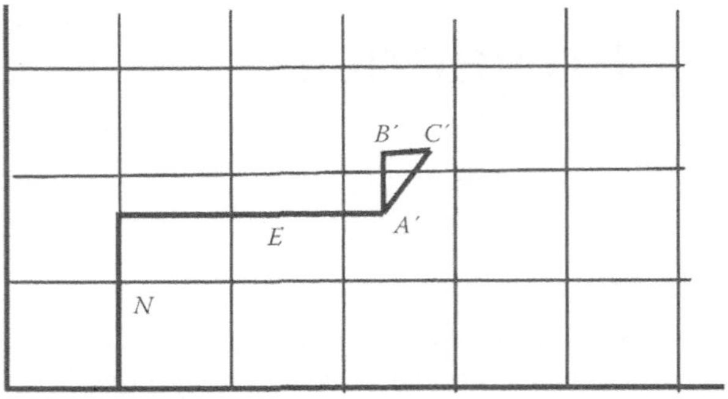

FIGURE 12.2 The simple cylindrical projection.

the globe must be increased by a scale factor of (1 / cos ø) or sec ø in order to plot them on the rectangular grid.

Another approach might be to plot the northings correctly as $R\phi$ and the eastings as $(R \cos \phi)\ \lambda$ so that both north–south and east–west had a scale factor of 1. But consider what happens to triangle *ABC*. If for the sake of argument we set the latitude of $A = 30°$ and $C = 40°$ so $\Delta\phi = 10° = 0.1745$ radians and we set $\Delta\lambda = 15° = 0.2618$ radians and we have a globe of radius $R = 100$ units then $A'B'$ on the flat $= R\Delta\phi = 17.45$ and $B'C' = R\Delta\lambda * \cos 40 = 20.05$. Tan $(\angle B'A'C') = 20.05 / 17.45$ or $\angle B'A'C' = 49°$ while the distance $A'C' = \sqrt{(17.45^2 + 20.05^2)} = 26.58$.

On the globe, however, we have to use the spherical triangle formulae on the triangle *ANC* in Figure 12.1. In this case, $NC = 50°$, $NA = 60°$, angle $ANC = 15°$, and if we call the angle $AC = b$, then by using the cosine formula for spherical triangles

$$\cos b = \cos 50\ \cos 60 + \sin 50 \sin 60 \cos 15 = 0.9622, \text{ or}$$
$$b = 15.8° = 0.2758 \text{ radians}$$

Thus, on the sphere, the distance is 27.58 but on the grid it is 26.58. Likewise, using the sine formula the angle *NAC* on the globe = 53.7° rather than 49° as previously derived. Thus, although we have preserved the scale in both the north–south and east–west directions we have changed the scale along the diagonal and the angle at *A*. We would also have the meridians bending toward the center and not crossing the parallels at right angles. Some distortion is inevitable because a curved surface must be stretched or shrunk in order to represent it on a plane.

Now let us consider the triangle *ABC* in Figure 12.1 as being very small, so small as to allow us to use differential calculus. We will call this small triangle the *elemental triangle* with sides *AB* and *BC* on the map (i.e., on a flat surface) being represented by δN and δE, while on the sphere they are $R\delta\phi$ and $R \cos \phi\ \delta\lambda$ (Figure 12.3). The question is, "How are δN and δE related to $R\delta\phi$ and R

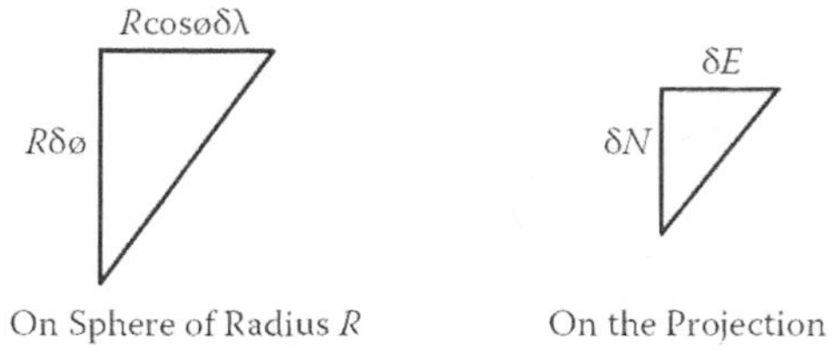

FIGURE 12.3 The elemental triangle on a sphere and on a plane surface.

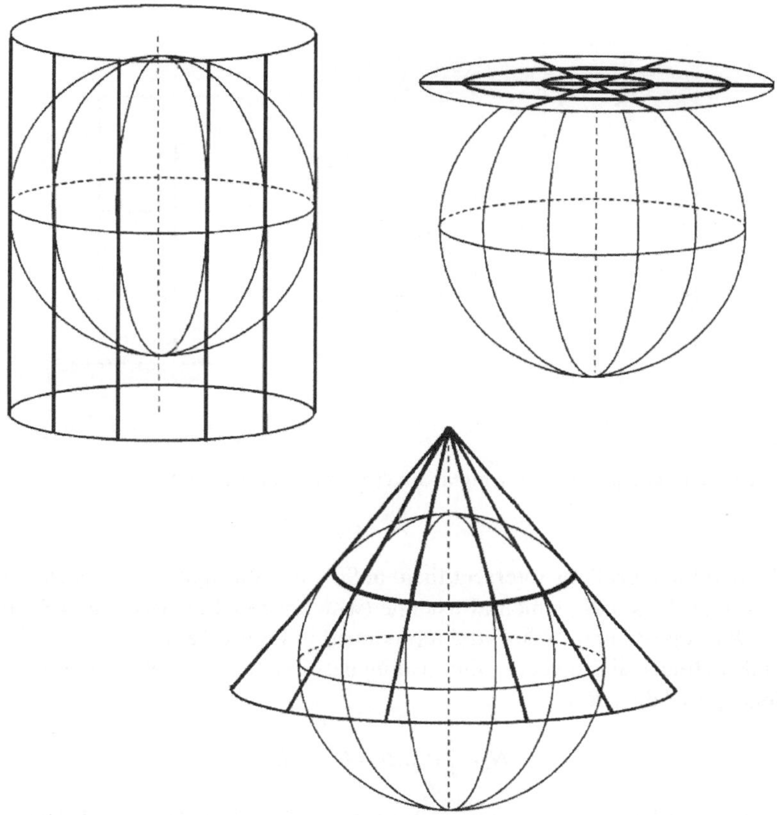

FIGURE 12.4 Cylinder, cone, and plane.

cos ø δλ?" There are many possible answers, each one trying to preserve the angles, the distances, or the area, or somehow compromise between these three quantities. There are in general three approaches, one based on a cylinder wrapped around the sphere, another based on a cone touching or cutting it, and a third on a plane. Both the cylinder and cone can be cut along a line and opened up into a flat surface (Figure 12.4).

Projections based around these three approaches are called *cylindrical*, *conical*, and *azimuthal* (or *zenithal*). They are normally considered to have the axis of rotation of the Earth as vertical but they can be applied obliquely. In much of what follows, we will consider the Earth as a sphere. The calculations need to be slightly modified when we treat the Earth as an ellipsoid.

12.2 CYLINDRICAL PROJECTIONS

First, let us consider the cylinder that wraps around the equator so that when it is cut along a vertical line and unfolded it creates a flat surface on which the meridians become parallel lines. The parallels of latitude cut the cylinder in a series of parallel lines that are perpendicular to the meridian lines. Three examples are shown in Figure 12.5, the first preserving distances north–south, the second preserving area, and the third scale around any point.

On the globe, the meridians are convergent, that is, they come together at the poles but always cross the parallels of latitude at right angles. In the ordinary cylindrical projections, the parallels

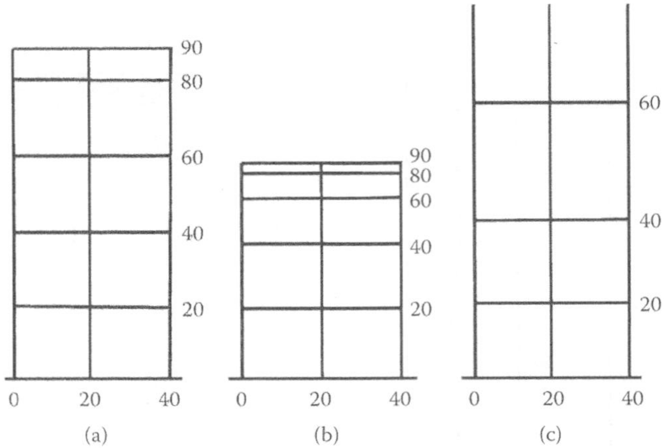

FIGURE 12.5 Cylindrical equidistant (a), equal area (b), and conformal (c).

remain parallel and the meridians intersect them at 90°. It is the spacing between the parallels that is adjusted. Consider the small elemental triangle (which we had enlarged as *ABC* in Figure 12.1) with two sides "$R \cos \phi\ \delta\lambda$" and "$R\ \delta\phi$" and reproduced in Figure 12.3.

For the simple cylindrical projection, the spacing between the parallels is retained and, therefore, $\delta N = R\ \delta\phi$. Hence, $\mathrm{d}N / \mathrm{d}\phi = R$, or

$$N = \int R\ \mathrm{d}\varnothing = R\varnothing + C_n$$

where C_n is some constant. For differences in longitude, $\delta E = R\ \delta\lambda$. Hence, $\mathrm{d}E / \mathrm{d}\lambda = R$, or

$$E = \int R\ \mathrm{d}\lambda = R\lambda + C_e$$

C_n and C_e are the constants that must be introduced on integration, and ϕ and λ must be measured in radians. Since $N = 0$ when $\phi = 0$ and $E = 0$ when $\lambda = 0$, then $C_n = C_e = 0$.

Hence, for the simple cylindrical projection

$$N = R\phi \quad \text{and} \quad E = R\lambda$$

To preserve the area of the small element, if we retain $E = R\ \lambda$ when it really ought to be $(R \cos \phi)\ \lambda$, then we have, in effect, increased the scale west–east by sec ϕ. Thus, to preserve area we must reduce the south–north distances by an amount cos ϕ. Put another way, since the area of a triangle is half its base * height, the area of the elemental triangle on the sphere is

$$(1/2)R^2 cos\ \varnothing\ \delta\varnothing\ \delta\lambda$$

The area on the projection is $(1 / 2)\ \delta N\ \delta E$. But,

$$\delta E = R\ \delta\lambda$$

hence,

$$\delta N = R \cos \varnothing\ \delta\varnothing \quad \text{and} \quad N = \int R \cos \varnothing\ \mathrm{d}\varnothing = R \sin \varnothing$$

Thus, for the cylindrical equal area projection,

$$E = R\lambda \text{ and } N = R \sin \varnothing$$

Projections that preserve area are called *equal-area* or *equivalent* or *authalic.*

If we want to maintain the shape of the elemental triangle, then since the west–east scale factor is "sec ø," then we must do the same for the south–north and thus, $\delta N = R \sec \phi\ \delta\phi$.

The integral of sec ø was quoted in Box 10.4 to be $\log_e(\sec \phi + \tan \phi)$. This can also be expressed as $\log_e \tan(\pi / 4 + \phi / 2)$. Hence, if $dN = R \sec \phi\ d\phi$, then ignoring the constant term,

$$N = \int R \sec \varnothing \ d\phi = R \log_e \tan(\pi / 4 + \varnothing / 2)$$

By maintaining the scale around any point, the shape of small areas is preserved. Projections that maintain this property are known as *orthomorphic* or *conformal.* They have the advantage that angles around any point are preserved so that angles measured in the field with a theodolite can be used directly in the projection. This is particularly important in topographic mapping.

One consequence is that at the North or South Pole, the value for N then becomes $\log_e \tan(90°)$ and that is infinitely large. Recalling that this is based on a cylinder wrapped around the equator, there is no way that the position of the poles can be plotted on a map when using this form of cylindrical projection.

The cylindrical orthomorphic projection is also known as *Mercator's projection.* Since angles around any point are preserved and since any straight line drawn on the map cuts all the meridians at the same angle, it means that the bearing along the line is always the same. A line of constant bearing is known as a *rhumb line* or *loxodrome.* Such lines have been particularly important to navigators who can draw a straight line on a chart using the Mercator's projection and know that by following the bearing of that line they will reach their destination. This line may not be the shortest route; but it may be the easiest to navigate. The scale at any point on the Mercator's projection is given by "sec ø." This becomes large at high latitudes; but around the equator it is close to 1.

By wrapping a cylinder around a meridian rather than the equator, we have a *transverse cylindrical projection.*

In Figure 12.6, the cylinder is wrapped around the meridian *NPQS*, known as the *central meridian*, with points *E* and *W* being 90° east and west of the central meridian, in effect becoming the poles in the standard case. For any point *A*, *EAPW* is a great circle and the coordinates of *A* can be

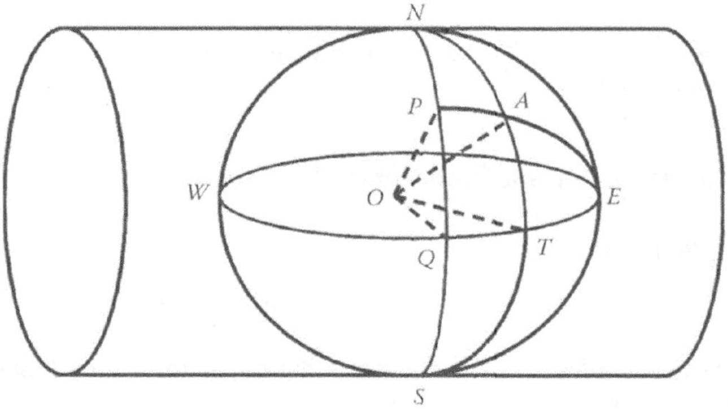

FIGURE 12.6 The Transverse Mercator.

expressed in terms of *QP* (the equivalent of λ in the normal cylindrical case but here being ϕ_P) and *PA* (the equivalent of the latitude in the nontransverse case). Let us assume that its value is α so angle $POA = \alpha$. *NPQS* is the central meridian to which we will give the value λ_M (= λ_Q).

Let *NA* cut the equator at *T*. *QT* is the difference in longitude between the central meridian and the meridian through *A*. AT = latitude of $A = \phi_A$; $NA = 90° - \phi_A$; angle *PNA* = difference in longitude between *A* and the central meridian = $\lambda_A - \lambda_M$. Let us call this λ_A so that angle $ANE = 90° - \lambda_A$.

In triangle *ATE*, angle $ATE = 90°$; side $AT = \phi_A$; side TE = angle $ANE = 90° - \lambda_A$; $AE = 90° - \alpha$. Using the cosine formula on triangle *ATE*

$$\cos(AE) = \cos(AT)\cos(TE) + \sin(AT)\sin(TE)\cos(90)$$

or, since $\cos(90) = 0$,

$$\sin\alpha = \cos\varnothing_A \sin\lambda_A = \cos\varnothing_A \sin(\lambda_A - \lambda_M) = \cos\varnothing_A \sin(\lambda_A - \lambda_Q)$$

Also, using the sine formula on triangle *ATE* where $ATE = 90°$

$$\sin(AET)/\sin(AT) = \sin(ATE)/\sin(AE)$$

or, since angle $AET = PQ$ = the latitude of $P = \phi_P$

$$\sin\varnothing_P = \sin\varnothing_A / \cos\alpha, \text{ or}$$

$$\cos\alpha = \sin\varnothing_A / \sin\varnothing_P$$

Using the cosine formula in triangle *PNA*,

$$\cos PA = \cos PN \cos NA + \sin PN \sin NA \cos(\lambda_A - \lambda_Q)$$

$$\cos\alpha = \sin\varnothing_P \sin\varnothing_A + \cos\varnothing_P \cos\varnothing_A \cos(\lambda_A - \lambda_Q) \text{ and from above,}$$

$$= \sin\varnothing_A / \sin\varnothing_P$$

Thus,

$$\sin\varnothing_A = \sin\varnothing_A \sin^2\varnothing_P + \sin\varnothing_P \cos\varnothing_P \cos\varnothing_A \cos(\lambda_A - \lambda_Q)$$

$$\sin\varnothing_P \cos\varnothing_P \cos\varnothing_A \cos(\lambda_A - \lambda_Q) = \sin\varnothing_A(1 - \sin^2\varnothing_P) = \sin\varnothing_A(\cos^2\varnothing_P)$$

Rearranging, $\tan\phi_P = \tan\phi_A \sec(\lambda_A - \lambda_Q)$, which with $\sin\alpha = \cos\phi_A \sin(\lambda_A - \lambda_Q)$ means that we can calculate the values of *QP* (= ϕ_P) and *PA* (= α).

Example 12.1 The Mercator and Transverse Mercator

With the Mercator projection,

$$E = R\lambda \quad \text{and} \quad N = R\log_e \tan(\pi/4 + \varnothing/2)$$

using the 100° west meridian and the equator as the origin, for a point *A* at latitude 40° north and longitude 103° west and a globe of radius 100 units

$$\lambda = 3° = 0.05236 \text{ radians;} \quad \tan(\pi/4 + \varnothing/2) = 2.1445$$

Hence, the coordinates of A would be

$$E = 5.236 \quad \text{and} \quad N = 76.291$$

The scale at A would be sec (40°) = 1.3054.

With the Transverse Mercator, in Figure 12.6, angle $AOT = ø_A = 40°$; angle $PNA = (\lambda_A - \lambda_P) = 3°$. Let us call angle $POA = \alpha$.

We have shown that

$$\sin\alpha = \cos\varnothing_A \sin\lambda = 0.04009. \text{ Hence, } \alpha = 2.3°$$

Also,

$$\tan\varnothing_P = \tan\varnothing_A \sec\left(\lambda_A - \lambda_P\right) = 0.84025. \text{ Hence, } \varnothing_P \approx 40.04°$$

From Box 12.1,

$$E = R\log_e\left((\sec)\ \alpha + (\tan)\ \alpha\right) \quad \text{and} \quad N = R\,\varnothing_P$$

Hence, the coordinates of A would be

$$E = 4.011 \quad \text{and} \quad N = 69.881$$

The scale at $A = \sec\alpha = 1.0008$.

Note: Along the central meridian the scale is sec (0) = 1 exactly. Our example shows that at 3° east or west the scale is 1.0008. If we apply an overall scale factor of 0.9995 to all our measurements, this reduces to 1.0003. In the Universal Transverse Mercator (UTM) where the Earth is mapped in strips 6° wide, this results in scales always being close to unity.

For the transverse cylindrical equidistant projection (also known as the *Cassini projection*), then the northings of point A are plotted as $N = R\, ø_P$ and while for the eastings, $E = R\,\alpha$. The Cassini projection was much used in early days for topographic mapping. Similarly, if we let $E = R\sin\alpha$, then we have the transverse cylindrical equal area. If, however, we make

$$E = R\log_e(\sec\alpha + \tan\alpha) = R\log_e\tan(\pi/4 + \alpha/2)$$

then we have an orthomorphic projection, known as the *Transverse Mercator* (see Example 12.1).

By extending this principle to an ellipsoid rather than to a sphere and then mapping the globe in strips 3° either side of the central meridians that are chosen every 6° of longitude, a picture of the globe can be built up in what is known as the *Universal Transverse Mercator* or UTM. This was originally developed in the 1940s by the U.S. military and has been widely adopted around the world, especially for topographic mapping.

The Transverse Mercator is the most commonly used projection in surveying and is normally related to the spheroid, rather than the sphere, as this gives the best mathematical representation of the shape of the Earth. As we will see in Chapter 13, the geometry of the ellipse and ellipsoid is more complex than that of the circle and sphere. In essence, there are two radii of curvature that must be considered namely, ρ and ν. Further discussions of this, and how they affect map transformations, is outside the scope of the present book. Likewise, the summary in Box 12.1 relates to spherical rather than ellipsoidal transformations.

BOX 12.1 CYLINDRICAL PROJECTIONS FOR A SPHERE

For the cylindrical equidistant or *plate carrée*

$$E = R\lambda \quad \text{and} \quad N = R\varnothing$$

For the cylindrical equal area

$$E = R\lambda \quad \text{and} \quad N = R \sin \varnothing$$

For the cylindrical orthomorphic (Mercator)

$$E = R\lambda \quad \text{and} \quad N = R \log_e(\sec \varnothing + \tan \varnothing)$$

For the transverse cylindrical, for point A and central meridian M

$$\tan \varnothing_P = \tan \varnothing_A \sec(\lambda_A - \lambda_M), \; \sin \alpha = \cos \varnothing_A \sin(\lambda_A - \lambda_M)$$

For the transverse cylindrical equidistant (the Cassini projection)

$$E = R\alpha \quad \text{and} \quad N = R\varnothing_P$$

For the transverse cylindrical orthomorphic (the Transverse Mercator)

$$E = R \log_e(\sec \alpha + \tan \alpha) \quad \text{and} \quad N = R\varnothing_P$$

12.3 AZIMUTHAL PROJECTIONS

An azimuthal projection is one where the surface onto which points are projected is a plane that is tangential to the sphere. One of the easiest azimuthal projections to imagine is the polar case (known as the *zenithal projection*, shown in Figure 12.7) in which either pole (here designated N) becomes the center, with the meridians radiating outward as straight lines. The parallels become circles centered around N, their radius depending on the characteristics that we wish to preserve.

On the sphere, the distance along the surface from the pole to the parallel = $R(\pi / 2 - \phi) = R\chi$ where ϕ is the latitude of point P and χ = colatitude (the angle as measured from the pole, rather than from the equator). This distance plots on the flat as the length r, the value of which depends on the characteristics of the projection. The angles λ representing the meridians are plotted as their values on the Earth's surface, that is, they are true to scale.

Consider a small triangle of width $\delta\lambda$ and height $\delta\phi$ on the sphere and on the projection. For the zenithal equidistant, circles are drawn around the pole with radius $r = R(\pi / 2 - \phi) = R\chi$. For the equal area projection, we have to make the two elemental triangles in Figure 12.7 the same area: hence,

$$(1/2)R^2 \sin \chi \; \delta\chi \; \delta\lambda = (1/2)r \; \delta r, \quad \text{or} \int R^2 \sin \chi \; d\chi = \int r \; dr$$

This gives

$$C - R^2 \cos \chi = (1/2)r^2$$

where C is some constant.

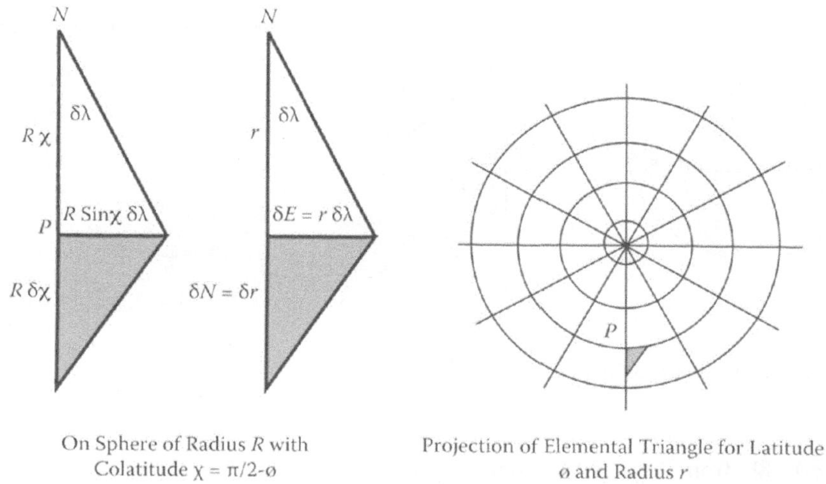

FIGURE 12.7 Zenithal projection.

Since $r = 0$ at the center where $\chi = 0$ and $\cos 0 = 1$, then $C = R^2$. Rearranging, $r^2 = 2R^2(1 - \cos\chi)$ and this equals $4R^2 \sin^2 (\chi / 2)$ as shown in Chapter 6: hence,

$$r = 2R \sin\left(\chi / 2\right) \text{ for the equal area projection.}$$

For the zenithal orthomorphic projection, in order to preserve shape then from Figure 12.7.

$$R \sin \chi \, \delta\lambda / R \, \delta\chi = r \, \delta\lambda / \delta r$$

or

$$(1 / \sin\chi)\mathrm{d}\chi = (1 / r)\mathrm{d}r$$

Integrating,

$$\log_e\left(\tan\chi / 2\right) + \text{constant} = \log_e r$$

If we call the constant $\log_e C$, then

$$\log_e\left(C \tan \chi / 2\right) = \log_e r$$

or

$$r = C \tan\left(\chi / 2\right)$$

The value of C determines the scale overall and is commonly set so that $C = 2R$. This projection is often known as the *stereographic projection.*

The relationships established above can also be applied to the projection onto a plane that touches the sphere at any point P, not just the pole N. The formulae modified as χ and λ used above take on different meanings. The angle NPA in Figure 12.8a and b represents the difference in longitude so we must replace λ by Ω (omega), while the angle PA is now what was the colatitude, so we must replace χ by Ψ (psi). The values can be calculated from the sine and cosine formulae for spherical triangles.

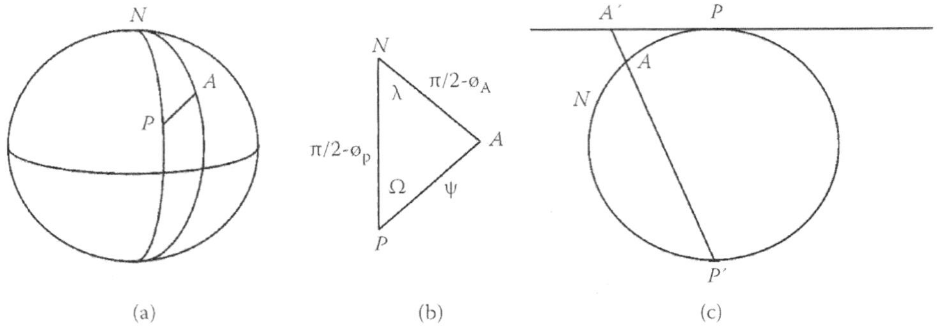

FIGURE 12.8 (a–c) Oblique azimuthal.

In Figure 12.8b, from the sine formula:

$$\sin\Omega / \sin(\pi/2 - \varnothing_A) = \sin\lambda / \sin\Psi \text{ or, since } \sin(\pi/2-\varnothing) = \cos\varnothing, \text{ then}$$

$$\sin\Omega \sin\Psi = \sin\lambda \cos\varnothing_A$$

From the cosine formula

$$\cos\Psi = \cos(\pi/2-\varnothing_A)\cos(\pi/2-\varnothing_P) + \sin(\pi/2-\varnothing_A)\sin(\pi/2-\varnothing_P)\cos\lambda$$

or

$$\cos\Psi = \sin\varnothing_A \sin\varnothing_P + \cos\varnothing_A \cos\varnothing_P \cos\lambda$$

BOX 12.2 ZENITHAL PROJECTIONS

If the colatitude is $\chi = 90 - \phi$, then

For the zenithal equidistant, $r = R\chi$
For the zenithal equal-area, $r = 2R \sin(\chi/2)$

For the zenithal orthomorphic (or stereographic), $r = 2R \tan(\chi/2)$.

Hence, we can find Ψ and from that we can find Ω. The results of all azimuthal projections for a sphere are summarized in Box 12.2.

The stereographic projection is equivalent to projecting onto a plane from the point on the sphere diametrically opposite point P shown as P' in Figure 12.8c. See Example 12.2. Although the plane onto which the projection is made is assumed in Example 12.2 to be tangential to the sphere at P, it can be any plane parallel to the tangent plane at P (for instance, through the center O) provided it is not through the point diametrically opposite to P. It has the characteristic that a circle on the surface of the Earth plots as a circle on the projection. Hence, when looking for the epicenter of an earthquake, for example, we can plot the time when shock waves were recorded and this will give an indication as to where the center of the earthquake lay.

Example 12.2 Oblique Stereographic Projection

For the stereographic projection of a sphere based on point P at 40° North and 100° West (see Figure 12.8a and b), then for a point A at 45° North and 103° West, using a sphere of radius $R = 100$

$$\varnothing_A = 45; \quad \varnothing_P = 40; \quad \lambda = 3$$

Using

$$\cos\Psi = \sin\varnothing_A \sin\varnothing_P + \cos\varnothing_A \cos\varnothing_P \cos\lambda = 0.99545, ° \text{ then } \Psi \approx 5.466°$$

Using

$$\sin\Omega \sin\Psi = \sin\lambda \cos\varnothing_A, \text{ then } \sin\Omega = 0.38848 \text{ and } \Omega \approx 22.86°$$

Thus, the distance $PA = 2R \tan(\Psi / 2) = 9.5478$.
Relative to the origin P,

$$E = PA\sin\Omega = 3.709; \quad N = PA\cos\Omega = 8.798$$

Now consider point P that is diametrically opposite P on a sphere in Figure 12.8c. Let the vector PA meet the tangent plane at P at point A. In geocentric coordinates

$$P \text{ is } \left(R\cos\varnothing_P, 0, R\sin\varnothing_P\right) = (76.60444,\ 0,\ 64.27876)$$

$$A \text{ is } \left(R\cos\varnothing_A \cos\lambda, R\cos\varnothing_A \sin\lambda, R\sin\varnothing_A\right) \approx (70.61377,\ 3.70071,\ 70.71068).$$

Now rotate about the whole configuration about the y-axis by an amount $(90° - ø_P)$ so that point P in effect becomes the North Pole. This means applying the matrix

$$\begin{pmatrix} \cos\varnothing_P & 0 & \sin\varnothing_P \\ 0 & 1 & 0 \\ -\sin\varnothing_P & 0 & \cos\varnothing_P \end{pmatrix}$$

The coordinates of P become $(R, 0, 0)$ while A becomes

$$x_A = R\cos\varnothing_A \cos\varnothing_P \cos\lambda + R\sin\varnothing_A \sin\varnothing_P; = 99.545$$

$$y_A = R\cos\varnothing_A \sin\lambda; = 3.701$$

$$z_A = -R\cos\varnothing_A \sin\varnothing_P \cos\lambda + R\sin\varnothing_A \cos\varnothing_P = 8.778$$

The center O remains (0, 0, 0) while P' becomes (−100, 0, 0).

We now project point A onto the tangent plane at P from P' so that the x value for A' becomes 100, as shown in Figure 12.8c. This means extending $P'A$ to $P'A'$ or from 199.545 to 200.

Thus,

$$y_A \text{ becomes } E = 3.701 * 200 / 199.545 = 3.709$$

$$z_A \text{ becomes } N = 8.778 * 200 / 199.545 = 8.798$$

These are the same values as we achieved earlier, showing that the stereographic projection is a view of the world from P', which is called the *antipode* of point P.

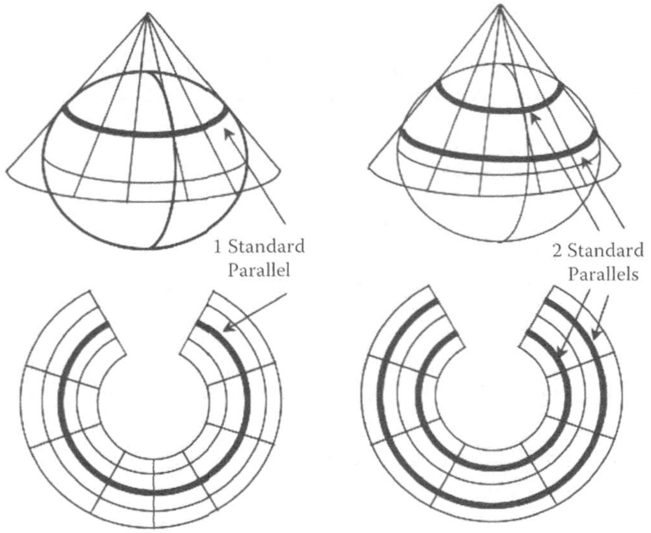

FIGURE 12.9 Conical projection with one or two standard parallels.

12.4 CONICAL PROJECTIONS

The third group of projections is known as *conical.* They are based on the idea of a cone wrapped around the midlatitudes, it being possible to cut a cone so that its surface can be uncurled onto a flat piece of paper. The cone may touch the sphere— in which case the parallel of latitude where this happens is known as the *standard parallel*—or cut the sphere in which case there will be *two standard parallels.*

When the cone is unwrapped and laid flat (Figure 12.9), each parallel of latitude becomes a part of a set of concentric arcs, while the meridians are lines that radiate from the center. The pole also becomes a circle, whose radius will depend upon the extent to which the cone is pointed.

In fact, the cylindrical and zenithal projections are special cases of the conical with the angle of the cone being zero for the cylindrical and 180° for the zenithal; if QN in Figure 12.10 is zero, the cone will have become a plane while if QN is infinite in length, then the cone will have become a cylinder.

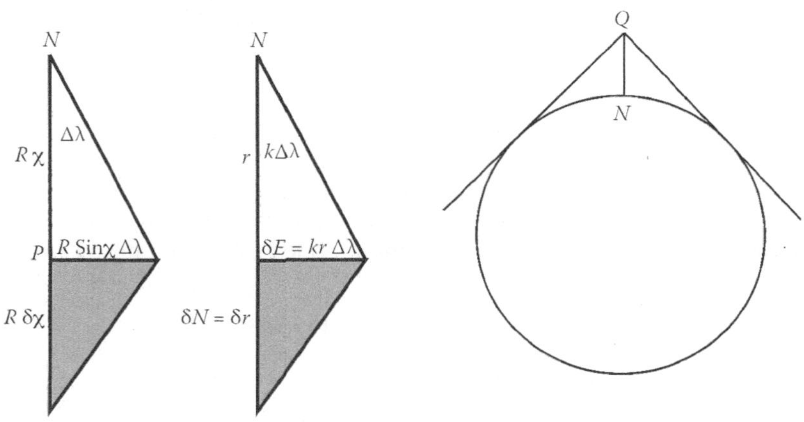

FIGURE 12.10 Elemental triangles for conical projections.

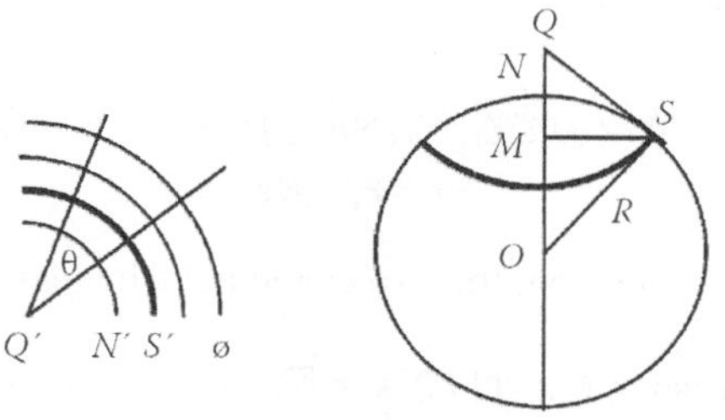

FIGURE 12.11 Conical equidistant with one standard parallel.

If we compare the conical case with the zenithal, the radius from the center has two elements, a fixed quantity (*Q′N′* in Figure 12.11) that represents the distance from the center of the circle plus an element that depends on the spacing of the parallels. The length *Q′N′* depends upon the shape of the cone and in addition upon whether there are one or two standard parallels.

In the case of any standard parallel (Figure 12.11), the North Pole becomes an arc of a circle of radius *Q′N′* (that we will call r_0) and the standard parallel has correct scale. The meridians become straight lines radiating out from the center, the angles between them being proportional to the true values by an amount called *k*, which is called the *constant of the cone*. The spacing out of the parallels depends on the characteristics of the projection.

If the angular separation between any two meridians is λ on the sphere, then on the projection this angle will have to be reduced by some proportion. For the standard parallel with latitude ϕ_s the distance along the parallel is the circumference of a circle of radius *SM*, which in Figure 12.11 = $R \cos \phi_S$ where *R* is the radius of the sphere. Thus, the whole length of the standard parallel on the sphere = $2\pi R \cos \phi_S$. On the flat this will be an arc of radius *QS*, which will equal $R \cot \phi_S$. The angle between the meridians will therefore have to be reduced by a factor *k* such that

$$K * 2\pi * R\cot \varnothing_S = 2\pi R\cos \varnothing_S = \text{the length of the standard parallel}$$

Thus, $k = \sin \phi_S = \cos \chi_S$ where χ is the colatitude or $(\pi / 2 - \phi)$. A difference in longitude of λ becomes $k\lambda$ on the projection so $\theta = k\,\delta\lambda$ (Figure 12.11).

For the conical equidistant with one standard parallel, we need to keep the parallels correctly spaced. Then for any parallel of colatitude χ, the radius of the circular arc that appears on the projection will be *QS* + the distance along the meridian from the standard parallel = $QS + R(\phi_S - \phi)$. But, $QS = R \cot \phi_s$. Thus, if the distance from the center $Q' = r$, then

$$r = R\cot \varnothing_S + R(\varnothing_S - \varnothing) = R\tan \chi_S + R(\chi - \chi_S)$$

In particular, at the pole, $\varnothing = \pi/2$, so $Q'N' = R\cot \varnothing_s + R(\varnothing_s - \pi/2)$

$$= R\tan\chi_s + R(\varnothing_S - \pi/2) = r_0$$

The radius at any other latitude = $r = r_0 + R(\pi / 2 - \phi)$.

For the case where there are two standard parallels ϕ_1 and ϕ_2, then the length of each parallel on the sphere will be such that $k\,r = R\cos\phi = R\sin\chi$ where *k* is the *constant of the cone* and χ is the colatitude. For the two standard parallels

$$k\{r_0 + R(\chi_1)\} = R\sin\chi_1$$

$$k\{r_0 + R(\chi_2)\} = R\sin\chi_2$$

After suitable manipulation, we obtain

$$r_o = R(\chi_2 \sin\chi_1 - \chi_1 \sin\chi_2)/(\sin\chi_2 - \sin\chi_1)$$
$$r = r_0 + R\chi$$

For the conical equal area projection, the two elemental triangles must be of equal area; hence,

$$(1/2)R\ \mathrm{d}\chi\ R\sin\chi\ \delta\lambda = (1/2)k\ r\ \delta\lambda\ \delta r \quad \text{or} \quad R^2 \sin\chi \mathrm{d}\chi = k\ r\ \mathrm{d}r$$

Integrating

$$C - 2R^2 \cos\chi = kr^2$$

where C is the constant of integration and k is not yet determined. Thus,

$$kC - 2kR^2 \cos\chi = k^2 r^2$$

Along a standard parallel, distances on the sphere are the same as on the projection. Hence, along a standard parallel

$$kr = R\sin\chi$$

If there are two standard parallels (1 and 2), then

$$kC - 2kR^2 \cos\chi_1 = R^2 \sin^2\chi_1 \quad \text{and} \quad kC - 2kR^2 \cos\chi_2 = R^2 \sin^2\chi_2$$

Subtracting the two equations, and dividing,

$$2k = (\sin^2\chi_2 - \sin^2\chi_1)/(\cos\chi_1 - \cos\chi_2)$$

But,

$$\sin^2 = 1 - \cos^2;\ \text{so}\ (\sin^2\chi_2 - \sin^2\chi_1) = (\cos^2\chi_1 - \cos^2\chi_2)$$
$$= (\cos\chi_1 + \cos\chi_2)(\cos\chi_1 - \cos\chi_2)$$

Hence,

$$k = (1/2)(\cos\chi_1 + \cos\chi_2) = (1/2)(\sin\varnothing_1 + \sin\varnothing_2)$$

where ϕ_1 and ϕ_2 are the latitudes rather than colatitudes of the standard parallels.

Substituting for k in the $C - 2R^2 \cos\chi = kr^2$, then

$$C = R^2(1 + \sin\varnothing_1 \sin\varnothing_2)/k = R^2(1 + \cos\chi_1\ \sin\chi_2)/k$$

From this, r^2 can be calculated and hence r derived.

The derivation of the values of C and r for the conical orthomorphic follows the same principles as those for the stereographic or zenithal orthomorphic projection. From the elemental triangles in Figure 12.10 the angles must be preserved: hence,

$$R\ \mathrm{d}\chi/(R\sin\chi\ \delta\lambda) = \delta r/(k\ r\ \delta\lambda)$$

or

$$k \text{ cosec } \chi \text{ d}\chi = (1/r) \text{ d}r$$

On integrating $k \log_e \{\tan(\chi/2)\} + C = \log_e r$.

After suitable manipulation and the insertion of the values for the standard parallels χ_1 and χ_2, it can be shown that

$$k = (\log_e \sin \chi_1 - \log_e \sin\chi_2)/\{\log_e \tan(\chi_1/2) - \log_e \tan(\chi_2/2)\}$$

$$r = R\left(\frac{\sin \chi_1}{k}\right) * \left(\tan\left(\frac{\chi}{2}\right) / \tan\left(\frac{\chi_1}{2}\right)\right)^k$$

The results are summarized in Box 12.3.

BOX 12.3 CONICAL PROJECTIONS OF THE SPHERE

For the conical equidistant with one standard parallel $\varnothing_s$, the radius of each arc of the projection = r where

$$r = r_0 + R(\pi/2 - \varnothing)$$

where

$$r_0 = R\{\cot \varnothing_s + (\varnothing_s - \pi/2)\}$$

The meridians converge at a point and if the angle of longitude between two meridians is λ on the sphere and θ on the projection:

$$\theta = (\sin \varnothing_s)\lambda$$

For the conical equidistant, with two standard parallels whose colatitudes are χ_1 and χ_2,

$$r = R(\chi_2 \sin\chi_1 - \chi_1 \sin\chi_2)/(\sin \chi_2 - \sin\chi_1) + R\chi$$

For the conical equal-area,

$$k = (1/2)(\cos\chi_1 + \cos \chi_2);\ C = R^2(1 + \cos \chi_1 \sin\chi_2)/k;\ r = \surd\{C/k - 2R^2 \cos\chi/k\}$$

For the conical orthomorphic,

$$k = (\log_e \sin\chi_1 - \log_e \sin\chi_2)/(\log_e \tan(\chi_1/2) - \log_e \tan(\chi_2/2))$$

$$r = R\{(\sin \chi_1)/k\}\{\tan(\chi/2)/\tan(\chi_1/2)\}^k$$

$$= R\left(\frac{\sin \chi_1}{k}\right) * \left(\tan\left(\frac{\chi}{2}\right)/\tan\left(\frac{\chi_2}{2}\right)\right)^k$$

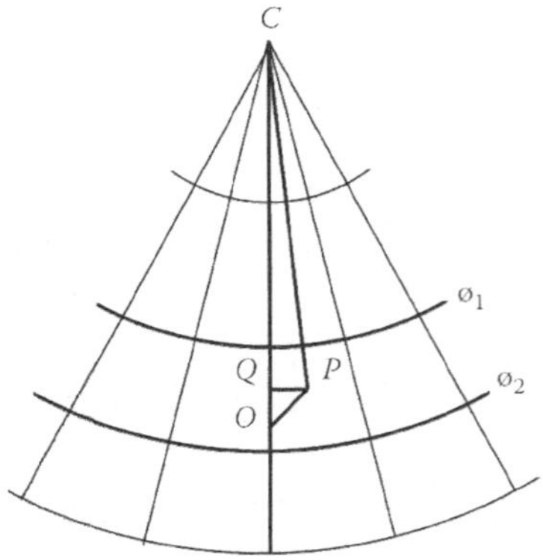

FIGURE 12.12 A conical projection with two standard parallels.

When working on the ellipsoid, the principles remain the same although the value of R, the radius of the Earth, will be replaced by ν, derived from the normal to the ellipse (see Figure 13.4 in Chapter 13) and the latitudes will be geodetic latitudes rather than geocentric.

In Figure 12.12, O is the origin for the coordinates of the projection, here located between the two standard parallels ϕ_1 and ϕ_2 (or colatitude χ_1 and χ_2). On the map, let the coordinates of O be (x_O, y_O) and of point P be (x_P, y_P), where P is any point to be plotted on the map. C is the point where the meridians converge in a conical projection.

On the ellipsoid, let O have latitude ϕ_O and longitude λ_O, and P have latitude ϕ_P and longitude λ_P.

On the projection, let $CP = r$; $CO = r_O$; and angle $OCP = k(\lambda - \lambda_O)$ where λ is the longitude of P relative to the origin (here OC). If Q is the foot of the perpendicular from P onto the line OC in the projection, then

$$x_p = x_O + QP = x_O + r\sin\{k(\lambda - \lambda_o)\}$$

$$y_P = y_O + OQ = y_O + r_O - r\cos\{k(\lambda - \lambda_O)\}$$

To plot our point P on the conical projection we need to know the formulae for r and k. If we want our projection to be conformal (i.e., orthomorphic), then if the Earth were a sphere:

$$k = (\log_e \sin\chi_1 - \log_e \sin\chi_2) / (\log_e \tan(\chi_1/2) - \log_e \tan(\chi_2/2))$$

$$r = R\left(\frac{\sin\chi_1}{k}\right) * \left(\tan\left(\frac{\chi}{2}\right) / \tan\left(\frac{\chi_2}{2}\right)\right)^k$$

for which R is the radius of the Earth to which a suitable scale factor must be applied.

If we are working on the best-fit ellipsoid that represents the surface of the Earth, then the formulae need some modification. The actual values will depend on which ellipsoid is chosen and that tends to be a matter of national preference. The key elements that need defining are the radius at the equator R and the flattening that is either defined as the ratio (a / b) or as the value of e (see Chapter 10, Section 10.6). Here we will use e with $R = a$ on the ellipsoid.

The formulae for the conformal conical with two standard parallels, known as *Lambert's projection*, then become:

For k, instead of $\sin\chi_1$ we need $\cos\varnothing_1 / \sqrt{(1 - e^2\sin^2\varnothing_1)}$

Call this $= n_1 = \cos\varnothing_1 / \sqrt{(1 - e^2\sin^2\varnothing_1)}$

$$\sin\chi_2 \qquad n_2 = \cos \varnothing_2 / \sqrt{(1 - e^2\sin^2\varnothing_2)}$$

$$\tan(\chi_1/2) \quad n_3 = \tan(\chi_1/2)/\{(1 - e\sin \varnothing_1)/(1 + e\sin \varnothing_1)\}^{e/2}$$

$$\tan(\chi_2/2) \quad n_4 = \tan(\chi_2/2)/\{(1 - e\sin \varnothing_2)/(1 + e\sin \varnothing_2)\}^{e/2}$$

$$k = (\ln n_1 - \ln n_2)/(\ln n_3 - \ln n_4)$$

(Here, we write "ln" to mean the natural logarithm, which is the same as $\log_e$ which we avoid, so as not to confuse the "e" of natural logarithms with the "*e*" for the ellipsoid.)

The formula for *r* must be modified in a similar way:

$$\sin \chi_2 \text{ is still} \quad n_1 = \cos \varnothing_1 / \sqrt{(1 - e^2\sin^2\varnothing_2)}$$

$$\tan(\chi_1/2) \quad n_3 = \tan(\chi_1/2)/\{(1 - e\sin\varnothing_1)/(1 + e\sin\varnothing_1)\}^{e/2}$$

$$\tan(\chi_2/2) \quad n = \tan(\chi/2)/\{(1 - e\sin\varnothing)/(1 + e\sin\varnothing)\}^{e/2}$$

$$r = R * (n_1/k) * \{n/n_3\}^k$$

The formulae are summarized in Box 12.4. For a worked example, see Example 12.3.

BOX 12.4 LAMBERT'S PROJECTION OF THE ELLIPSOID

Let the radius at the equator be *R*, the radius of the arcs for the conical projection of the parallels of latitude by *r* from the central point, and let the longitudes be modified by the factor *k*. Let the standard parallels be ϕ_1 and ϕ_2 and their colatitudes be χ_1 and χ_2.

$$\text{Let } n_1 = \cos \varnothing_1 / \sqrt{(1 - e^2\sin^2\varnothing_1)};\ n_3 = \tan(\chi_1/2)/\{(1 - e\sin \varnothing_1)/(1 + e\sin \varnothing_1)\}^{e/2}$$

$$n_2 = \cos\varnothing_2 / \sqrt{(1 - e^2\sin^2\varnothing_2)};\ n_4 = \tan(\chi_2/2)/\{(1 - e\sin \varnothing_2)/(1 + e\sin\varnothing_2)\}^{e/2}$$

Then

$$k = (\ln n_1 - \ln n_2)/(\ln n_3 - \ln n_4)$$

Let $n = \tan(\chi / 2) / \{(1 - e \sin \phi) / (1 + e \sin \phi)\}^{e/2}$

Then

$$r = R * (n_1/k) * \{n/n_3\}^k$$

Example 12.3 Lambert's Conical Orthomorphic with Two Standard Parallels

Consider an ellipsoid with $a = R = 6378206.400$ meters or 20925832.16 U.S. geodetic feet and $e = 0.08227185$. (This is known as the *Clarke 1866 ellipsoid.*) Let the origin *O* be at latitude 45° N and 100° W, and the standard parallels be 40° N and 50° N. Let point *P* be at 47° N and 95° W.

Then using the formulae $k = (\ln n_1 - \ln n_2) / (\ln n_3 - \ln n_4)$

$$r = R * (n_1 / k) * \{n / n_3\}^k$$

$$n_1 = 0.64406800504 \qquad n_2 = 0.76711787277$$
$$n_3 = 0.36586487239 \qquad n_4 = 0.46834279637$$
$$n_P = 0.39586764826 \qquad n_O = 0.41620305983$$

From these values

$$k = 0.708020879$$

$$r_P = 20128131.243 \text{ feet} \qquad r_o = 20854828.952 \text{ feet}$$

$$\{k(\lambda - \lambda_o)\} = 0.06178648 \text{ radians}$$

Let the coordinates of the origin in Figure 12.12 be (2000000, 0) using U.S. geodetic feet. Then,

$$x_P = 3242855.245 \qquad y_P = 765105.750$$

In addition to these basic projections, there are various alternatives that compromise between the preservation of various angles, small shapes, areas, and distances. The reader should consult other texts for the details (see "Further Reading").

12.5 MEASURING PROJECTION ERRORS

GIS professionals need to know the distortion of each map projection they encounter in order to correctly interpret and process the data stored using each map projection's coordinate system, and also to use their understanding of the pros and cons of different map projections to make better decisions on how to perform quantitative data analysis to the required accuracy levels. In the profession of land surveying each geographic area such as a country or a subdivision of a country has one or more designated "official" map projection for land measurement work in the locality. However, GIS professionals can custom create their own map projections for localized areas that achieve even higher levels of accuracy than official projections. As a general rule, an azimuthal conformal projection with its center set at the exact location of the locality would have the highest fidelity when representing the local land areas on a 2D map. Being able to measure and visualize the type and size of distortions generated by specific map projections is of immense value to the GIS professional.

Tissot's indicatrix is a useful method to visualize the distortion caused by any map projection in different parts of the Earth's surface, be it equatorial, sub-arctic, etc. Distortion is visualized by drawing realistic circular disk shapes of the same size at different locations on the earth's space in GIS, then visualize the shape of these disks when the map's coordinate system is set to various map projections. Where no spatial distortion happens, the disks will stay as perfect circles of equal size. In places where distortions occur, the shape of the projected circular disk becomes a variation of an ellipse, which visually shows the type and the amount of distortion that was applied to the disk, and hence to any spatial data at that location.

Note that since a map's scale is a number that can be easily changed to correct discrepancies, Tissot's graphs measure the relative distortion between locations in different parts of the earth's surface rather than the absolute discrepancy in distances. Since bearing angles are not affected by map scales, the distortions of bearing angles can be deducted from the projected shapes of these circles.

Below are the distorting effects of different kinds of map projection that can be visualized by Tissot's indicatrix:

1. Conformal projections: these projections will show circles of different sizes on the graphs, an indication that both the x (horizontal) and y (vertical) dimensions are scaled equally at each location, which means that local bearing angles are preserved. For example, in the Mercator projection, points along the equator have the smallest circles, while those near the North and South Poles have the largest, as illustrated in Figure 12.17. A conformal projection of the conic type is shown in Figure 12.14. Conformal projections for small areas (i.e. areas whose east-west and north-south dimensions are small compared to the earth's radius) will appear very similar whether they belong to cylindrical, conical, or azimuthal types of projections, and often the only differences are a rotation of the North angle and an adjustment of local map scale.
2. Equal Area projections: these projections show ellipses of different shapes on the Tissot's indicatrix graphs but the ellipses located in different parts of the Earth's surface have the same area sizes. These projections are illustrated in Figures 12.13, 12.15, and 12.16. As equal area projections, in order to achieve preservation of area measurements map scale changes in the north-south direction must compensate for changes in the east-west direction, or vice versa, at each location. Therefore the Tissot circles become ellipses, and at the most distorted locations in the periphery of the map projection areas they become extremely flattened. This characteristic also has the consequence of distorting bearing angles. For example, a 45-degree bearing whose tangent is 1 (i.e. dy / dx = 1) will be changed to the new tangent value of dy / dx of the Tissot ellipse, which varies by location.

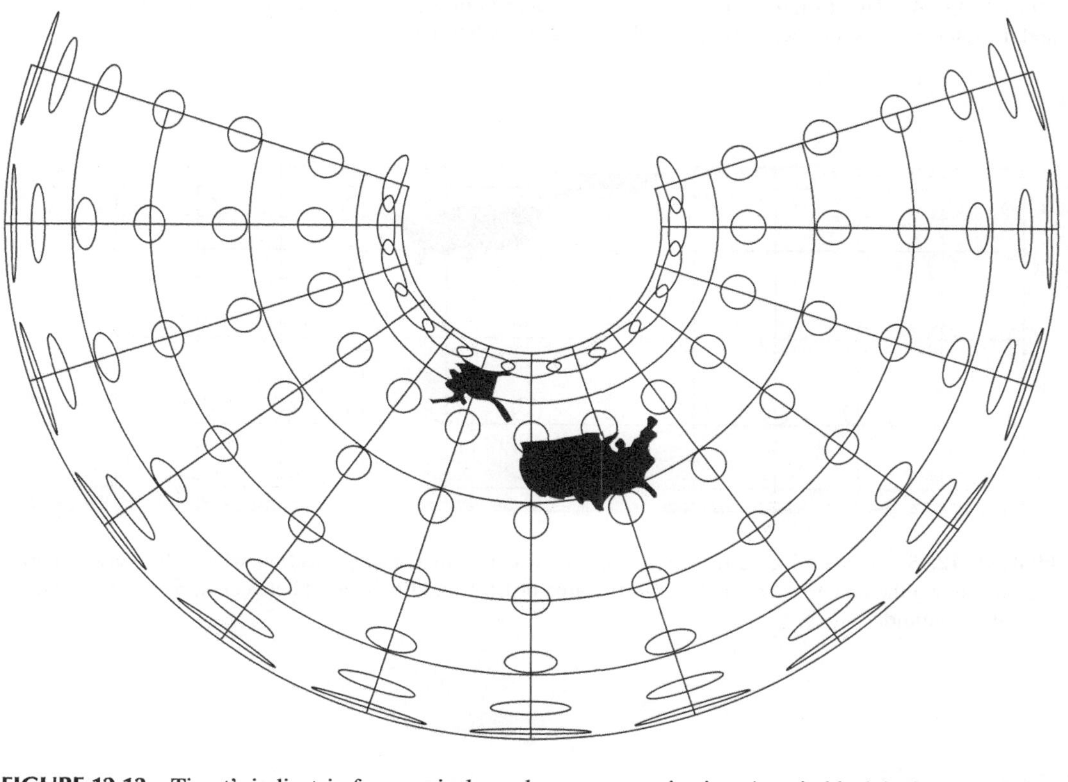

FIGURE 12.13 Tissot's indicatrix for a conical equal-area map projection. Area in black is the area of USA and the interval between successive parallels is 30 degrees latitude.

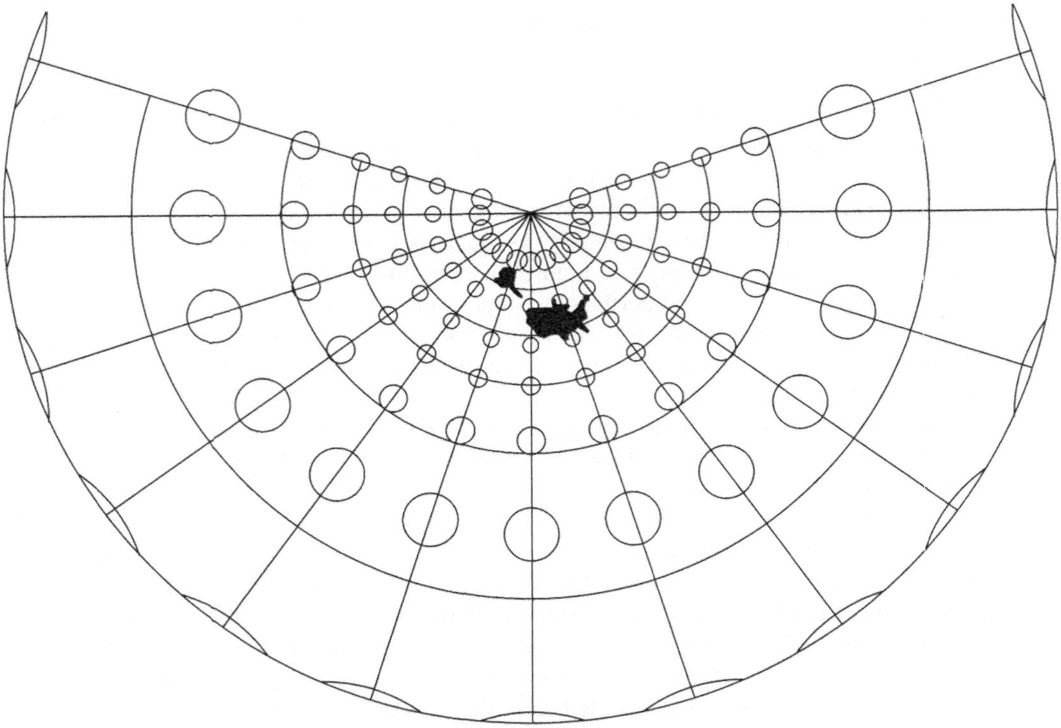

FIGURE 12.14 Tissot's indicatrix for a conical conformal map projection. Area in black is the area of USA and the interval between successive parallels is 30 degrees latitude.

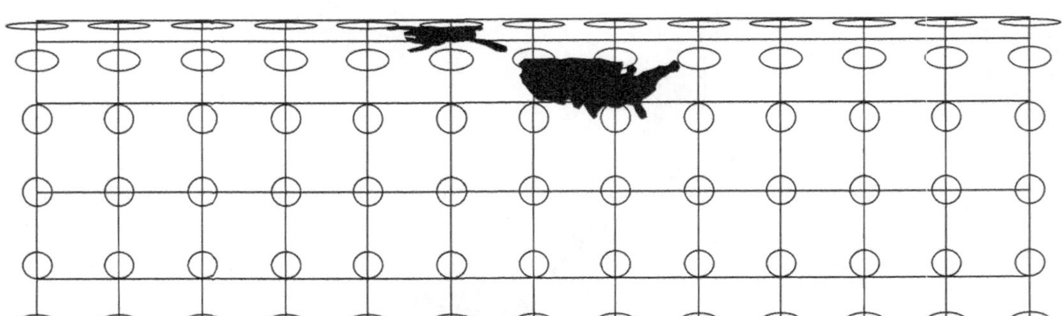

FIGURE 12.15 Tissot's indicatrix for a cylindrical equal-area map projection where the least distorted region is along the equator. Area in black is the area of USA and the interval between successive parallels is 30 degrees latitude.

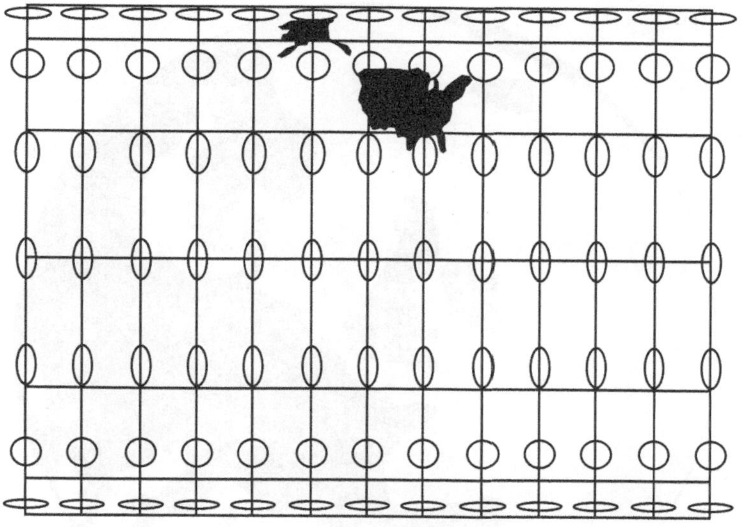

FIGURE 12.16 Tissot's indicatrix for a cylindrical equal-area map projection where the least distorted regions are at latitudes 45 degrees North and 45 degrees South. Area in black is the area of USA and the interval between successive parallels is 30 degrees latitude.

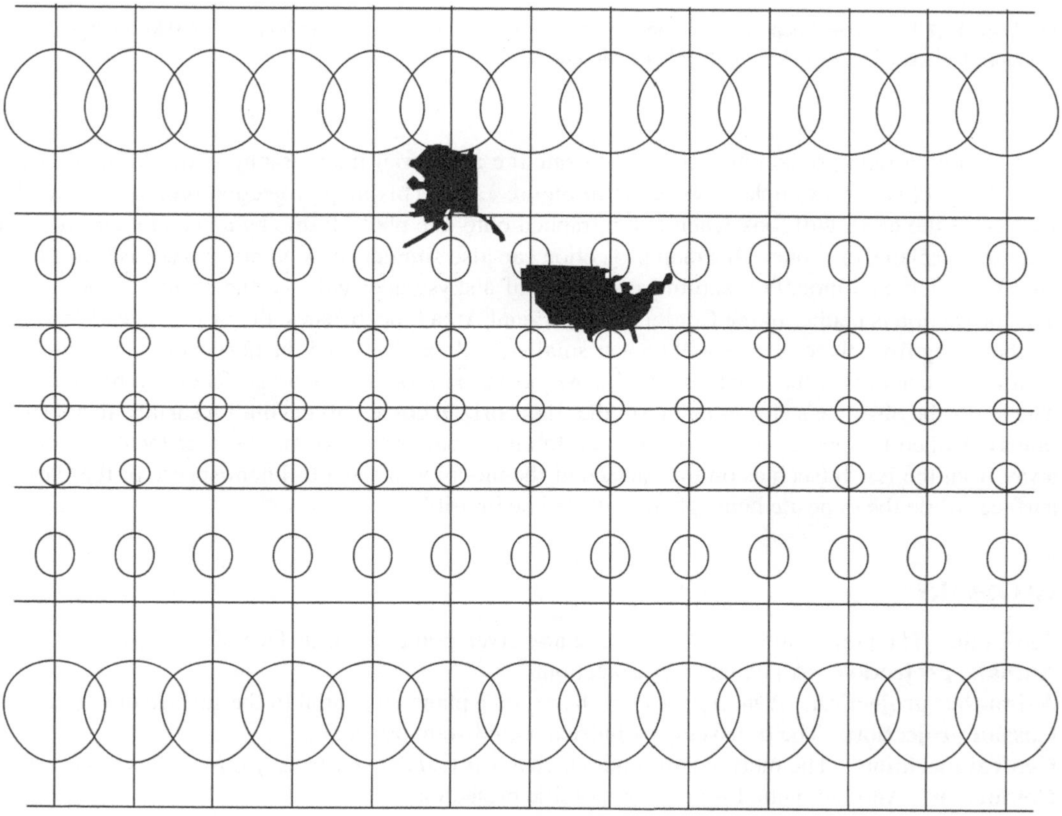

FIGURE 12.17 Tissot's indicatrix for the Mercator map projection, which is a conformal cylindrical map projection.

FIGURE 12.18 Vertical near-side perspective map projection for an observer located 1000 kilometers above London, United Kingdom. Area in black are land areas.

A common map projection type used in satellite and aerial photography is the Vertical Near-Side Perspective, an example of which is in Figure 12.18. This map projection represents how the surface of the earth will look when photographed onto one photo frame by a camera carried by a satellite, airplane, or drone. This map projection can also simulate how a person who rises to a high elevation such as a mountain summit or the top of a skyscraper will see the earth beneath them. This projection is neither of the Conformal nor Equal Area type, because the visible size of features further away from the observer will appear smaller, and those beyond the horizon circle (a circular area whose center is at the point directly below the observer on the earth's surface) will be invisible. The lower the observer's elevation the smaller this horizon circle will be, until it shrinks to approximately 0 when the person is at ground level. In an extreme scenario, an observer located very far away from the Earth (such as on the surface of the moon) will see a full hemisphere of the Earth's surface, while the opposite hemisphere will still be invisible.

GLOSSARY

Antipode The point diametrically opposite any given point on the surface of the Earth.
Authalic projection An equal area projection.
Azimuthal projection A *map projection* based on a plane tangential to the surface of the Earth.
Cassini projection The transverse cylindrical equidistant projection.
Central meridian The meridian around which a *transverse cylindrical projection* is wrapped.
Conformal Another name for an orthomorphic projection.
Conical projection A *map projection* based on a cone that touches or cuts the surface of the Earth and whose vertex is usually taken as being above the North or South Pole.

Constant of the cone The factor used in *conical projections* to determine the radius of circles representing the parallels of latitude.

Cylindrical projection A *map projection* based on a cylinder wrapped around the equator of the Earth.

Elemental triangle A small triangle on the surface of the globe defined by the increments $\delta\lambda$ and $\delta\phi$, both of which tend to zero.

Equal-area projection Also known as *equivalent* or *authalic*. A projection that preserves area by contracting distances in one direction and stretching them in the direction at right angles.

Equidistant projection Also known as *equirectangular*, a projection that retains the correct lengths along the two axes, which are usually the parallels and meridians.

Equivalent Another name for the *equal area* projection.

Loxodrome Also known as a *rhumb line*. A line of constant bearing.

Map projection A method for transforming the location of features on the curved surface of the Earth onto a flat or plane surface.

Mercator's projection The *cylindrical orthomorphic* projection. It has the characteristic that straight lines on the map are lines of constant bearing.

Orthomorphic projection Also known as a *conformal* projection. A projection that preserves the shape of small areas by maintaining scale along short distances around that point. Scale may, however, be different at different points.

Rhumb line Also known as a *loxodrome*. A line of constant bearing.

Simple cylindrical The simplest projection based on plotting north–south distances and east–west distances correctly. It introduces very large distortions east–west when nearing the North or South Pole.

Standard parallel The parallel of latitude where the encircling cone touches the sphere or, in the case of two standard parallels, it intersects the sphere (or ellipsoid).

Stereographic projection The *orthomorphic* version of the *azimuthal* projection. It has the characteristic that circles on the Earth plot as circles on the projection.

Tissot's indicatrix A method invented by French mathematician Nicolas Tissot to visualize the distortions of a map projections.

Transverse cylindrical projection A projection based on a cylinder wrapped around a meridian of longitude.

Transverse Mercator An *orthomorphic* projection based on a cylinder wrapped around a meridian of longitude. It is often used in topographic mapping.

UTM The Universal Transverse Mercator projection. It is based on a series of central meridians at six-degree intervals around the globe, extending three degrees either side.

Zenithal projection An *azimuthal projection* based on one or other pole. The meridians radiate out from the center as straight lines and the parallels are circles around that center.

13 Curves and Surfaces

13.1 PARAMETRIC FORMS

So far, we have seen that for points on a plane we can express a straight line as a series of values (x, y) such that

$$ax + by + c = 0$$

We can also express it in the form that

$$y = mx + n$$

where m is the slope or gradient of the line and n is a constant. These two expressions represent the same line if

$$m = -a / b \text{ and } n = -c / b$$

We also saw in Chapter 10 and can see from Figure 13.1 that if the slope of the line AB is θ (measured from the horizontal axis anticlockwise), then

$$y = x \tan \theta + c$$

Here, c is the value of y when x is zero. We can also express this as $\mathrm{d}y / \mathrm{d}x = \tan \theta = m$, which is constant for a straight line since

$$\int \tan \theta \, \mathrm{d}x = x \tan \theta + c$$

The differential of a constant is zero; hence, for a line, $\mathrm{d}^2y / \mathrm{d}x^2 = 0$.

Also note that if we use clockwise bearings from the vertical rather than anticlockwise angles from the horizontal, the bearing of line $AB = (90 - \theta)$ and

$$y = x \cot \left(\text{Bearing } AB\right) + c$$

If the line CD is perpendicular to AB, its slope will be $(90 + \theta)$. Hence, for the line CD

$$y = m'x + c'$$

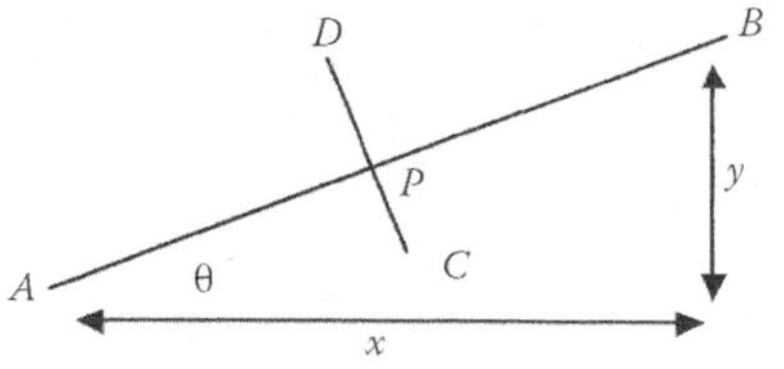

FIGURE 13.1 Orthogonal lines.

DOI: 10.1201/9781003464402-17

where

$$m' = -\cot(\theta) = -\tan(\text{Bearing } AB)$$

We have also seen that when two lines AB and CD are at right angles, their slopes are such that

$$m\,m' = \tan\theta * (-\cot\theta) = -1$$

If the product of the slopes of two lines = −1 they are said to be *orthogonal*. Thus, in Figure 13.1, the lines AB and CD are orthogonal. In Chapter 9, we also used the term *orthogonal* for a matrix where $\mathbf{A}\,\mathbf{A}^{-1} = \mathbf{I}$.

In the form $y = mx + n$, y is the dependent variable and x the independent. We can however express a straight line in *parametric* form as

$$x = p + t\cos\theta;\ y = q + t\sin\theta$$

where (p, q) are the coordinates of a point on the line, θ is the fixed slope angle, and "t" is the parameter. Here, t represents the distance along the line from point (p, q). There is then only one independent variable (t) and two dependent variables (x and y).

In the case of curves, we have seen that the slope at any point on the curve is dy / dx and the normal to it will have a slope $(-dx / dy)$. The straight line that represents the slope at any point is called the *tangent*. In the case of a point on a circle, the tangent is normal to the radius at that point. In the case of the tangent to an ellipse, the normal does not pass through the center except when the tangent is parallel to one of the axes (Figure 13.2).

As we have already seen, the equation for a circle with center at point (x_c, y_c) can be expressed in the form $(x - x_c)^2 + (y - y_c)^2 = r^2$.

An alternative, in parametric form, would be

$$x = r\cos t + x_c \quad y = r\sin t + y_c$$

where the variable t (the angle measured from the horizontal) is such that $0 \le t \le 2\pi$.

Consider the ellipse in Figure 13.3 with center O, semimajor axis a, and semiminor axis b. As we saw in Chapter 5, the circle of radius a centered on O is called the *auxiliary circle*. In effect the auxiliary circle is squashed or scaled down in the vertical or y-direction to give the ellipse. Thus, if N is a point on the auxiliary circle, then we can regard it as having coordinates

$$x_n = a\cos t + x_c, y_n = a\sin t + y_c$$

The length $MN = a\sin t$. By squashing the auxiliary circle so that OM remains the same but MN becomes scaled down to MP in the ratio $MP / MN = b / a$, then

$$MP = (b / a) * a\sin t = b\sin t$$

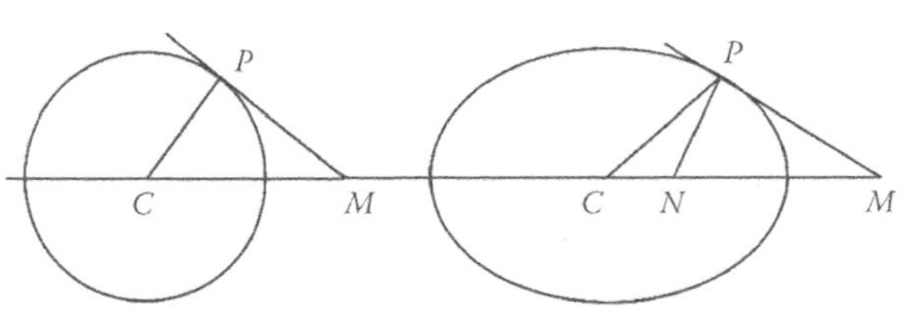

FIGURE 13.2 Tangents to a circle and an ellipse.

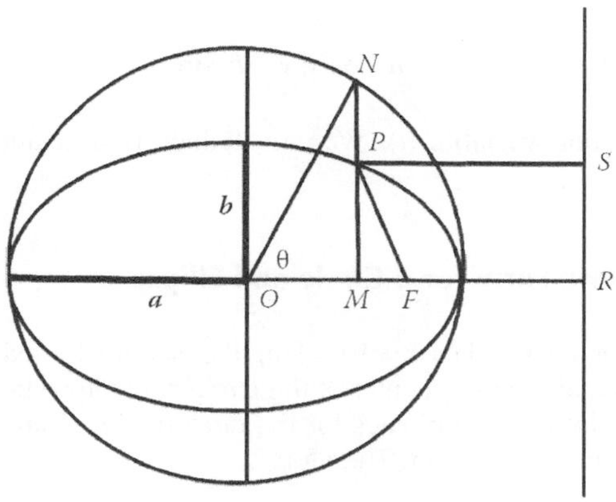

FIGURE 13.3 The ellipse and auxiliary circle.

Hence, for point P (x_P, y_P) on the ellipse with center O at (x_c, y_c)

$$x_p = a \cos t + x_c$$
$$y_p = b \sin t + y_c$$

Then, since $\sin^2 t + \cos^2 t = 1$, we have $(x_p - x_c)^2 / a^2 + (y_p - y_c)^2 / b^2 = 1$.

To pursue the ellipse a little further, the distance OR to the directrix (the line RS) was stated in Chapter 5 as (a / e) with the value of e being such that $0 < e < 1$. As for any conic with center O and focus F, the distance $OF = a * e$, $OR = a / e$, and $FP = e * PS$.

For convenience, we will assume that the center of the ellipse is taken as the origin. Since in Figure 13.3, the lengths $PS = MR$ and $MR + OM = a / e$, then $PS = a / e - x$. Hence,

$$\mathrm{FP} = e * PS = a - ex.FM = OF - OM = ae - x \text{ while } MP = y$$

Also, using Pythagoras, $FP^2 = FM^2 + MP^2$. So

$$(a - ex)^2 = (ae - x)^2 + y^2$$

or

$$a^2 - 2aex + e^2x^2 = a^2e^2 - 2aex + x^2 + y^2$$

We can write this as

$$\left(1 - e^2\right)x^2 + y^2 = a^2\left(1 - e^2\right) = b^2$$

where

$$e = \sqrt{\left(1 - b^2 / a^2\right)}$$

Once again this confirms that for an ellipse, $x^2 / a^2 + y^2 / b^2 = 1$.

Expressing the equation of a circle in the parametric form

$$x = r \cos t;\ y = r \sin t$$

or of an ellipse as

$$x = a \cos t; y = b \sin t$$

results in our having only one variable (t). We can deduce both x and y from that variable (see Example 13.1).

Example 13.1 Parametric Form for a Circle and Ellipse

In Figure 13.3, let the circle have radius $a = 10$ = length of semimajor axis of ellipse. Let the ellipse have semiminor axis $b = 8$ and let the center of the circle and ellipse be (0, 0).

For the first quadrant, let the coordinates for the circle be (x, y) and for the ellipse be (x, y'). Using $x = a \cos t$; $y = a \sin t$; $y' = b \sin t$. Then for:

$t =$	0°	10°	20°	30°	40°	50°	60°	70°	80°	90°
cos t	1	0.98	0.94	0.87	0.77	0.64	0.5	0.34	0.17	0
sin t	0	0.17	0.34	0.5	0.64	0.77	0.87	0.94	0.98	1

Given that $a = 10$ and $b = 8$, then

$x =$	10	9.8	9.4	8.7	7.7	6.4	5.0	3.4	1.7	0
$y =$	0	1.7	3.4	5.0	6.4	7.7	8.7	9.4	9.8	10
$y' =$	0	1.4	2.7	4.0	5.1	6.2	7.0	7.5	7.8	8

All intermediate values and the values for the other three quadrants can be calculated in a similar fashion.

The parabola $y^2 = 4ax$ can be expressed in parametric form as:

$$x = at^2 \quad \text{and} \quad y = 2at$$

The hyperbola can be expressed as $x^2 / a^2 - y^2 / b^2 = 1$. Its parametric form uses the fact that $(1 + \tan^2 \theta) = (1 + \sin^2 \theta / \cos^2 \theta) = (\cos^2 \theta + \sin^2 \theta) / \cos^2 \theta = \sec^2 \theta$. Thus,

$$\sec^2 \theta - \tan^2 \theta = 1$$

The hyperbola can therefore be expressed in the form

$$x = a \sec \theta, y = b \tan \theta$$

or using the parameter t

$$x = a \sec t; y = b \tan t$$

Parametric forms (see Box 13.1) can be used to determine the slope at any point on a curve. If, for example, we consider the circle with center (x_c, y_c) and radius r, then its parametric form is:

$$x = x_c + r \cos t \quad y = y_c + r \sin t$$

Differentiating with respect to t

$$\mathrm{d}x / \mathrm{d}t = -r \sin t; \quad \mathrm{d}y / \mathrm{d}t = r \cos t$$

$$(\mathrm{d}y / \mathrm{d}t) / (\mathrm{d}x / \mathrm{d}t) = -(\cos t / \sin t) = -\cot t$$

BOX 13.1 PARAMETRIC EQUATIONS FOR LINES AND CONIC SECTIONS (SECOND-DEGREE CURVES)

p, q, and θ are constant, that is, fixed quantities; t is the independent variable.

For a line:

$$x = p + t\cos\theta; y = q + t\sin\theta$$

For a circle:

$$x = p + r\cos t; y = q + r\sin t$$

For an ellipse:

$$x = p + a\cos t; y = q + b\sin t$$

For a parabola:

$$x = p + at^2; y = q + 2at$$

For a hyperbola:

$$x = p + a\sec t; y = q + b\tan t$$

or

$$dy / dx = -\cot t$$

This is the rate of change of the curve, or put another way, it is the measure of the slope of the curve at the point defined by t.

13.2 THE ELLIPSE

Figure 13.4 shows the auxiliary circle of radius a and the ellipse that has semiminor axis of length b. A point P on the ellipse has coordinates ($a\cos\theta$, $b\sin\theta$) referred to the center. PT is the tangent to the ellipse at point P with T being on the major axis. PQ is the normal at P, with Q being on the minor axis. The line QP makes an angle ø with the major axis and is known as the *geodetic* or *spheroidal latitude* of P. It is essentially the direction of the vertical at P. In practice, a plumb bob on the Earth's surface may not point exactly in this direction because of the gravitational attraction of nearby mountains, giving rise to what is known in geodesy as the *deviation of the vertical*. Given an ellipsoid of uniform density, the line PQ would be the vertical at P and, except along the equator and at the poles (D' and E'), it would not pass through the center O.

From the parametric form of a circle we can calculate the slope of the curve at any point on the circle, for example, NT in Figure 13.4. (Later, we show why the tangent at N and the tangent at P both intersect the major axis at T.) This is the slope of the tangent at that point.

Similarly for the ellipse, the tangent is PT where

$$\begin{aligned} dy / dx &= (dy / d\theta) / (d\theta / dx) = (b\cos\theta) / (-a\sin\theta) \\ &= -(b / a)\cot\theta \end{aligned}$$

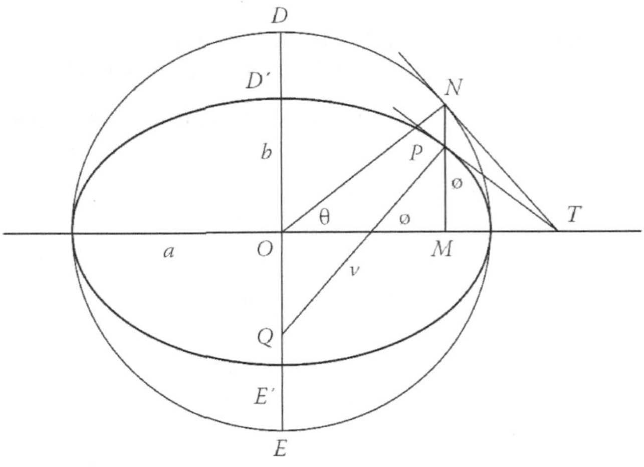

FIGURE 13.4 Normal to an ellipse.

Circles and ellipses are important figures in geomatics and geographic information system since, when rotated about their north / south axis, they create volumes that represent good approximations to the shape of the Earth, depending on the scale at which one is working.

The angle *POT* (not shown in the diagram) would be the *geocentric latitude* but is not normally used in geodetic calculations, although it is used in satellite position fixing (see Chapter 16) and in astronomy. The angle θ is sometimes known as the *reduced latitude*. The length *PQ* is usually referred to by the Greek letter ν (pronounced "nu" as in "new"). On an ellipsoid that is obtained by rotating the ellipse about its minor axis, the length of ν is the same for all points along a given parallel of latitude but has different values at different latitudes. The *x*-coordinate of $P = a \cos \theta$ from before but also $= \nu \cos \phi$ from the projection of *QP* onto the *x*-axis. Hence (see Example 13.2),

$$a \cos \theta = \nu \cos \varnothing$$

We mentioned earlier that in Figure 13.4, the tangents at *N* and *P* meet at *T* on the major axis. *TN* is tangential to the auxiliary circle and by squashing the auxiliary circle down to the ellipse, tangency will not be affected even though angles change. In fact, we can show this using the parametric forms of the auxiliary circle and ellipse.

For point *N* on the auxiliary circle:

$$x = a \cos t; y = a \sin t; \mathrm{d}y / \mathrm{d}x = (\mathrm{d}y / \mathrm{d}t) / (\mathrm{d}x / \mathrm{d}t) = -\cot t$$

If we assume that the tangent to the ellipse at *N* cuts the *x*-axis at *T*, then the equation of the line *NT* is $(y - a \sin t) = -\cot t\,(x - a \cos t)$.

This intersects the *x*-axis where $y = 0$, and $x = a \cos t + a \sin^2 t / \cos t = a \sec t$. Hence, $OT = a \sec t$ or $a \sec \theta$ in Figure 13.4.

For point *P* on the ellipse:

$$x = a \cos t; y = b \sin t; \mathrm{d}y / \mathrm{d}x = (\mathrm{d}y / \mathrm{d}t) / (\mathrm{d}x / \mathrm{d}t) = -(b / a)\cot t$$

Example 13.2 The Earth's Radii

The equatorial radius of the Earth = a = 6378.1370 km. The polar radius = b = 6356.7523 km.
For the Earth as an ellipsoid, $e^2 = 1 - (b / a)^2 = 0.00669438$ or e = 0.081818.

$$\text{At } \varnothing = 50° \text{ north, } \tan\theta = b/a \tan\varnothing = 1.1877578657775.$$

$$\text{Hence, } \theta = 49.9052218256 \text{ degrees} = 49°54'18.8''.$$

Thus, there is a difference of 5 minutes 41.2 seconds of arc between the auxiliary circle latitude and the geodetic latitude. Since 1 second of arc ≈ 30 meters on the Earth's surface, this represents a difference of approximately 10.55 km on the ground.

Given that $a \cos\theta = \nu \cos ø$, ν = 6390.7021 km at 50° north.

If we assume that the tangent to the ellipse at P cuts the x-axis at T', then the equation of the line PT' *is* $(y - b \sin t) = -(b / a) \cot t\, (x - a \cos t)$. This intersects the x-axis where $y = 0$, and

$$x = a \cos t + b(a/b) \sin^2 t / \cos t = a \sec t$$

Hence, $OT' = a \sec t$ demonstrating that T and T' are the same point.

Thus in Figure 13.4, if $\angle MNT = \theta$, and $\angle MPT = ø$, then $NM = a \sin\theta$, $OM = x$ value of $P = a \cos\theta$, and $PM = y$ value of $P = b \sin\theta$.

In triangle MNT, $MT = NM \tan\theta = a \sin\theta \tan\theta$ while in triangle MPT,

$$MT = PM \tan\varnothing = b \sin\theta \tan\varnothing$$

Hence, $a \tan\theta = b \tan ø$. We can express this as:

$$\tan\theta = (b \sin\varnothing)/(a \cos\varnothing)$$

Since angles are just ratios, we can represent this relationship by the right-angled triangle HIJ in Figure 13.5 in which $IJ = b \sin ø$ and $HI = a \cos ø$ and $\angle JHI = \theta$. Then by suitable manipulation as shown in the Box 13.2, we can show that:

$$x = \nu \cos\phi;\ y = \nu(1 - e^2) \sin\phi;\ \nu = a/\sqrt{(1 - e^2 \sin^2\varnothing)}$$

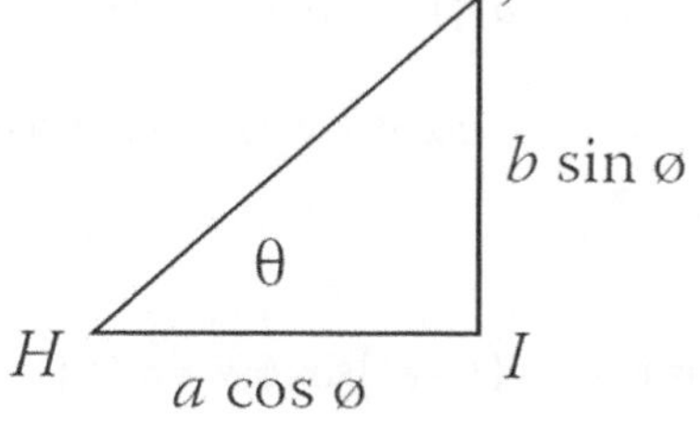

FIGURE 13.5 Theta and phi.

BOX 13.2 THE RADIUS ν

From Figure 13.5

$$HI = a\ \cos\ \varnothing,\quad IJ = b\ \sin\ \varnothing,\quad \text{and}\quad \angle JHI = \theta$$

Thus,

$$IJ\ /\ HI = \tan\ \theta = (b\ /\ a)\ \tan\ \varnothing.\ \text{Using Pythagoras, }HJ^2 = a^2\cos^2\varnothing + b^2\sin^2\varnothing$$

Since $\cos^2\phi + \sin^2\phi = 1$, we can write this as

$$HJ^2 = a^2 - a^2\sin^2\varnothing + b^2\sin^2\varnothing = a^2\left(1 - \frac{(a^2 - b^2)}{a^2}\sin^2\varnothing\right)$$
$$= a^2\left(1 - e^2\sin^2\varnothing\right)$$

where

$$e^2 = \left(1 - b^2\ /\ a^2\right)\ \text{as before. Hence,}\quad HJ = a\sqrt{1 - e^2\sin^2\varnothing}$$

In triangle HIJ,

$$\cos\ \theta = HI\ /\ HJ = a\ \cos\ \varnothing\ /\left\{a\sqrt{1 - e^2\sin^2\varnothing}\right\}$$

or

$$a\ \cos\ \theta = \nu\ \cos\ \varnothing$$

We have previously shown that:

$$a\ \cos\ \theta = \nu\ \cos\ \varnothing$$

So $a\ \cos\ \theta = \nu\ \cos\ \phi = a\ \cos\ \phi\ /\ \sqrt{(1 - e^2\sin^2\phi)}$ and $\nu = a\ /\ \sqrt{(1 - e^2\sin^2\phi)}$, or

$$a = \nu\ \sqrt{\left(1 - e^2\sin^2\varnothing\right)}$$

But,

$$x^2\ /\ a^2 + y^2\ /\ b^2 = 1\ \text{and}\ x = \nu\ \cos\ \varnothing$$

Thus,

$$y^2 = b^2 - \left(b^2\ /\ a^2\right)x^2 = a^2\left(1 - e^2\right) - \left(1 - e^2\right)x^2 = \left(1 - e^2\right)\left(a^2 - x^2\right)$$

Since $a^2 = \nu^2\,(1 - e^2\sin^2\phi)$ and $x^2 = \nu^2\cos^2\phi = \nu^2\,(1 - \sin^2\phi)$, then $\left(a^2 - x^2\right) = \nu^2\left(1 - e^2\right)\sin^2\varnothing$. As a result $y^2 = \nu^2\left(1 - e^2\right)^2\sin^2\varnothing$.

Thus,

$$x = \nu\ \cos\ \phi;\ y = \nu\left(1 - e^2\right)\sin\ \phi;\ \nu = a\ /\ \sqrt{\left(1 - e^2\sin^2\phi\right)}$$

These relationships are used in geodetic computations.

The quantities v and ø are of particular importance in the field of geodesy, the scientific study of the size and shape of the Earth and, for example, in calculations of position using the satellite global positioning system (GPS).

13.3 THE RADIUS OF CURVATURE

Before leaving the ellipse, there is one further quantity that is required. It is known as the *radius of curvature* at *P* along the line of the ellipse and is normally referred to as "ρ," the Greek letter "rho" (pronounced "row" as in a row of beans). *Curvature* is the rate of change of the direction of a tangent in relation to the length of arc; the *radius of curvature* at *P* is the radius of a circle with curvature equal to that of the curve at *P* or, put another way, the radius of the circle that just touches the curve. At the point of contact, the tangent to the circle is also the tangent to the curve. This applies to all curves, not just ellipses.

Consider the curve *QPQ′* in Figure 13.6 with the circle center *C* just touching the curve at *P*. At the point of contact the radius *CP* (= ρ) will be normal to the tangent *PT* that has a slope value that is here called θ. Consider also the two nearby points *P* and *P′* shown on the inset. Their coordinates will differ by δx and δy, while the length along the curve is denoted by δs. If the slope at *P* is θ and at *P′* is θ + δθ, then the angle *PCP′* = δθ (exaggerated in Figure 13.6).

By measuring the angle θ in radians in triangle *PCP′*, then δs = *CP*δθ = ρ δθ or (δθ / δs) = 1 / ρ. The slope of the line *PP′* = θ, therefore $\delta y / \delta x = \tan\theta$ or

$$dy / dx = \tan\theta$$

The second derivative or rate of change of the slope is the curvature = $d^2y / dx^2 = d(dy / dx) / dx = d(\tan\theta) / dx = \{d\theta / dx\} * \{d(\tan\theta) / d\theta\} = (d\theta / dx)\sec^2\theta$ since the differential of $\tan\theta = \sec^2\theta$. Also, $\delta\theta / \delta x = (\delta\theta / \delta s) * (\delta s / \delta x)$ and $\delta s / \delta x = \sec\theta$ as can be seen from Figure 13.6. So

$$d^2y / dx^2 = (1/\rho)\sec^3\theta$$

Since $\sec^2 = 1 + \tan^2$ and $\tan\theta = dy / dx$, therefore

$$d^2y / dx^2 = (1/\rho)\left(1 + (dy/dx)^2\right)^{3/2}$$

On rearranging, the radius of curvature of any curve is given by

$$\rho = \left\{\left(1 + (dy/dx)^2\right)^{3/2}\right\} / \left(d^2y / dx^2\right)$$

If this is applied to an ellipse, then it can be shown that

$$\rho = v\,(1 - e^2)(1 - e^2\sin^2\varnothing)^{-1} = a(1 - e^2)(1 - e^2\sin^2\varnothing)^{-3/2}$$

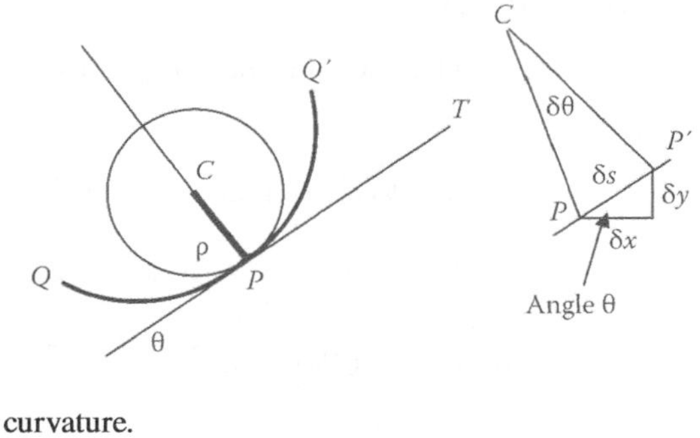

FIGURE 13.6 Radius of curvature.

BOX 13.3 RADII OF CURVATURE

For any curve the radius of curvature

$$\rho=\left\{\left(1+(\mathrm{d}y/\mathrm{d}x)^2\right)^{3/2}\right\}/\left(\mathrm{d}^2y/\mathrm{d}x^2\right)$$

For an ellipse

$$\rho=\mathrm{v}\left(1-e^2\right)\left(1-e^2\sin^2\varnothing\right)^{-1}=a\left(1-e^2\right)\left(1-e^2\sin^2\varnothing\right)^{-3/2}$$

$$x=\mathrm{v}\cos\varnothing$$
$$y=\mathrm{v}\left(1-e^2\right)\sin\varnothing$$
$$\mathrm{v}=a/\sqrt{\left(1-e^2\sin^2\varnothing\right)}$$

where ϕ is the geodetic latitude.

The value "ρ" (Box 13.3) is used extensively in geodetic computations; but these are beyond the scope of the present text. For now, we will consider more about curvature. Every second-degree (quadratic) curve bends only one way. The circle and ellipse, for example, bend all the way around until they close back on themselves, while the parabola and hyperbola go off to infinity before turning back.

From Figure 13.6 we can also note that $\delta s^2=\delta x^2+\delta y^2$ *or* $\delta s/\delta x=\sqrt{1+(\delta y/\delta x)^2}$. From this, we obtain

$$\mathrm{d}s/\mathrm{d}x=\sqrt{1+(dy/dx)^2}$$

or

$$s=\int\sqrt{1+(dy/dx)^2}\,\mathrm{d}x$$

This provides a method for calculating the length along a curve. In the particular case of the ellipse, the integration is not straightforward; but for some functions it provides an elegant way to determine such lengths.

13.4 FITTING CURVES TO POINTS

The simplest curve that bends forward and backward is the cubic. If we take the general case of the cubic

$$y=a+bx+cx^2+dx^3$$

Then

$$\mathrm{d}y/\mathrm{d}x=b+2cx+3dx^2$$

There can be two values for which $dy / dx = 0$. These are where the maxima and minima for the curve are located. Also,

$$d^2y / dx^2 = 2c + 6dx$$

There is only one value for which $d^2y / dx^2 = 0$, namely, where $x = -c / (3d)$. This is the point of inflection. (See Figure 10.3 in Chapter 10.)

Because there are four unknown constants (a, b, c, d) in the cubic equation for y, we need four independent equations to calculate them. Put another way, a cubic can be made to pass through four points just as a quadratic curve such as a circle can be made to pass through any three points (albeit with infinite radius for three points on a straight line). Five points are needed to define a quartic (fourth degree), six for a quintic (fifth degree), and so forth.

Given a cubic in the form $y = f(x)$ with four known points (x_1, y_1), (x_2, y_2), (x_3, y_3), and (x_4, y_4), then the equation:

$$y = y_1 \frac{(x-x_2)(x-x_3)(x-x_4)}{(x_1-x_2)(x_1-x_3)(x_1-x_4)} + y_2 \frac{(x-x_1)(x-x_3)(x-x_4)}{(x_2-x_1)(x_2-x_3)(x_2-x_4)}$$
$$+ y_3 \frac{(x-x_1)(x-x_2)(x-x_4)}{(x_3-x_1)(x_3-x_2)(x_3-x_4)} + y_4 \frac{(x-x_1)(x-x_2)(x-x_3)}{(x_4-x_1)(x_4-x_2)(x_4-x_3)}$$

is a cubic in x and passes through all four points. Note that none of the elements in the denominator can equal zero because there is only one value for y for a given value of x in the equation that we are considering. Note also that the sequence can be extended to quartics, quintics, and any other order and can be expressed in mathematical shorthand as

$$y = \sum_{i=1}^{n} y_i \prod_{i \neq j} \frac{(x - x_j)}{(x_i - x_j)}$$

where $\sum_{i=1}^{n} y_i$ means the sum of all the y_i from $i = 1$ to $i = n$, and $\prod_{i \neq j}$ means the product of the following expression, subject to the constraint that i does not equal j $(i \neq j)$.

But what if we want to fit a cubic to five points? One solution would be to make the best-fit approximation to all five. Another would be to ensure that the curve goes exactly through all five points but that is normally not possible with only one cubic. The solution is to fit the curve step by step in sections, or what is called *piecewise*. This means fitting several cubic curves so that if we have points $A, B, C, D, E, \ldots, N$ then we fit the curve to AB, then BC, then CD, then DE, and so on, until the last point N is reached. The result is what is called a *spline* curve, the points A, B, C, and so forth, being known as *nodes* or *knots*.

Before considering piecewise polynomials, let us return to the way in which we express a straight line between two points in parametric form, t being the independent variable.

$$x = a + bt \quad y = c + dt$$

For a quadratic

$$x = a + bt + ct^2 \quad y = d + et + ft^2$$

For a cubic

$$x = a + bt + ct^2 + dt^3 \quad y = e + ft + gt^2 + ht^3$$

The reason for expressing (x, y) in this form is that with a computer plotter we can increment t from 0 to 1 in, for example, 10 steps of 0.1. The computer calculates (x, y) for $t = 0$, then for $t = 0.1$, then for $t = 0.2$, and so on, until $t = 0.9$, and then $t = 1.0$. We would then have 11 values for (x, y) from the start point (x_s, y_s) where $t = 0$ to the end point (x_e, y_e) where $t = 1$. A computer-driven plotter can draw a series of straight lines joining these points consecutively so as to portray the curve from its beginning to its end. By taking finer increments of t (for example, every 0.01 or 0.05 rather than 0.1) we can create the appearance of a smoother curve but at a cost of longer processing time.

If we have two points A (x_1, y_1) and B (x_2, y_2), then any point on the line AB can be expressed in the form

$$x = x_1 + (x_2 - x_1)t \quad \text{and} \quad y = y_1 + (y_2 - y_1)t$$

(where $t = 0$ at point A and $t = 1$ at B).

Consider three points A (x_A, y_A), B (x_B, y_B), and C (x_C, y_C) in Figure 13.7. A second-degree curve can be made to pass exactly through three points but here we will divide it into two separate but continuous sections. For each section AB and BC there is a quadratic of the form:

$$x = a + bt + ct^2 \quad \mathrm{d}x/\mathrm{d}t = b + 2ct \quad \mathrm{d}^2x/\mathrm{d}t^2 = 2c$$
$$y = e + ft + gt^2 \quad \mathrm{d}y/\mathrm{d}t = f + 2gt \quad \mathrm{d}^2y/\mathrm{d}t^2 = 2g$$

Let us consider two sets of equations for the two sections of the curve from A to B and from B to C. Let any point on the first section of the curve be (x_1, y_1) and on the second section (x_2, y_2). For the first section, these coordinates must satisfy the quadratic expressions:

$$x_1 = a_1 + b_1t_1 + c_1t_1^2 \quad \mathrm{d}x/\mathrm{d}t = b_1 + 2c_1t_1 \quad \mathrm{d}^2x/\mathrm{d}t^2 = 2c_1$$
$$y_1 = e_1 + f_1t_1 + g_1t_1^2 \quad \mathrm{d}y/\mathrm{d}t = f_1 + 2g_1t_1 \quad \mathrm{d}^2y/\mathrm{d}t^2 = 2g_1$$

and for the second section:

$$x_2 = a_2 + b_2t_2 + c_2t_2{}^2 \quad \mathrm{d}x/\mathrm{d}t = b_2 + 2c_2t_2 \quad \mathrm{d}^2x/\mathrm{d}t^2 = 2c_2$$
$$y_2 = e_2 + f_2t_2 + g_2t_2{}^2 \quad \mathrm{d}y/\mathrm{d}t = f_2 + 2g_2t_2 \quad \mathrm{d}^2y/\mathrm{d}t^2 = 2g_2$$

For section 1 from A to B:

$$\text{At } A, t_1 = 0 \text{ and } x_1 = a_1 \text{ and } y_1 = e_1$$

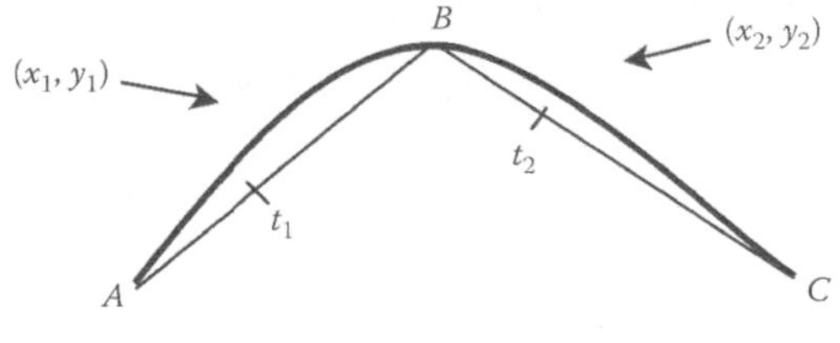

FIGURE 13.7 Fitting a second-degree (quadratic) curve.

Thus,

$$a_1 = x_A \text{ and } e_1 = y_A$$
$$At\ B, t_1 = 1 \text{ and } x_1 = x_B = a_1 + b_1 + c_1 \text{ or}$$

hence,

$$b_1 + c_1 = x_B - x_A$$

Likewise,

$$f_1 + g_1 = y_B - y_A$$

For section 2 from B to C:

$$\text{At } B, t_2 = 0 \text{ and } x_2 = a_2 \text{ and } y_2 = e_2$$

Since this is point B, $a_2 = x_B$ and $e_2 = y_B$

$$\text{At } C, t_2 = 1 \text{ and } x_2 = x_C = a_2 + b_2 + c_2 \text{ or } b_2 + c_2 = x_C - x_B$$

Likewise,

$$f_2 + g_2 = y_C - y_B$$

For the curve to be continuous at point B the first and second derivatives must be the same. For the first derivative this means that $b_1 + 2c_1 = b_2$, and $f_1 + 2g_1 = f_2$. For the second derivative, $c_1 = c_2$, and $g_1 = g_2$.

Putting this together

$$a_1 = x_A; a_2 = x_B; b_1 = (4x_B - 3x_A - x_C)/2; b_2 = (x_C - x_A)/2; c_1 = c_2 = (x_A - 2x_B + x_C)/2$$

Similarly,

$$e_1 = y_A; e_2 = y_B; f_1 = (4y_B - 3y_A - y_C)/2; f_2 = (y_C - y_A)/2; g_1 = g_2 = (y_A - 2y_B + y_C)/2$$

Example 13.3 Fitting a Quadratic Piecewise

As an example of a piecewise quadratic, consider the points A (3, 5), B (23, 25), and C (40, 15) shown in Figure 13.8. Using the equations derived in Section 13.4,

$$a_1 = x_A = 3; a_2 = x_B = 23; b_1 = (4x_B - 3x_A - x_C)/2 = 21.5; b_2 = (x_C - x_A)/2 = 18.5$$
$$c_1 = c_2 = (x_A - 2x_B + x_C)/2 = -1.5$$
$$e_1 = y_A = 5; e_2 = y_B = 25;$$
$$f_1 = (4y_B - 3y_A - y_C)/2 = 35; f_2 = (y_C - y_A)/2 = 5$$
$$g_1 = g_2 = (y_A - 2y_B + y_C)/2 = -15$$

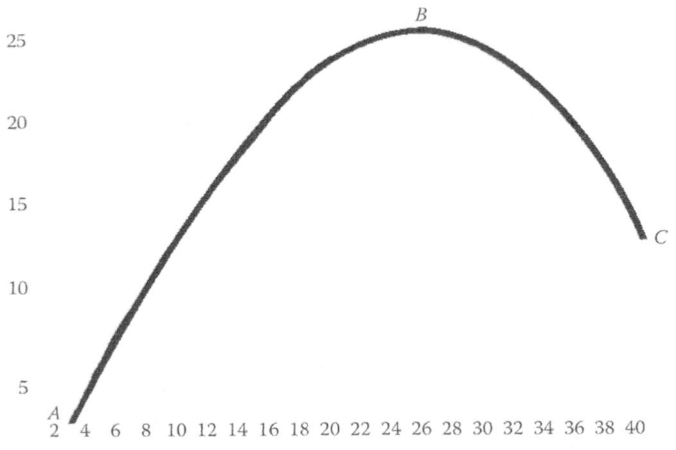

FIGURE 13.8 Two quadratics fitted to three points.

We can now compile a table based on $x_1 = a_1 + b_1t_1 + c_1t_1^2$, and so forth, using x_1, y_1 for section *AB* and x_2, y_2 for section *BC*:

$t =$	0	0.1	0.2	0.3	0.4	0.5	0.6	0.7	0.8	0.9	1
x_1	3	5.1	7.2	9.3	11.4	13.4	15.5	17.3	19.2	21.1	23
y_1	5	8.4	11.4	14.2	16.6	18.8	20.6	22.2	23.4	24.4	25
x_2	23	24.8	26.6	28.4	30.2	31.9	33.6	35.2	36.8	38.4	40
y_2	25	25.3	25.4	25.2	24.6	23.8	22.6	21.2	19.4	17.4	15

The (x, y) values can then be plotted as shown in Figure 13.8.

With these values, we can calculate $x = a_1 + b_1t + c_1t_2$ and $y = e_1 + f_1t + g_1t_2$ for t incremented from 0 to 1. From these values we can then plot the first section of the curve; and similarly for the second section as illustrated in Example 13.3.

The trouble with a quadratic is that it bends only one way—it cannot twist left and then right, clockwise then counterclockwise, or vice versa. A cubic can do this since it can have a maximum and minimum and a point of inflection in between. As a result we can fit a set of cubic curves piecewise through any number of points.

Consider the series of points *A*, *B*, *C*, *D*, …, *M*, *N* in Figure 13.9. The upper part of Figure 13.9 shows a series of points joined by lines. Let these be:

$$A(x_0, y_0), B(x_1, y_1), C(x_2, y_2), D(x_3, y_3), \ldots, M(x_{n-1}, y_{n-1}), N(x_n, y_n)$$

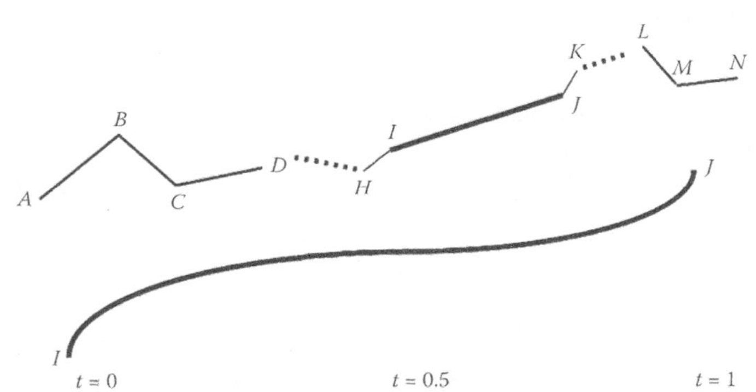

FIGURE 13.9 A piecewise cubic.

The series of lines $A - B$, $B - C$, and so forth, is said to form a *string*.

Now consider the two successive points along this string, I (x_i, y_i) and J (x_{i+1}, y_{i+1}).

For a cubic curve between I and J, we can let

$$x = a_i + b_i t + c_i t^2 + d_i t^3$$
$$y = e_i + f_i t + g_i t^2 + h_i t^3$$

where t is made to take a series of values from 0 to 1.

We then have eight unknowns $(a_i, b_i, c_i, d_i, e_i, f_i, g_i, h_i)$ for the section between I and J. But when $t = 0$, we are at I, hence,

$$x_i = a_i \text{ and } y_i = e_i$$

When $t = 1$, we are at J, hence,

$$x_j = a_j + b_j + c_j + d_j \quad \text{and} \quad y_j = e_j + f_j + g_j + h_j$$

or

$$x_j - x_i = b_j + c_j + d_j \quad \text{and} \quad y_j - y_i = f_j + g_j + h_j$$

Also,

$$\mathrm{d}x / \mathrm{d}t = b_i + 2c_i t + 3d_i t^2 \quad \mathrm{d}y / \mathrm{d}t = f_i + 2g_i t + 3h_i t^2$$
$$\mathrm{d}^2 x / \mathrm{d}t^2 = 2c_i + 6d_i t \qquad \mathrm{d}^2 y / \mathrm{d}t^2 = 2g_i + 6h_i t$$

To obtain a smooth curve at I, there must be continuity in both the first and second differentials so that both the slope and rate of change of the slope, namely the curvature are continuous. That means that at I (where $t = 0$), $\mathrm{d}x / \mathrm{d}t = b_i$ and must have the same value as the previous section HI where $t = 1$. This would have been $b_h + 2c_h + 3d_h$ using the same notation. Hence,

$$b_i = b_h + 2c_h + 3d_h$$

Similarly,

$$f_i = f_h + 2g_h + 3h_h$$

At point I on the section IJ, where $t = 0$, $\mathrm{d}^2x / \mathrm{d}t^2 = 2c_i$ and this must equal the value from the previous section of the curve when $t = 1$, namely, $2c_h + 6d_h$. Hence, we can calculate c_i and similarly g_i and since we have

$$d_i = (x_j - x_i) - b_j - c_j \quad \text{and} \quad h_i = (y_j - y_i) - f_j - g_j$$

we have all the parameters for the section IJ provided that we have established the parameters for HI. Thus, for all the sections of the curve BC, CD, …, IJ, …, MN we can fit a cubic in parametric form, based on values obtained from the previous section. The trouble is how to get started, since we need to know the initial slope and curvature in order to calculate the section AB. There are an infinite number of possibilities.

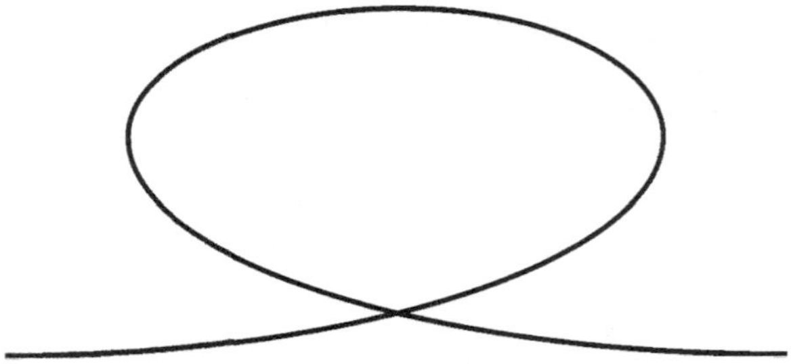

FIGURE 13.10 Looping curve.

Of the many practical solutions, the most common is to set the second derivative to zero at the very start and very end of the string. An alternative is to use the values from the quadratic described above as the start for the line *AB* and carry through the values at *B* as the initial values for fitting a cubic to *BC,* and so on; or fit a cubic exactly through the first four points but use only the first section.

There is, however, a fundamental problem with the approach described in that although it works in many circumstances, it can lead to a curve that crosses over itself as the angle of slope increases from zero to 2π—see Figure 13.10.

Such a loop is unacceptable if it is meant to represent a contour line through a set of interpolated points. It is not the intention of this text to discuss curve-fitting algorithms but rather to explain the basic mathematics behind them. Readers should consult relevant books on computer graphics for more details.

13.5 THE BEZIER CURVE

There is, however, another approach that will be mentioned, since it is commonly referred to in computer drawing packages. Known as the *family of Bezier curves*, they use Bernstein polynomials (named after the Russian mathematician, Sergei Bernstein) and take the form:

$$x = f_x(t) = \sum_{i=0}^{i=n} {}_n C_i t^i (1-t)^{n-i} x_i$$

$$y = f_y(t) = \sum_{i=0}^{i=n} {}_n C_i t^i (1-t)^{n-i} y_i$$

where *f*(*t*) means the function for which *t* takes a series of steps from 0 to 1 calculated separately for each increment of *x* and *y* and

$${}_n C_i = n! / \{i!(n-1)!\}$$

${}_n C_i$ is a symbol often used for the number of distinct *combinations* of *i* elements chosen from a set of *n* objects. It is sometimes written as $\binom{n}{i}$. Thus, for $n = 3$ then, given that $0! = 1$,

$${}_3 C_0 = 3!/\{0!3!\} = 1 = {}_3 C_3 \quad \text{and} \quad {}_3 C_1 = 3!/\{1!2!\} = 3 = {}_3 C_2$$

Thus,

$$\sum_{i=0}^{i=3} {}_3 C_i t^i (1-t)^{3-i} = (1-t)^3 + 3t(1-t)^2 + 3t^2(1-t) + t^3$$

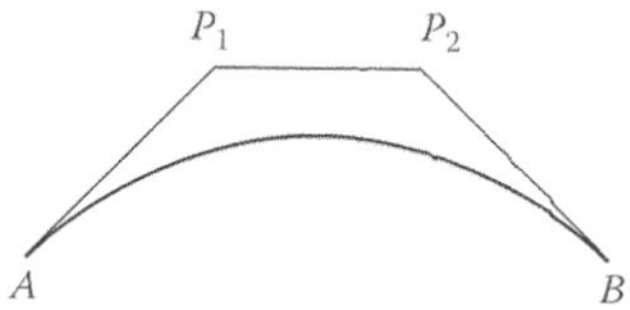

FIGURE 13.11 A Bezier curve with two control points.

The curve starts at A (x_o, y_o) and finishes at B (x_n, y_n) with $(n - 1)$ guiding points P_1 (x_1, y_1), P_2 (x_2, y_2) through to $P_{(n-1)}$ (x_{n-1}, y_{n-1}). With a quadratic ($n = 2$), there would be one guiding point; with a cubic ($n = 3$) there would be two guiding points as in Figure 13.11. A and B can be regarded as P_0 and P_n.

The curve starts tangential to the line P_0P_1 (AP_1 in Figure 13.11) and finishes tangential to the line P_{n-1} P_n (here P_2B). For any point t along a Bezier curve of the third degree, we have

$$x_t = (1-t)^3 x_0 + 3t(1-t)^2 x_1 + 3t^2(1-t)x_2 + t^3 x_3$$
$$y_t = (1-t)^1 y_0 + 3t(1-t)^2 y_1 + 3t^2(1-t)y_2 + t^3 y_3$$

Example 13.4 Bezier Curves

Figure 13.11 shows two versions of a cubic curve from A to B with two intermediate points P_1 and P_2. In the upper curve we have A (20, 20), B (40, 25), P_1 (25, 25), and P_2 (35, 35). In the lower curve, the same values are used except that the y value of P_2 has been changed to $P_2'(35,28)$.

$$x_t = (1-t)^3 x_0 + 3t(1-t)^2 x_1 + 3t^2(1-t)x_2 + t^3x_3$$
$$y_t = (1-t)^3 y_0 + 3t(1-t)^2 y_1 + 3t^2(1-t)y_2 + t^3y_3$$

T	$1-t$	$(1-t)^3$	$3t(1-t)^2$	$3t^2(1-t)$	t^3	x_t	y_t	y_t'
0	1	1	0	0	0	20	20	20
0.1	0.9	0.729	0.243	0.027	0.001	21.64	21.625	21.409
0.2	0.8	0.512	0.384	0.096	0.008	23.52	23.4	22.632
0.3	0.7	0.343	0.441	0.189	0.027	25.58	25.175	23.663
0.4	0.6	0.216	0.432	0.288	0.064	27.76	26.8	24.496
0.5	0.5	0.125	0.375	0.375	0.125	30	28.125	25.125
0.6	0.4	0.064	0.288	0.432	0.216	32.24	29	25.544
0.7	0.3	0.027	0.189	0.441	0.343	34.42	29.275	25.747
0.8	0.2	0.008	0.096	0.384	0.512	36.48	28.8	25.728
0.9	0.1	0.001	0.027	0.243	0.729	38.36	27.425	25.481
1	0	0	0	0	1	40	25	25

In the case of P' the curve is closer to the straight line AB. In both cases, the curve starts off tangential to AP_1 and finishes tangential to P_2B (or $P_2'B$) (Figure 13.12).

We then increment t in small steps from 0 to 1 and plot the points (x_t, y_t) to obtain the curve as in Example 13.4.

Thus, the selection of the control points P_1, P_2, and so forth, is critical to the shape of the curve. By choosing various controlling points, a graphic designer can adjust a curve to meet any required design. In general, in matrix notation, if

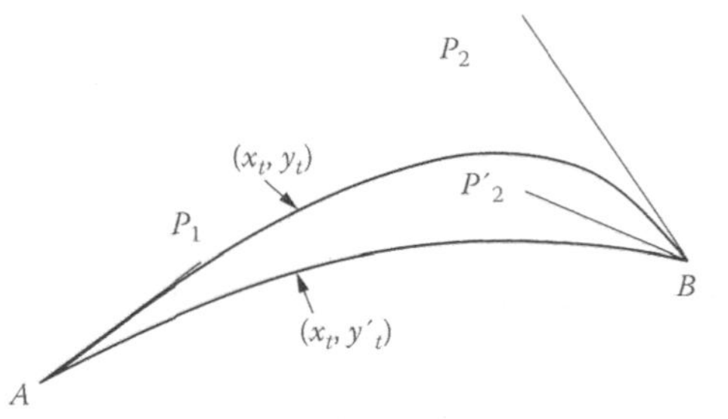

FIGURE 13.12 Two versions of a Bezier curve.

$\mathbf{P}_i = \begin{bmatrix} x_i \\ y_i \end{bmatrix}$ and $\mathbf{P}(t) = \begin{bmatrix} p_x(t) \\ p_y(t) \end{bmatrix}$ where $p_x(t)$ means the x value of **P** for the value t.

Then

$$\mathbf{P}(t) = \sum_{r=0}^{r=n} {}_nC_r t^r (1-t)^{n-r} \mathrm{P}_r$$

The idea of curve fitting can be extended to three dimensions. Thus, if we are fitting the cubic in the form

$$x = a_i + b_i t + c_i t^2 + d_i t^3$$

and

$$y = e_i + f_i t + g_i t^2 + h_i t^3$$

then we can add

$$z = l_i + m_i t + n_i t^2 + o_i t^3$$

Similarly, the Bezier function can be extended so that

$$\mathbf{P}(t) = \begin{bmatrix} p_x(t) \\ p_y(t) \\ p_z(t) \end{bmatrix}$$

Curves and surfaces may be made to go exactly through a series of fixed points or can be constructed as the best mean fit. The latter requires a statistical approach and is discussed in Chapters 14 and 15.

13.6 B-SPLINES

In Section 13.4, we considered a set of cubic curves that passed through a series of points while in Section 13.5 we fitted a cubic to end points using points off the curve as guides to the shape. We can combine both ideas by drawing a smooth curve piecewise, that is, section by section, that uses

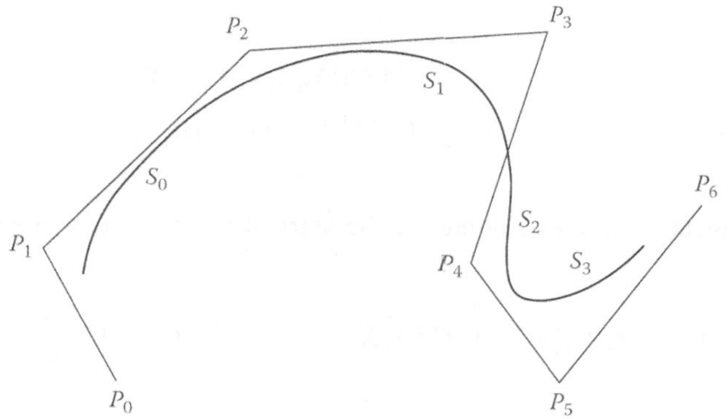

FIGURE 13.13 B-spline and knots.

both approaches in what is known as a set of *B-spline curves*. B-splines have the advantage that they can be used to describe regular figures such as circles, ellipses, parabolas, and hyperbolas and their shapes are invariant under the rotation of axes, translations, and scale changes.

Consider a series of "$n + 1$" control points P_0, P_1, P_2, …, P_i, …, P_n whose coordinates are (X_0, Y_0), (X_1, Y_1), and so forth (Figure 13.13). These points are sometimes referred to as *knots*. The curve will be made up from "$n - 2$" sections S_0, S_1, S_2, …, S_i, …, $S_{(n-3)}$ with S_i being influenced by the four points P_i, $P_{(i+1)}$, $P_{(i+2)}$, and $P_{(i+3)}$. Thus, when $i = n - 3$ we will have reached the final point P_n.

Each section is then defined as

$$S_i(t) = \sum_{k=0}^{k=3} P_{i+k} Q_k(t)$$

where t is incremented from 0 to 1 in an appropriate number of steps (for example, $t = 0, 0.01, 0.02, \ldots, 1$) and the function Q is given by

$$\begin{aligned}
Q_0(t) &= (1-t)^3/6 \\
Q_1(t) &= \left(3t^3 - 6t^2 + 4\right)/6 \\
Q_2(t) &= \left(-3t^3 + 3t^2 + 3t + 1\right)/6 \\
Q_3(t) &= t^3/6
\end{aligned}$$

This means that for the section S_i:

$$x_i(t) = \left\{(1-t)^3/6\right\}X_i + \left\{\left(3t^3 - 6t^2 + 4\right)/6\right\}X_{(i+1)} + \left\{\left(-3t^3 + 3t^2 + 3t + 1\right)/6\right\}X_{(i+2)} + \left\{t^3/6\right\}X_{(i+3)}$$

$$y_i(t) = \left\{(1-t)^3/6\right\}Y_i + \left\{\left(3t^3 - 6t^2 + 4\right)/6\right\}Y_{(i+1)} + \left\{\left(-3t^3 + 3t^2 + 3t + 1\right)/6\right\}Y_{(i+2)} + \left\{t^3/6\right\}Y_{(i+3)}$$

Note that the section starts where

$$\begin{aligned}
x &= (1/6)X_i + (4/6)X_{(i+1)} + (1/6)X_{(i+2)} \\
y &= (1/6)Y_i + (4/6)Y_{(i+1)} + (1/6)Y_{(i+2)}
\end{aligned}$$

It ends where

$$x = (1/6)X_{(i+1)} + (4/6)X_{(i+2)} + (1/6)X_{(i+3)}$$
$$y = (1/6)Y_{(i+1)} + (4/6)Y_{(i+2)} + (1/6)Y_{(i+3)}$$

Thus, the coordinates of the end point are the start of the next section so there is continuity. Furthermore:

$$\mathrm{d}x/\mathrm{d}t = \left\{-3(1-t)^2/6\right\}X_i + \left\{(9t^2 - 12t)/6\right\}X_{(i+1)} + \left\{(-9t^2 + 6t + 3)/6\right\}X_{(i+2)} + \left\{3t^2/6\right\}X_{(i+3)}$$

When $t = 0$,

$$\mathrm{d}x/\mathrm{d}t = -(1/2)X_i + (1/2)X_{(i+2)}$$

Similarly,

$$\mathrm{d}y/\mathrm{d}t = -(1/2)Y_i + (1/2)Y_{(i+2)}$$

Hence,

$$\text{at } t = 0,\ \mathrm{d}y/\mathrm{d}x = \left\{Y_i - Y_{(i+2)}\right\}/\left\{X_i - X_{(i+2)}\right\}$$

When $t = 1$,

$$\mathrm{d}x/\mathrm{d}t = -(1/2)X_{(i+1)} + (1/2)X_{(i+3)}$$
$$\mathrm{d}y/\mathrm{d}t = -(1/2)Y_{(i+1)} + (1/2)Y_{(i+3)}$$

Hence,

$$\text{at } t = 1, \mathrm{d}y/\mathrm{d}x = \left\{Y_{(i+1)} - Y_{(i+3)}\right\}/\left\{X_{(i+1)} - X_{(i+3)}\right\}$$

The slope at the end of the section is the same as the slope at the start of the next section. Similarly,

$$\mathrm{d}^2x/\mathrm{d}t^2 = (1-t)X_i + (3t-2)X_{(i+1)} + (-3t+1)X_{(i+2)} + tX_{(i+3)}$$
$$\text{at } t = 0,\ \mathrm{d}^2x/\mathrm{d}t^2 = X_i - 2X_{(i+1)} + X_{(i+2)}$$
$$\text{at } t = 1,\ \mathrm{d}^2x/\mathrm{d}t^2 = X_{(i+1)} - 2X_{(i+2)} + X_{(i+3)}$$

and similarly for y. Thus, the rate of change of the slope at the end of one section is the same as the rate of change at the start of the next section. There is therefore continuity for both the first and second derivatives and hence a smooth curve throughout.

The ideas behind the B-spline can be extended to surfaces and beyond and can also be generated by more complex curves than the cubic. Such procedures are beyond the present text. Readers who are interested should consult textbooks on computer graphics for more information.

GLOSSARY

Bezier curve A curve built from the family of polynomials of the form

$$x = f_x(t) = \sum_{i=0}^{i=n} {}_nC_i t^i (1-t)^{n-i} x_i; \quad y = f_y(t) = \sum_{r=0}^{r=n} {}_nC_r t^r (1-t)^{n-r} y_i$$

Combination The number of distinct ways in which *r* objects can be chosen from a set of *n* objects. Written as ${}_nC_r$

Curvature The rate of change of a tangent to a curve.

Geocentric latitude The latitude as measured from the notional center of the Earth.

Geodetic latitude Also known as the *spheroidal latitude.* The latitude determined by the direction of the normal to an ellipsoid at any point.

Nodes A point at which two or more branches of a graph meet.

Orthogonal lines Lines which are perpendicular to each other.

Parametric form A set of equations that relate the variables such as coordinates to a set of parameters, for instance,

For a line: $x = p + lt; y = q + mt$
For a circle: $x = p + r \cos t; y = q + r \sin t$
For an ellipse: $x = p + a \cos t; y = q + b \sin t$
For a hyperbola: $x = p + a \sec t; y = q + b \tan t$

Piecewise The fitting of a curve section by section using lower-order polynomial equations.

Radius of curvature The radius of the circle that has the same *curvature* as that of a point on a curve. Known as ρ (rho). $\rho = \{(1 + (dy / dx)^2)^{3/2}\} / (d^2y / dx^2)$

Reduced latitude The latitude determined by the normal to the auxiliary circle to an ellipse.

Spline A simplified way of representing a complex curve using low-order polynomials.

String A series of elements connected in a specific order, as with points along a line.

Unit V

Finding Patterns and Predicting

As larger and larger quantities of data have become available in the digital world nowadays, GIS practitioners are increasingly tasked with building models that simulate real world processes or make predictions about real life events. Making predictive models and assessing confidence of their results often requires using statistical methods. Statistics embodies a different kind of logical and mathematical approach from algebra, geometry, and calculus. Unit V in the book is about understanding the fundamental concepts in statistics, and exploring the different methods used to model real world processes. These methods include the more traditional regression and Monte Carlo simulation, as well as various machine learning methods that fall under the umbrella of artificial intelligence, or AI.

In Chapter 14, we introduce the fundamental concepts of statistics including probability, distributions, and methods to simulate "random" events. In Chapter 15, we introduce commonly used statistical modeling methods such as correlation and regression. In Chapter 16, we discuss basic principles of machine learning and the common methods used in different applications to draw patterns and predictions from large quantities of data.

DOI: 10.1201/9781003464402-18

14 Basic Statistics

14.1 PROBABILITIES

So far we have dealt with matters for which there is at least in theory an exact answer—curves and surfaces have been fitted through points and values such as sin ø have been calculated to as many significant figures as were necessary. In many circumstances, however, things are not exact. Every measure can be subject to a small amount of error, which means that we must look for a "best-fit" solution.

Many measures are of phenomena for which there are no precise answers, such as measures of human preference as in estimates of how people may vote in an election prior to the actual event. We need some way of estimating the reliability of any measurement and some way of handling inconsistencies. In other words, we need statistical measures where a *statistic* is some function of random variables that can be used as an estimate of a population. A *population* (sometimes referred to as a *universe*) is the complete set of individual components or events from which *samples* are drawn (Figure 14.1). The sample should be chosen so that its characteristics mimic those of the whole population. If it does not, then the sample will be biased.

Descriptive statistics are sets of numbers used to summarize a set of known data in a clear and concise way, while *inferential statistics* result from the theory and practice of using statistical data to draw conclusions from random samples. Both processes are *heuristic,* meaning that they are guided by experience and experiment rather than by rigorous logical argument from precisely defined axioms.

The word *random* means that the value of the item under consideration cannot be predetermined other than in terms of the probability that it may occur; in particular, it cannot be determined by what has happened before. Thus, a *random number* is one that cannot be determined from any of the previous numbers that have been selected and, therefore, it does not follow any particular regular or repetitive pattern. In a series of random numbers, for example, each number is as likely to occur as often as any other in the series. If this does not happen, then the data are biased.

Processes that can be described by random variables are said to be *stochastic* and are described in terms of *probability,* which is a measure of the degree of confidence that can be had in any event.

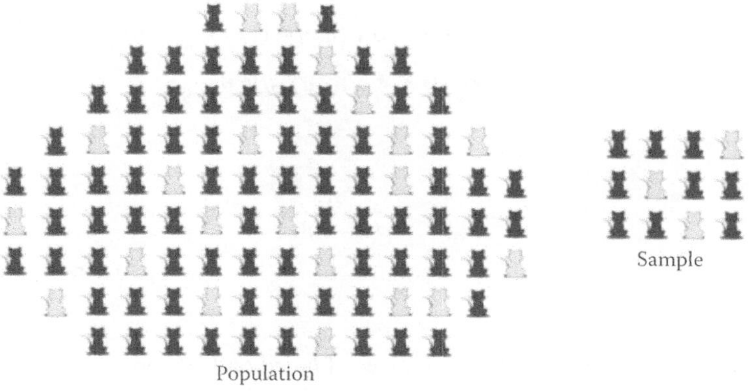

FIGURE 14.1 Population and sample.

DOI: 10.1201/9781003464402-19

Thus, when tossing an unbiased coin, there is a 50% probability that it will come down head-side up and 50% that it will be "tails."

If we consider the toss of a coin, we can describe the probability of the outcome as $(ph + qt)$ where h is simply a dummy term meaning "heads" and t is a dummy meaning "tails"; p is the chances out of 1 that the main event will happen and q is the chances that it will not happen. $q = (1 - p)$ since it is 100% certain that either it will or will not happen. In the case of tossing an unbiased coin, $p = q = (1 / 2) = 0.5$.

Thus, after one throw, we have $(0.5h + 0.5t) = (h + t) / 2$. If we repeat the experiment a second time, we can multiply the probabilities and obtain

$$\left\{\frac{h+t}{2}\right\} * \left\{\frac{h+t}{2}\right\} = \left\{\frac{h^2+2ht+t^2}{4}\right\}$$

This tells us that out of four possible outcomes (the denominator) there is one chance that we have heads twice (h^2), two chances of having either a head and a tail or a tail and a head ($2ht$), and one chance of having two tails (t^2). If we repeat the process a third time, we obtain

$$\left(\frac{h+t}{2}\right)^3 = \left(\frac{h^3+3h^2t+3ht^2+t^3}{8}\right) = \left(\frac{h^3t^0+3h^2t^1+3h^1t^2+h^0t^3}{8}\right)$$

Writing this as $(h^3t^0 + 3h^2t^1 + 3h^1t^2 + h^0t^3) / 8$ tells us that after three throws, there are eight possible outcomes, there being one chance in eight of obtaining three heads (h^3) with no tails (t^0), three chances of there being two heads and one tail ($3 * h^2t^1$), three chances of there being one head and two tails ($3 * h^1t^2$), and one chance out of eight of obtaining three tails (t^3) with no heads (h^0).

Note that in the expression $(h^3t^0 + 3h^2t^1 + 3h^1t^2 + h^0t^3) / 8$, the sum of the coefficients of the terms h^mt^n is $(1 + 3 + 3 + 1) = 8$, which is the same as the denominator. This confirms that the total probability $= 8 / 8$ which is precisely one. In each component h^mt^n the sum of the indices $= m + n$ is the same as the number of events under consideration; here, after three throws, it equals 3.

For four throws,

$$\left(\frac{h+t}{2}\right)^4 = \left(h^4t^0 + 4h^3t^1 + 6h^2t^2 + 4h^1t^3 + h^0t^4\right)/16$$

BOX 14.1 THE BINOMIAL EXPANSION

$$(p+q)^n = p^n + np^{(n-1)}q + \{n(n-1)/2)\}p^{(n-2)}q^2 + \cdots +_n C_r p^{(n-r)}q^r + \cdots + npq^{(n-1)} + q^n$$

where

$${}_nC_r = \frac{n!}{(n-r)!r!}$$

and again $(1 + 4 + 6 + 4 + 1) / 16 = 1$. The indices of h and t all add up to 4. There is one chance in 16 of obtaining four heads, four chances of obtaining three heads, six of obtaining two heads, four of obtaining only one head, and one chance in 16 of there being no heads at all.

The expansion of $\left(\frac{h+t}{2}\right)^n$ is known as the *binomial expansion* (Box 14.1) and results in

$$\left\{h^n t^0 +_n C_1 h^{(n-1)} t^1 +_n C_2 h^{(n-2)} t^2 + \cdots +_n C_{(n-1)} h^1 t^{(n-1)} + h^0 t^n\right\} / 2^n$$

where ${}_nC_r$ is a shorthand way of writing the number $\frac{n!}{(n-r)!r!}$ and n! = 1 * 2 * 3…* $(n - 1)$ n. For instance, ${}_5C_2 = \frac{5*4*3*2*1}{(3*2*1)*(2*1)} = 10$.

The sum of the indices of h and t or p and $q = n$. The sum of the coefficients in the numerator divided by the denominator = 1. If, for example, n = 100, we can calculate the chances that on a purely random basis all the 100 throws of the coin will be heads. The chances of this are 1 / 2^{100}—pretty unlikely! The chances of getting 99 heads will be 100 / 2^{100} while for 98 heads they will be 4950 / 2^{100} where $4950 = \frac{100*99}{1*2}$.

We can extend the use of the binomial expansion where the probabilities are not 50–50. As an example, all other things being equal, the chances that a Wednesday will be the most rainy day in the week will be one in seven. The probability is $(pw + q)$ where p = 1 / 7 and q = 6 / 7 and w is a dummy to represent "Wednesday." For two weeks we have $(pw + q)$ * $(pw + q)$ or for three weeks $(pw + q)^3$. Hence, the chance that a Wednesday is the most rainy day in the week for three weeks running is

$$\left(\frac{(w+6)}{7}\right)^3 = \left(\frac{w^3}{343} + \frac{18w^2}{343} + \frac{108w}{343} + \frac{216}{343}\right)$$

This means there is one chance in 343 that Wednesday will be the most rainy day of the week for three weeks, 18 / 343 chances that it happens twice, 108 / 343 chances that it happens once, and 216 / 343 chances that no Wednesdays are the most rainy. (Note that 1 + 18 + 108 + 216 = 343.)

The coefficients in the expansion of $(p + q)^n$ are given in Box 14.2. For instance, the second line, (1 2 1) indicates 1 * p^2 + 2 * p^1q^1 + 1 * q^2, while in the fifth line, (1 5 10 10 5 1) means that

$$(p+q)^5 = p^5 + 5p^4q^1 + 10p^3q^2 + 10p^2q^3 + 5p^1q^4 + q^5$$

BOX 14.2 THE BINOMIAL COEFFICIENTS

n =	Coefficients in $(p + q)^n$
1	1 1
2	1 2 1
3	1 3 3 1
4	1 4 6 4 1
5	1 5 10 10 5 1
6	1 6 15 20 15 6 1
7	1 7 21 35 35 21 7 1
8	1 8 28 56 70 56 28 8 1
9	1 9 36 84 126 126 84 36 9 1
10	1 10 45 120 210 252 210 120 45 10 1
11	1 11 55 165 330 462 462 330 165 55 11 1
12	1 12 66 220 495 792 924 792 495 220 66 12 1
13	1 13 78 286 715 1287 1716 1716 1287 715 286 78 13 1
14	1 14 91 364 1001 2002 **3003 3432** 3003 2002 1001 364 91 14 1
15	1 15 105 455 1365 3003 5005 **6435** 6435 5005 3003 1365 455 105 15 1

There is, of course, a pattern in all this whereby the numbers in the lower row are the sum of the two numbers directly above them—for example, the 8th number in row 15 is 6435 = 3003 + 3432, the seventh and eighth numbers in row 14. The number 3003 is itself the sum of the two numbers above it, namely, 1287 + 1716; likewise, 3432 is the sum of 1716 + 1716, and so on. The triangular array of integers shown in Box 14.2 is known as a *Pascal triangle*, named after the French mathematician, Blaise Pascal, who lived in the 17th century.

To obtain the probabilities assuming $p = q = ½$, for row 15 we must divide all the numbers by $2^{15} = 32768$ so that the whole probability adds up to 1. Hence, dividing through, we obtain 0.00003, 0.00047, 0.0032, 0.0139, 0.0417, 0.0916, 0.1527, 0.1964 with the numbers repeated in reverse order for the remaining eight coefficients, all adding up to 1.0.

14.2 MEASURES OF CENTRAL TENDENCY

In Figure 14.2, the 16 values of the coefficients of p and q have been plotted as a *histogram,* which is a set of contiguous rectangles with width proportional to the size of the class interval and the area proportional to its frequency. They have also been traced in the form of a continuous curve that interpolates the frequency. The chances of there being no heads or 15 heads in 15 throws are 1 in 32,768 or about 0.00003. This means that it is probable that if you carried out 32,768 experiments with tossing a coin 15 times, then you would expect on average one case where there were 15 consecutive heads. But it might not happen or it might happen several times. Nothing is certain. What you would expect is that if you carried out a set of 32,768 experiments a large number of times, then on average you would find 15 consecutive heads once in each set of experiments.

In Figure 14.2, almost all the area under the curve lies between 3 and 12. The area under the whole curve is made up of all the separate probabilities and must equal 1 or 100%. We achieved this by dividing the original coefficients by 2^{15}. The areas represented by rectangles in Figure 14.2 are proportional to the overall area and hence represent graphically the probability of the number shown in the bottom line occurring; the total area of these boxes (and under the curve) must be exactly 1.

Furthermore, the middle or average value is 7.5. Of course, there cannot be 7.5 heads in a single experiment, since we have specified that the outcome is either heads or tails with a 50% probability of either. But with a series of experiments throwing 15 coins one would expect that in 7.5 cases there would be 8 or more heads and in 7.5 cases there would be 7 or less heads. In general, if there are n attempts at something for which the probability of it happening is p, then the average number of times that it will happen will be np – n times p. Here, $n = 15$, $p = ½$, so $np = 7.5$.

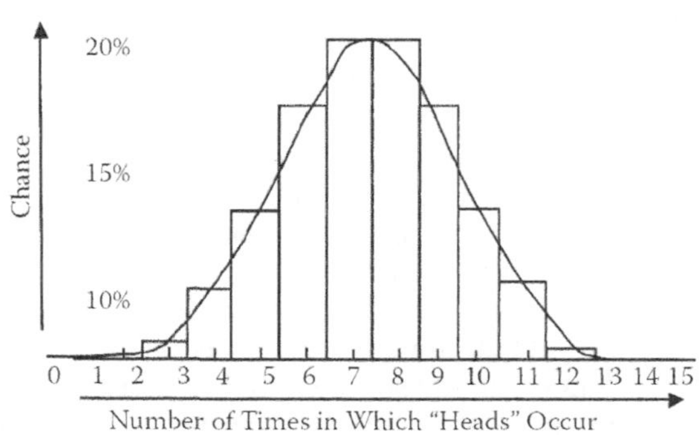

FIGURE 14.2 A plot of equal probability after 15 events.

In producing a general description of the shape of the curve in Figure 14.2, there is a need to know not only the mean value, but also how spread out or dispersed the individual answers are likely to be. There are three parameters that are used to express the central point—the *mean*, the *median*, and the *mode*. The mean is the average value of a set of n observations x_i, calculated as $\sum x_i / n$, meaning "add all the x values together and divide by n." The median value is the middle number, while the mode is the one that most frequently occurs.

For example, with two sets of numbers A (1, 5, 24) and B (1, 2, 2, 2, 8, 12, 15, 22), both have a median value of 5 since in the case of B where there is an even number of entities and 5 lies half way between 2 and 8. In B, the mode is 2, while the mean is 8. For data set A, the mean is 10. If the distribution were symmetrical about the middle, then all the values of the mean, median, and mode would coincide.

The average or mean value calculated from $\sum x_i / n$ assumes that each value of x is comparable with every other value. In practice this may not be the case—some observations may be dodgy while others may be more reliable. We may therefore need to introduce a weighting factor w_i to allow for the fact that, for example, one observation is twice as reliable as another. The weight is in effect a relative value of the importance of any individual observation. The weighted mean then becomes $\sum w_i x_i / \sum w_i$.

If x is part of a very large, in theory, infinite, population of random variables, then its weighted average is known as the *expected value* (E). If a random variable X can take the value x_i with probability p_i, then $E(X) = \sum p_i x_i / \sum p_i = \sum p_i x_i$ since the sum of all probabilities $= \sum p_i = 1$. Hence, $w_i / \sum w_i = p_i$. $E(X)$ is sometimes referred to as the *first product moment* or *first moment* for the set X.

For the present, we will treat all observations as being of equal weight unless we declare otherwise (see Chapter 15, for example). The parameter that is most commonly used to measure how spread out or dispersed is a set of data is known as the *variance*. This is defined as the average sum of the squares of the amount by which each observation differs from the mean. If an observation has a value x_i and if the Greek letter for m, which is "μ" (pronounced "mu") is used to denote the mean value of the population, then $(x_i - \mu)$ is said to be the *residual* value for x_i.

For n observations, the average value of the sum of the squares of all the residuals is equal to $\sum(x_i - \mu)^2 / n$ where $\sum$ (the Greek capital letter "sigma") means "the sum for all the values of x_i." The variance is usually represented by σ^2 where the symbol "σ" represents the Greek lowercase letter "sigma." So, $\sigma^2 = \sum(x_i - \mu)^2 / n$.

The square root of the variance, σ, is known as the *standard deviation*. Thus,

$$\sigma = \sqrt{\left(\sum (x_i - \mu)^2 / n\right)} \text{ for the whole population.}$$

The amount by which any observation differs from the mean when measured in units of standard deviation is called the *Z-score* where $z_i = (x_i - \mu) / \sigma$. The mean of a set of z-scores $= 0$ while the variance and the standard deviation for the set of z-scores $= 1$.

We have summarized the important statistics in any given data set in Box 14.3. In Example 14.1, we give a simple example with two sets of measurements and their means and standard deviations.

BOX 14.3 VARIANCE AND STANDARD DEVIATION

For any set of n numbers of equal weight $x_1, x_2, \ldots x_i, \ldots x_n$: The mean value is

$$\left(x_1 + x_2 + \cdots + x_i + \cdots + x_n\right) / n = \sum x_i / n = \mu$$

The variance is

$$\sigma^2 = \sum (x_i - \mu)^2 / n$$

The standard deviation is

$$\sigma = \sqrt{\left(\sum (x_i - \mu)^2 / n\right)}$$

For each number x_i the Z value is

$$z_i = (x_i - \mu) / \sigma$$

Example 14.1 Example of Variances

Consider two groups of 10 people and their heights in meters.

Group A has heights of:

1.66, 1.66, 1.67, 1.69, 1.70, 1.71, 1.71, 1.72, 1.73, 1.75

Group B has heights of:

1.46, 1.52, 1.58, 1.60, 1.68, 1.74, 1.78, 1.84, 1.89, 1.91

The average or mean height in both groups is 1.70 meters. The residuals are:

For group A: −0.04, −0.04, −0.03, −0.01, 0, 0.01, 0.01, 0.02, 0.03, 0.05

For group B: −0.24, −0.18, −0.12, −0.10, −0.02, 0.04, 0.08, 0.14, 0.19, 0.21

For group A the variance = 0.00082 and the standard deviation ≈ 0.03. For group B the variance = 0.02226 and the standard deviation ≈ 0.15.

As demonstrated in Box 14.4, on average, there will be $A = np$ successful outcomes from n experiments where the probability of success = p. Put another way, the mean value $\mu = np$ as illustrated in Example 14.2, which demonstrates what commonsense should indicate, namely, that if each event has a probability of p, then thereafter n tries, there are likely to be np successful outcomes. The derivation of the standard deviation is perhaps less obvious and is shown in Box 14.5 to be $\sqrt{(npq)}$ where $q = 1 - p$.

To summarize, with a set of probabilities, $\mu = np$ and $\sigma = \sqrt{(npq)}$.

In Box 14.5, we show that the mean and variance and standard deviation can also be calculated for any set of numbers whatever they represent if we assume that the numbers are part of a binomial distribution. Normally, however, we simply add the numbers all together and divide by the number n to obtain the mean. Sum the squares of the differences between each number and the mean, and again divide by n and you have the variance σ^2; take the square root of the variance to obtain the standard deviation σ.

The smaller the standard deviation, the more bunched are the observations around the mean, and therefore the greater confidence there is that the mean represents the best possible estimate of the quantity being measured. The variance and standard deviation are sometimes referred to as *measures of central tendency*.

The binomial function allows precise probabilities that can be calculated, given certain basic assumptions such as heads occurring one time out of two, or rain falling more on a Wednesday one time out of seven. The function becomes difficult to handle when the numbers become large and, of course, it assumes a precise number of outcomes. Where there is more uncertainty, then a slightly different approach is needed.

BOX 14.4 MEAN OUTCOME FROM BINOMIAL EXPANSION

In Box 14.1, we quoted the binomial expansion in the generalized form

$$(q+p)^n = \left(q^n p^0 +_n C_1 q^{(n-1)} p^1 + \cdots +_n C_r q^{(n-r)} p^r + \cdots + q^0 p^n\right)$$

where $(p + q) = 1$, p being the probability that an event will happen, q that it will not happen. $p^0 = q^0 = 1$ while $p^1 = p$ and $q^1 = q$. In the expansion of $(q + p)^n$, we defined $_nC_r$ as $n! / \{r!(n - r)!\}$ where r is an integer between 0 and n. It follows that:

$$_{(n-1)}C_{(r-1)} = \frac{(n-1)!}{\{(n-1)-(r-1)\}!(r-1)!} = \frac{(n-1)!}{\{(n-1)-(r-1)\}!(r-1)!} * \frac{n*r}{n*r} = \frac{n!}{n} * \frac{r}{(n-r)!r!}$$
$$= (r/n)_n C_r$$

Hence,

$$_nC_r = (n/r)_{(n-1)} C_{(r-1)}$$

On average, the probability of there being no successes $= q^n p^0$. The probability of one successful outcome $= {_nC_1} q^{(n-1)} p^1$. The probability of r successful outcomes $= {_nC_r} q^{(n-r)} p^r$.
If the total number of expected successful outcomes after n attempts $= A$, then

$$A = \left\{q^{(n-0)}0 +_n C_1 q^{(n-1)} p * 1 + \cdots +_n C_r q^{(n-r)} p^r * r + \cdots +_n C_n p^n n\right\}$$

with 0, 1, …, r, …, n representing the number of possible outcomes. We can write this as:

$$A = \left\{q^n * 0 +_n C_1 q^{(n-1)} p^0 * (np/n) + \cdots +_n C_r q^{(n-r)} p^{(r-1)} * (npr/n) + \cdots +_n C_n p^{(n-1)} * (np)\right\}$$

and hence as:

$$A = \left\{ {_nC_1} q^{(n-1)} p^0 * (np/n) + \cdots +_n C_r q^{(n-r)} p^{(r-1)} * (npr/n) + \cdots +_n C_n p^{(n-1)} * (np)\right\}$$

Since $_nC_1 = n$ and $_nC_n = 1$, we have

$$A = (np) * \left\{q^{(n-1)} p^0 + \cdots +_n C_r q^{(n-r)} p^{(r-1)} * (r/n) + \cdots + p^{(n-1)}\right\}$$

But we have just shown that $_nC_r = (n / r)\ {_{(n-1)}C_{(r-1)}}$; hence,

$$A = np\left\{q^{(n-1)} +_{n-1} C_1 q^{(n-2)} p + \cdots +_{n-1} C_r q^{(n-1-r)} p^r + \cdots + p^{(n-1)}\right\}$$
$$= np(q+p)^{(n-1)} = np \text{ since } (q+p) = 1$$

Thus, on average, there will be ($A =$) np successful outcomes or the mean value from the binomial distribution $= np$.

Example 14.2 Mean Outcome Under the Binomial Distribution

We show in Box 14.4 that the mean value of a set of n observations with probability p of a particular outcome will on average $= np$. For instance, after 20 throws of a dice with $p = 1 / 2$ then one would expect 20 * 1 / 2 = 10 heads. With rain on a Wednesday ($p = 1 / 7$) then over 14 weeks it is probable that there will be 14 * (1 / 7) = 2 occasions on which Wednesday was the most rainy day, which is what one would expect.

**BOX 14.5 THE BINOMIAL DISTRIBUTION—
VARIANCE AND STANDARD DEVIATION**

Given a mean of np then the original binomial equation shows the probability of the event occurring $(np-1)$, $(np-2)$, $(np-3)$, … $(np-r)$, … $(np-n)$ times. These numbers will be positive or negative, depending on whether the event is less or more common than the average. The variance σ^2 = the sum of the squares of these differences from the mean multiplied by their relative frequency, or

$$\sigma^2 = q^n p^0 (np-0)^2 +{}_n C_1 q^{n-1} p^1 (np-1)^2 + \cdots {}_n C_r q^{(n-r)} p^r (np-r)^2 + \cdots + p^n (np-n)^2$$

where ${}_nC_r\, q^{(n-r)}\, p^r$ is the relative frequency of $(np-r)$.

This can be expanded to $L+M+N$ where

$$\begin{aligned}
L &= n^2p^2\left(q^n p^0 +{}_n C_1 q^{n-1} p^1 +{}_n C_2 q^{n-2} p^2 + \cdots +{}_n C_r q^{n-r} p^r + \cdots + p^n\right)\\
M &= -2np\left(q^n p^{0*} 0 +{}_n C_1 q^{n-1} p^{1*} 1 +{}_n C_2 q^{n-2} p^2 * 2 + \cdots +{}_n C_r q^{n-r} p^{r*} r + \cdots + p^{n*} n\right)\\
N &= +\left(q^n p^0 * 0 +{}_n C_1 q^{n-1} p^1 * 1^2 +{}_n C_2 q^{n-2} p^2 * 2^2 + \cdots +{}_n C_r q^{n-r} p^r * r^2 + \cdots + p^{n*} n^2\right)
\end{aligned}$$

The first line, L, gives

$$n^2p^2\left(q^n p^0 +{}_n C_1 q^{n-1} p^1 +{}_n C_2 q^{n-2} p^2 + \cdots +{}_n C_r q^{n-r} p^r + \cdots + p^n\right) = n^2p^2 (q+p)^n$$

But $(q+p)=1$, hence the first line $= n^2p^2$.

By expanding and rearranging the terms in the second and third lines M and N we obtain $M = -2n^2p^2$ and $N = np + n(n-1)p^2$. Hence,

$$L+M+N = \sigma^2 = n^2p^2 - 2n^2p^2 + np + n(n-1)p^2 = np - np^2 = np(1-p) = npq$$

Hence, $\sigma^2 = npq$. This represents the variance in the probability. The standard deviation = $\sigma = \sqrt{(npq)}$.

14.3 THE NORMAL DISTRIBUTION

The curve that has been drawn in Figure 14.2 passes through discrete points but implies intermediate values. If we accept the possibility of an observation taking any value and still retain the area under the curve to be 1 in order to give 100% probability, then we need a more complex function than the binomial distribution.

The fundamental assumption behind much of statistical analysis is known as the *central limit theorem*, which states that if a sequence of independent identically distributed random variables each has a finite variance, then as the number of observed values increases, they tend to mirror the probability distribution known as the *normal distribution*. Randomness means a lack of bias with the result of each observation being totally independent of any other observation.

In Figure 14.2, we showed a series of rectangles that represented the probabilities or proportion of events such as 1, 2, 3, and so forth, heads occurring after a given number of throws. The rectangles were 1 unit wide with heights derived from the binomial expansion. The total area made up by all the rectangles was 1.0 when every eventuality was considered. In the case shown, there were 15 throws. In Box 14.6, we extend this further to n events and then to what in theory is an infinite number so that the curves in Figures 14.2 and 14.3 become smooth. Such a curve is then known as the *normal curve*.

BOX 14.6 THE CURVE OF THE NORMAL DISTRIBUTION

Let us consider the proportions as being the result of n events as shown in Figure 14.3. Each proportional representation is separated by an amount w. Let the height of each rectangle be y_r so that the area of rectangle $r = w * y_r$. This represents the probability of an event occurring. According to the binomial equation, the frequency of an event at point r is ${}_nC_r\, p^r\, q^{n-r}$. To ensure that the total overall probabilities add up to 1, we must scale all the values y accordingly. Hence, we need to let $y_r = K * {}_nC_r\, p^r\, q^{n-r}$ where K is some appropriate scaling factor.

As shown in Boxes 14.4 and 14.5, the mean value $= n * pw$ and the variance $= \sigma^2 = n * pw * qw$. Let

$$x_r = \text{distance from the mean value to rectangle } r = (r - np)w$$

As the number n gets larger, the widths of the rectangles get narrower and n becomes infinite and w becomes zero and we finish up with a smooth curve. The probability of an error

$$x_r + \delta x_r \text{ is } y_r + \delta y_r = K *_n C_{r+1} q^{n-r-1} p^{r+1}$$

Hence,

$$\delta y_r = K * \{ {}_nC_{r+1} q^{n-r-1} p^{r+1} - {}_nC_r q^{n-r} p^r \} = K * \{q^{n-r-1} p^r\}\{p *_n C_{r+1} - q *_n C_r\}$$

Since

$${}_nC_{r+1} = \frac{n!}{(n-r-1)!(r+1)!} = \frac{(n-r)}{(r+1)}\,{}_nC_r \text{ then:}$$

$$\delta y_r = K\{ {}_nC_r\}\{q^{n-r} p^r\}\{q^{-1}\}\{[(n-r)p - q]/(r+1)\} = \{y_r\}\{(np-r)/q(r+1)\}$$

or

$$\delta y_r / y_r = (np-r)/\{q(r+1)\} = (np-r)w/\{q(r+1)w\} \text{ based on strips of width } w$$

$$r \text{ represents the value } (npw + x_r) \text{ since } (r-np)w = x_r$$

Thus,

$$(np-r)w = -x_r \text{ while } (r+1)w = npw + x_r + w$$

Hence,

$$\delta y_r / y_r = -x_r / q(npw + x_r + w)$$

Again multiplying top and bottom by w

$$\delta y_r / y_r = -x_r w / \{npwqw + qw(x_r + w)\} = -x_r w / \{\sigma^2 + q(x_r + w)w\}$$

As the widths get smaller we can treat w as δx_r. Hence,

$$\delta y_r / y_r = -x_r \delta x_r / \{\sigma^2 + q(x_r + \delta x_r)\delta x_r\};$$

or

$$\delta y_r / \delta x_r = -y_r x_r / \{\sigma^2 + q(x_r + \delta x_r)\delta x_r\}$$

In the limit as we increase n and reduce w ($= \delta x$), then $q(x_r + \delta x_r)\delta x_r$ tends to zero, so

$$\mathrm{d}y / \mathrm{d}x = -xy / \sigma^2\text{; and}$$
$$y = \int(-xy / \sigma^2)\mathrm{d}x\text{; or}$$
$$\int(1/y)\mathrm{d}y = \int(-x / \sigma^2)\mathrm{d}x\text{; or}$$

$\log_e y = -x^2 / 2\sigma^2 + c$ where c is the constant of integration.
If we measure x from the mean value "μ," we can rearrange this as

$$y = k\ e^{-\{(x-\mu)^2/2\sigma^2\}}$$

where k is some constant and "μ" is the mean value. The curve is symmetrical about the mean value "$x = \mu$."

In order that the total area under the curve is unity and since $\int_{\infty}^{\infty} e^{-t^2}\,dt$ can be shown (though is not demonstrated in this text) to equal $\sqrt{\pi}$, then $k = 1/\{\sigma\sqrt{(2\pi)}\}$. The equation of the smooth curve, called the *normal curve*, becomes

$$y = \left[1/\{\sigma\sqrt{(2\pi)}\}\right]e^{-\{(x-\mu)^2/2\sigma^2\}}$$

Figure 14.3 shows the plot of probability of n events that are normally distributed. Figure 14.4 shows the percentage of a distribution that lies within 1, 2, 3, or more standard deviations of the mean.

For those who are interested, Box 14.6 outlines how the formula for representing the normal distribution is derived. What is important is that the normal distribution has the form

$$y = \left[1/\{\sigma\sqrt{(2\pi)}\}\right]e^{-\{x^2/2\sigma^2\}}$$

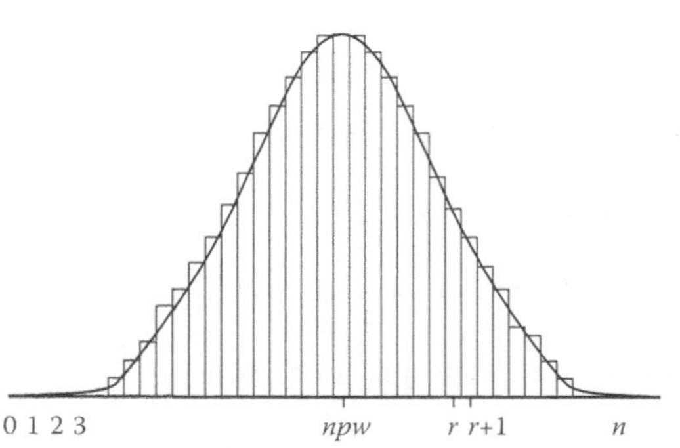

FIGURE 14.3 A plot of probability for n events.

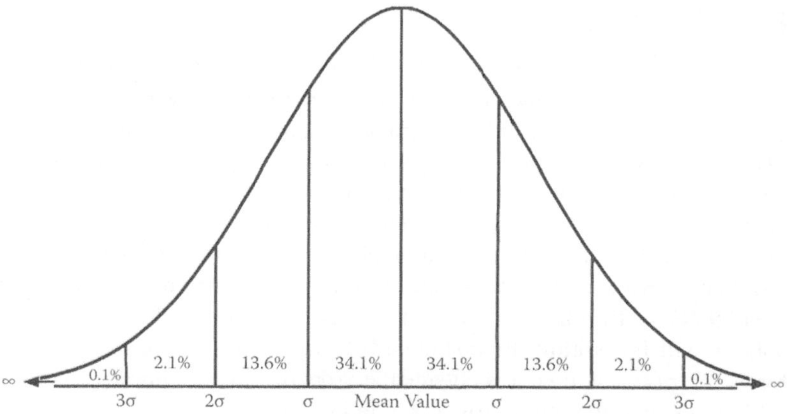

FIGURE 14.4 Percentage of distribution within one, two, and three standard deviations from the mean.

where e is the base of natural logarithms.

Or, using the z statistics

$$y = \frac{1}{\sigma\sqrt{2\pi}} e^{-(z^2/2)}$$

Those who are prepared to take Box 14.6 as read should note that:

a. The total area under the curve in each of the Figures 14.3 and 14.4 is 1.
b. The curve is symmetrical about the value $x = \mu$.
c. Although the curve goes off to infinity either side of the mean, in practice over 99% of the observations lie within about three standard deviations of the mean.

If we choose to record our measurements about a mean of zero, then $\mu = 0$ and we have

$$y = ke^{-\{x^2/2\sigma^2\}} \quad \text{with} \quad k = 1/\left\{\sigma\sqrt{(2\pi)}\right\}$$

This is symmetrical about $x = 0$. The area under the curve = 1 and the variance = σ^2. If we assume that x is a measure of the error in an observation, then the area under the curve between $x = 0$ and $x = r$ represents the total probability of an error equal to or less than r. This is given by $\left[1/\left\{\sigma\sqrt{(2\pi)}\right\}\right]\int_0^r e^{-\{x^2/2\sigma^2\}}dx\cdot$

Unfortunately, there is no easy way in which to calculate this when using elementary techniques. Suffice it to say that, using various numerical methods, the sums have been worked out and are available in statistical tables. The ratio between the size of the residual and the standard deviation is illustrated in Table 14.1 where values of r / σ are tabulated against the area under the curve in Figure 14.3. A summary of results is shown in Figure 14.4.

TABLE 14.1
Partial Areas under the Normal Curve

$Z = (x - \mu) / \sigma$	**+0.0**	**0.2**	**0.4**	**0.6**	**0.8**
0.0	0%	7.93%	15.54%	22.57%	28.81%
1.0	34.13%	38.49%	41.92%	44.52%	46.41%
2.0	47.72%	48.61%	49.18%	49.53%	49.74%
3.0	49.87%	49.93%	49.97%	49.98%	49.99%

14.4 LEVELS OF SIGNIFICANCE

The normal distribution shows that almost all observations should fall within the limits of three standard deviations away from the mean. If, for example, we take a series of measurements then it is easy enough to calculate the mean and standard deviation of the set of measurements. If we try to determine any quantity very precisely, then we would expect to have slight differences each time we make the measurement. Any observation that was more than three standard deviations away from the mean should be suspect and it is probably wise to reject it.

Table 14.1 shows that if we measure how far an observation is away from the mean value in units of σ (i.e., r / σ) then, if errors in the measurement are random, the chances are that 0.4987 of 0.5 (half the area) or 99.74% of the observations will be within the range $-3 < r / \sigma < +3$ or that 99.74% of the observations will fall within three standard deviations of the mean. Hence, if we have an observation for which $r / \sigma = 4$, then it is outside the probable range. This suggests that, in practical terms, it is highly likely that there is a gross error and the observation does not belong to the group under investigation. Example 14.3 illustrates a set of observations where one member of the set may or may not be included.

Before carrying out any statistical analysis we need to create what is called the *null hypothesis*. A hypothesis is essentially a supposition or an unproved theory; the null hypothesis is the assumption that there is no significant difference between any particular measure and any other unless we can demonstrate to the contrary. The object of a statistical test is to prove that something is significantly different from what we would otherwise expect. Statistics cannot prove that anything is true, but they can indicate the probability that something is significantly different from what might otherwise be anticipated. The key questions then become how to phrase the null hypothesis and how to judge what is significant.

It is highly likely that if an observed quantity x is such that $(x - \mu) / \sigma$ is greater than 3, there has been an error and the observation can be rejected as very unlikely to have happened by chance. There is, of course, a very remote possibility that this was a random event since the normal curve extends to infinity. If we have a hypothesis that an event is so rare that it cannot belong to the population under consideration and an observation that really should be included is rejected, then a *Type I Error* is said to occur; if, however, an observation is included that should be rejected, then the reverse is the case and a *Type II Error* occurs.

Example 14.3 Error Detection

In a series of 10 observations, the measurement of an angle was found to be 28° 33′ plus 24.8″, 23.9″, 25.8″, 32.2″, 25.0″, 24.5″, 24.2″, 23.7″, 24.1″, and 23.8″.

The mean value = 28° 33′ 25.2″.
The residuals are 0.4, 1.3, −0.6, −7.0, 0.2, 0.7, 1.0, 1.5, 1.1, and 1.4.

The sum of the squares of the residuals $= 58.16$, so $\sigma = \sqrt{\sum (x_i - \mu)^2 / n} = 2.4$.

If we set a threshold of three standard deviations away from the mean as being our limit, (here being 3 * 2.4 or 7.2), then all the observations lie just within it. If we set two standard deviations as the limit, then the observation with 32.2″ should be rejected and our mean would become 28° 33′ 24.4″ and the standard deviation would be 0.6″.

Type I errors can be reduced by increasing the threshold for rejecting observations (known as the *significance level*) but in so doing the chances of making a Type II error are increased. In practice, it is usual to set significance levels at either 95% (the equivalent approximately of $r / \sigma = 2$) or 99% (r / σ is approximately equal to 3).

Consider a data set A that is a sample set of measurements of a particular phenomenon for which there is potentially an infinite possible number of observations. Let the mean and variance of the sample data set A be $\underline{x}$ and S_A^2. In practice it is very likely that any small sample will have a slightly different mean and variance from that derived from the infinite series.

Let us assume that we have n observations in data set A and let these be $x_1, x_2 \ldots. x_n$ with a mean $\underline{x}$. Then for the variance of the sample set A, we have

$$\begin{aligned} S_A{}^2 &= (1/n)\sum(x_i - \bar{x})^2 = (1/n)\sum(x_i^2 - 2\bar{x}x_i + \bar{x}^2) \\ &= (1/n)\sum x_i^2 - (2\bar{x}/n)\sum x_i + (n/n)\bar{x}^2 \\ &= (1/n)\sum x_i^2 - \bar{x}^2 \text{ since } (1/n)\sum x_i = \bar{x} \end{aligned}$$

If the whole population has a variance σ^2 and mean μ we would expect that

$$\begin{aligned} \sigma^2 &= (1/n)\sum(x_i - \mu)^2 = (1/n)\sum x_i^2 - (2\mu/n)\sum x_i + \mu^2 \\ &= (1/n)\sum x_i^2 - 2\mu\bar{x} + \mu^2 \end{aligned}$$

Subtracting S_A^2 from this we obtain $(\sigma^2 - S_A^2) = (\underline{x} - \mu)^2$.

Thus, the sample variance is smaller than the variance of the infinite set. The mean values of the population and the sample will also differ. The question then arises as to how reliable is the value $\underline{x}$ as an estimate of μ, which is the mean of the whole population.

In fact, if we have N data sets A, B, C, and so on, each of size n_A, and so on, then they in turn form a set of values with means $\underline{x}_A$, $\underline{x}_B$, $\underline{x}_C$, and so on, the mean of all of which should be close to the mean of the overall population μ. The estimated variance for the combined N data sets will be

$$S^2 = (S_A{}^2 + S_B{}^2 + \ldots.)/N = \sigma^2/N, \text{ so } S = \sigma(\sqrt{N})$$

What this shows is that a sample of mean $\underline{x}$ is itself part of a wider distribution arising from many possible samples that could be taken. This larger distribution has the same mean as the total population (μ) and a standard deviation of $\sigma / (\sqrt{n})$ where σ is the standard deviation of the whole population and n is the total size of all the samples that have been taken. Put another way, the estimated standard deviation of the mean of a sample A of size n_A will be $(1 / \sqrt{n_A})$ times the standard deviation of the sample itself.

We showed above that $(\sigma^2 - S_A{}^2) = (\underline{x} - \mu)^2$ where σ^2 and μ relate to the whole population and S_A^2 and $\underline{x}$ to the sample. The average value of $(\underline{x} - \mu)^2$ over n samples will be σ^2 / n so for any general sample $(\sigma^2 - S^2) = \sigma^2 / n$ or $S^2 = (n - 1)\sigma^2 / n$. Thus, if we have a sample with standard deviation S_A the better estimate for the standard deviation for the population as a whole is

$$S = S_A\sqrt{\left(\frac{n}{(n-1)}\right)}$$

Box 14.7 summarizes all these relationships.

Putting $\sigma^2 = \{1/(n-1)\}\Sigma(x_i - \underline{x})^2$[rather than $\sigma^2 = \{1/n\}\Sigma(x_i - \underline{x})^2$] gives what is called the *unbiased estimate* of the population variance, although it makes little practical difference when the number n is large. The estimate is based on $(n - 1)$ *degrees of freedom*, the term referring to the minimum number of parameters necessary to describe completely the state of a system. In statistics the number of degrees of freedom is equal to the number of independent unrestricted random variables that constitute a particular statistic. Hence, for example, if we have a mean value of 10 observations, the mean is fixed, and once nine of the 10 observations have been chosen, the final 10th observation is already determined. There are thus nine degrees of freedom.

BOX 14.7 MEAN, VARIANCE, AND STANDARD DEVIATION

For any sample set of n numbers $x_1, x_2, \dots x_i, \dots x_n$ taken from a large population with variance σ^2, standard error σ and mean μ:

The mean value of the sample is

$$\left(x_1 + x_2 + \cdots + x_i + \cdots + x_n\right) / n = \sum x_i / n = \bar{x}$$

The variance of the sample is

$$S^2 = \sum \left(x_i - \bar{x}\right)^2 / n = (1/n) \sum x_i^2 - \bar{x}^2 = \sigma^2 * (n-1) / n$$

where σ^2 is the estimate for the population as a whole.

The standard deviation of the sample $A = S_A = \sqrt{\frac{\Sigma(x_i - \underline{x})^2}{n}}$. A better estimate for the standard deviation of the population $= S = S_A \sqrt{\left(\frac{n}{(n-1)}\right)}$.

Also: The standard deviation of the mean $\left(\underline{x}\right) = \frac{\sigma}{\sqrt{n}} = \frac{s}{\sqrt{(n-1)}}$.

14.5 THE *t*-TEST

The normal curve is derived on the basis of a random selection of measures from an infinite population, although in practice it applies equally well so long as the size of the sample is large. When this is not the case, the normal distribution has to be modified, but we can still use units of standard deviation and calculate $\left(x - \underline{x}\right) / \sigma$.

$\left(x - \underline{x}\right)$ is the difference between an observation and the mean, and the division by σ means that we are measuring this difference in units of σ.

An English mathematician, William Gosset, writing under the pseudonym of "Student" developed a way in which to test whether a random sample of normally distributed observations of unknown parameters has the same mean as the population. His statistic is known as t and his test as the *Student's t-test* (Box 14.8).

The number $t = \left(\underline{x} - \mu\right) / \left\{S / \sqrt{n}\right\} = \left\{\left(\underline{x} - \mu\right)\sqrt{n}\right\} / S$, where n is the number of observations in a sample, $\underline{x}$ is the sample mean, S the standard deviation of the sample (and hence, $S / \sqrt{n}$ is the standard deviation of its mean). "μ" is the estimated population mean. Put another way, t is a measure of the difference in the means of two sets of data divided by the standard deviation of the mean. It is the measure of the disparity between the means of two data sets as measured in units of the standard deviation of their means.

The test can be used to compare a sample mean with the estimated mean of the whole population. Alternatively, it can be used to test two samples to see if their means differ significantly, in which case x would be the mean of one sample and μ the mean of the other, with S and n being taken either from the combination of the two samples or from one sample, depending on the hypothesis.

The various values of t can be found in statistical tables. These show the probability (p) or significance level that arises for various values of n and the calculated value of t, as illustrated in Example 14.4. Table 14.2 shows a sample set of values.

TABLE 14.2
Levels of Significance (*p*) for Values of *t* (Given 9 Degrees of Freedom)

p	0.9	0.8	0.7	0.6	0.5	0.4	0.3	0.2	0.1	0.05	0.01
t	0.13	0.26	0.40	0.54	0.70	0.88	1.10	1.38	1.83	2.26	3.25

BOX 14.8 THE *t*-TEST

1. Tests the significance of the difference in means between two data sets.
2. Assumes that the population is normally distributed and that the samples are random.
3. Takes the form $t = \{(\underline{x} - \mu)\sqrt{n}\} / S$. It measures the difference in means by measuring $(\underline{x} - \mu)$ in units of the standard deviation of the mean $\{S / \sqrt{n}\}$.
4. The level of significance or probability can be found from special tables.

Example 14.4 Using the *t*-Test

In Example 14.1, we had two groups A and B of 10 people whose heights had been measured. Both had mean values of 1.70 but their standard deviations were (A) 0.03 and (B) 0.15. Does either population differ significantly from an estimated population average of 1.75 meters?

$$\bar{x} = 1.70, \mu = 1.75, n = 10, \sqrt{n} = 3.16 \text{ in both cases.}$$

For Group A,

$$t = \frac{(\bar{x} - \mu)\sqrt{n}}{S} = 0.158 / 0.03 = 5.27$$

For Group B,

$$t = \frac{(\bar{x} - \mu)\sqrt{n}}{S} = 0.158 / 0.15 = 1.05$$

There are 9 degrees of freedom, and (from tables) the level of significance for group A is less than 1%. So one can expect this to happen in less than 1 of 100 cases (see Table 14.2).

For group B, the result can be expected in slightly more than 0.3 of 1 or more than 30% of cases and the result is insignificant. Thus, group A seems to have been a special case, while group B is not unusual.

14.6 ANALYSIS OF VARIANCE

The *t*-test allows one to compare sample means. The *F-test* (Box 14.9) may be used to compare the variances between two samples, where $F = S_1^2 / S_2^2$. F is the ratio between the variance of sample set "1" and the variance of sample set "2." For very large data sets from the same population, the value of F would be expected to be ≈ 1 if the two samples were from the same population. It is usual to place the set with the larger variance as set "1" and the smaller as set "2." In Example 14.1, we had set A with a standard deviation of 0.03 and set B with 0.15. It is clear that since the two sets used in this example share the same mean, the smaller standard deviation fits easily within the range for set B. The reverse is not true.

For smaller data sets, the significance of the value of F will depend on the number of degrees of freedom in A and B, for instance, $(n_A - 1)$ and $(n_B - 1)$. It will also depend on whether one is looking for the variance in one of the data sets, such as A being greater than B or whether one is looking at the probability of their difference being significant regardless of whether A is greater or less than B. If the difference is only relevant in one direction, the test is said to be *one-tailed*; if the absolute difference is important regardless of whether one sample is greater or less than the other, then the test is *two-tailed.*

BOX 14.9 THE *F*-TEST

1. Is applied to sample variances.
2. Assumes that the samples A and B are normally distributed.
3. Assumes that the variance of $A\left(=S_A^2\right)$ and $B\left(=S_B^2\right)$ are independent estimates of the population variance σ^2 and $S_A^2 > S_B^2$.
4. Takes the form $F = S_A^2 / S_B^2$.
5. If there are n_A measures in A and n_B in B, then the degrees of freedom are $(n_A - 1)$ and $(n_B - 1)$.

The F-distribution is of particular use in the tests known as the analysis of variance.

In practical terms, the importance lies in where the level of significance is set and the difference only affects observations that are around the limit of acceptability. Roughly speaking, if 95% of observations are to be within the limit, then for the one-tailed test, the 5% that are outside the limit all occur at one end of the left-hand, bell-shaped probability curve in Figure 14.5. For a two-tailed test (the right-hand, bell-shaped curve), the two black areas are each 2.5% of the total area below the curve.

The idea of comparing variances leads to a powerful series of tests known as *analysis of variance* or *ANOVA*. If a series of observations are taken, for instance, of the yield of some crop when treated with different types of fertilizer, and if there is no significant difference between any of the fertilizers, then one would expect that the variance of each sample would be much the same. If, however, one fertilizer gave rise to significantly greater yields, then the samples taken from the crop treated with it would consistently have a greater mean. The difference in the mean values could be tested using the t-test, while the analysis of variance could help to discriminate between possible factors that are less obvious in a simple test of the mean values. There may be a whole variety of variables and the analysis of variance can be used to identify which of these is significant.

For example, if the results of a series of c different experiments are recorded for r different samples (see Table 14.3), then they represent $n = r * c$ different experiments for which the overall mean $(\underline{x})$ and standard deviation can be calculated. Let us arrange these in a table with r rows and c columns and label each observation x_{ij} as in a matrix.

The sum of the squares of the residuals from the mean of all the observations

$$= \sum\left(x_{ij} - \bar{x}\right)^2 = \sum\left\{x_{ij}{}^2 - 2\bar{x}\,x_{ij} + (\bar{x})^2\right\} = \sum x_{ij}{}^2 - 2\bar{x}\sum x_{ij} + n(\bar{x})^2\}$$

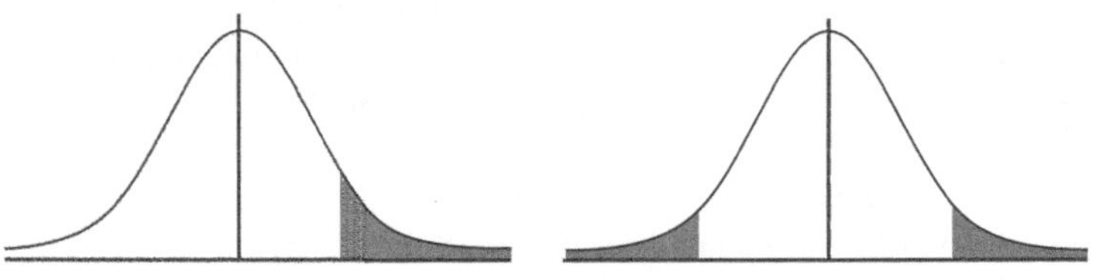

FIGURE 14.5 One- and two-tailed tests.

TABLE 14.3
Data Classified into Rows and Columns

	Col 1	Col 2	...	Col j	...	Col c	$\sum$ = Row Sum	Number in Row	Sum of Squares
Row 1	x_{11}	x_{12}	...	x_{1j}	...	x_{1c}	T_{r1}	n_{r1}	SS_{r1}
Row 2	x_{21}	x_{22}	...	x_{2j}	...	x_{2c}	T_{r2}	n_{r2}	SS_{r2}
Row 3	x_{31}	x_{32}	...	x_{3j}	...	x_{3c}	T_{r3}	n_{r3}	SS_{r3}
...	...	...	...	...	...	...	...	...	...
Row i	x_{i1}	x_{i2}	...	x_{ij}	...	x_{ic}	T_{ri}	n_{ri}	SS_{ri}
...	...	...	...	...	...	...	...	...	...
Row r	x_{r1}	x_{r2}	...	x_{rj}	...	x_{rc}	T_{rr}	n_{rr}	SS_{rr}
Column Sum	T_{c1}	T_{c2}	...	T_{cj}	...	T_{cc}	$T = \Sigma x_{ij}$		
No. in Column	n_{c1}	n_{c2}	...	n_{cj}	...	n_{cc}		n	
Sum of Squares	SS_{c1}	SS_{c2}	...	SS_{cj}	...	SS_{cc}			

Since $\underline{x} = \left(\Sigma x_{ij}\right)/n$ then the sum of the squares of the residuals $= \Sigma x_{ij}{}^2 - \left(\Sigma x_{ij}\right)^2/n$. This is sometimes called the *variation*. It equals n times the variance.

$$\text{Sample variation} = \sum x^2 - \left(\sum x\right)^2/n$$

$$\text{Sample variance} = \left(\sum x^2\right)/n - \left\{\left(\sum x\right)/n\right\}^2$$

The unbiased estimate of the variance of the population as a whole from which the observations have been drawn is therefore

$$\left(\sum x^2\right)/(n-1) - \left(\sum x\right)^2/\{n(n-1)\}$$

The reason for calculating the variation and variance in this way is that it is no longer necessary to calculate each of the individual residuals. Rather than having to work out all the residuals from the mean and then squaring each one in turn, we can calculate the sum of each observation squared and subtract the total sum squared divided by n. This gives us the variation from which we can obtain the variance and standard deviation by applying the appropriate number of degrees of freedom.

Referring to Table 14.3, we can calculate the overall variation or sum of the squares of the residuals as $\Sigma\left(x_{ij}{}^2\right) - \left(\Sigma x_{ij}\right)^2/n$ where x_{ij} is the observed value in row i and column j, and n = the total number of observations. If all elements are filled, then $n = r * c$ where r is the number of rows and c is the number of columns. The unbiased estimate of the variance for the population as a whole is therefore $\Sigma\left(x_{ij}{}^2\right)/(n-1) - \left(\Sigma x_{ij}\right)^2/\{n(n-1)\}$.

Let the sum of each column $j = T_{cj} = \Sigma\, x_{ij}$ for $i = 1$ to n_{cj} and the sum of each row $i = T_{ri} = \Sigma\, x_{ij}$ for $j = 1$ to n_{ri}. $\Sigma\, x_{ij}$ for the whole data set $= T = \Sigma T_{ri} = \Sigma T_{cj}$. If all elements are filled (that is, there are r observations in each column), then all $n_{ri} = c$ and all $n_{cj} = r$.

The mean value for each column $= T_{cj}/n_{cj}$. If we treat every element in the column as having the value of the mean for that column, then summing all c columns we have a total variation $= \Sigma\left\{n_{cj} * \left(\frac{T_{cj}}{n_{cj}}\right)^2\right\} - T^2/n = \Sigma\left(T_{cj}\right)^2/n_{cj} - T^2/n$. This is known as the *variation between columns.* If we assume that all elements were observed, then we can write this as $\Sigma\left(T_{cj}\right)^2/r - T^2/n$. If we divide this by the number of degrees of freedom ($c - 1$), we obtain an estimate of the overall population variance. Similarly, the variation between rows $= \Sigma(T_{ri})^2 / c - T^2 / n$. These variances are known as the *between-group variability* or *explained variance*.

Assuming that all the elements were observed, the unbiased estimate for the variance of the population obtained solely from column j is $SS_{cj} / (r-1) - (T_{cj})^2 / \{r(r-1)\}$. Let us call this V_{cj}.

The pooled estimate for all c columns will be the weighted mean of each such unbiased estimate, the weights being the degrees of freedom in each column. Call the result V_c. Since there are r items in each column, there will be ($r - 1$) degrees of freedom for each of the c columns. The sum of the weights with c lots of "($r - 1$)" or $c * (r - 1)$ will be $= (n - c)$. Hence,

$$V_c = \left\{\sum V_{cj} * (r-1)\right\} / (n-c) = \left\{\sum SS_{cj} - \sum (T_{cj})^2 / r\right\} / (n-c)$$

This is known as the *overall variance within columns*. It is also known as the *unexplained variance* or *within-group variability*. It is an estimate of the overall population variance based on ($n - c$) degrees of freedom. It can also be applied across the rows rather than across the columns, depending on the hypothesis being tested.

In order to determine whether there are any significant differences, we can apply the F-test, which in the context of analysis of variance takes the form:

$$F = (\text{between} - \text{samples estimate of variance}) / (\text{within} - \text{samples estimate of variance})$$

or

$$F = (\text{explained variance}) / (\text{unexplained variance})$$

The value of F can then be checked against the values given in statistical tables, based on the appropriate degrees of freedom.

The same calculations can be made with reference to rows rather than columns, resulting in different numbers. Their significance depends on the hypotheses being tested and whether one is attempting a one-way analysis of variance (in which only one result, for instance, the column differences, is being tested) or two-way analysis of variance in which more complex combinations are evaluated. The data are analyzed in terms of whether there is a significant difference between the circumstances giving rise to what is shown in rows or columns compared with the whole, or whether there are any anomalies.

In each case, the mean and standard deviation can be calculated and if there is no overall significant difference between any of the samples, both the variance between columns and the variance between rows should be about the same as the variance of the whole n experiments. This can then be checked by the F-test, see Example 14.5.

Example 14.5 Using Analysis of Variance

Consider a set of observations with five observations under four headings A, B, C, D with T referring to totals and SS to sums of squares.

	A	B	C	D	Tr	SS_r
	15	17	20	18	70	1238
	23	27	29	24	103	2675
	18	13	24	16	71	1325
	11	12	15	14	52	686
	17	16	21	14	68	1182
T_c	84	85	109	86	$\sum = 364$	
SS_c	1488	1587	2483	1548		$\sum = 7106$

$n = 4 * 5 = 20$. Total variation $= 7106 - 364^2 / 20 = 481.2$.
Between sample variation

$$= \sum (T_{cj})^2 / r - T^2 / n$$
$$= 84^2 / 5 + 85^2 / 5 + 109^2 / 5 + 86^2 / 5 - 364^2 / 20$$
$$= 6711.6 - 6624.8 = 86.8$$

With four columns there are three degrees of freedom; hence,

$$\text{Between sample variance} = 86.8 / 3 = 28.87 = \text{"explained variance"}$$

$$\text{Within sample variance} = \left\{ \sum SS_{cj} - \sum (T_{cj})^2 / r \right\} / (n - c)$$

$$= (7106 - 6711.6) / 16 = 394.4 / 16$$
$$= 24.65 = \text{"unexplained variance"}$$

$F = 28.87 / 24.65 = 1.17$. The between column differences are not significant.

14.7 THE CHI-SQUARED TEST

Analysis of variance compares the frequencies of certain events under differing conditions. A somewhat similar test is known as the chi-square or χ^2 test (from the Greek letter "chi" or χ) (Box 14.10). The test considers how often something is observed to happen (f_o) compared with the expected or calculated frequency (f_e) in which $\chi^2 = \sum \frac{(f_o - f_e)^2}{f_e}$.

The χ^2 test may be used, for example, if there is a predicted distribution of points on a map (such as cases of cancer near some electrical transmitter) so that the expected numbers can be compared with the observed numbers using the χ^2 test, as in Example 14.6.

The χ^2 test provides a way in which an observed distribution can be examined to see if it is probably part of a normal distribution—the estimated frequency being provided by the normal function, while the observed frequency is that which has been measured. The χ^2 test is an example of a *nonparametric test* in that it can be applied to ordinal and nominal data. All the other tests that have been discussed so far have been *parametric* tests in that they have had parameters such as the mean and standard deviation that can be subject to arithmetic operations.

Parametric tests are based on fairly stringent assumptions, for instance, about the nature of the distribution of the population, whereas nonparametric tests, sometimes referred to as *distribution free*, are much less demanding. For example, nonparametric tests can be used to compare two lists that have been ranked in order to see if they show similar characteristics, such as whether secondary or high school level examination results correlate well with final university degree awards. For further discussion on nonparametric tests the reader should consult specialist books on statistics.

BOX 14.10 THE χ^2 TEST

1. The test checks whether an observed distribution is probably consistent with estimated values based on a predictive model.
2. The data must be measured at least at a nominal scale or any higher level of measurement. There must be at least two mutually exclusive categories in which the data can be placed, each with a frequency greater than five.
3. If there are more than two categories, then at least 80% of these categories must have at least five expected outcomes.

4. If these restrictions cannot be met then a different test must be used or data categories amalgamated so that the criteria are met.
5. The test assumes an observed frequency (f_o) and expected frequency (f_e).

$$\chi^2 = \sum (f_o - f_e)^2 / f_e.$$

6. The significance level for the value of χ^2 can be found from statistical tables.

Example 14.6 χ^2 Test

According to evidence, 15% of properties are sold each year, on average across the whole of a country. In one town with 20,000 properties, only 1000 properties were sold last year. Is that significantly below average?

The expected frequency f_e would be 3000 of 20,000 or 3000 sold and 17,000 not sold. The observed quantities were 1000 sold and 19,000 not sold (we need to take the nonsales into account to ensure that the total probability {referred to before as ($p + q$)} is equal to 1 or 100%.

$$\chi^2 = (1000 = 3000)^2 / 3000 + (19{,}000 - 17{,}000)^2 / 17{,}000$$
$$= 4/3 + 4/17 = 1.57$$

The tables for χ^2 (not shown here) give the chances of this happening as greater than 20%; thus, at least one time in five this may happen. There is nothing particularly significant about the shortfall in house sales.

14.8 THE POISSON DISTRIBUTION

χ^2 can also be used to test other types of distribution, for example, that which bears the name of a French mathematician, Siméon Poisson. Although related to the normal distribution, it is particularly useful when dealing with small samples.

It takes the form

$$P = \sum \frac{m^r}{r!} e^{-m}$$

where p is the probability of an event, m is the mean value of a set of samples, r is the number of events, and e is the Euler number. Since

$$\frac{m^r}{r!} = \frac{m}{r} * \frac{m^{(r-1)}}{(r-1)!}$$

it means that each term except the first is $\frac{m}{r}$ times the previous term. An example is given in Examples 14.7 and 14.8.

Thus, if we lay a grid over an area and count the number of events that occur within each square, then we can calculate the probability of their being distributed with an expected frequency. This can be tested using the χ^2 test to see whether the predictions of the Poisson model are in conformity with what is found.

Example 14.7 Example of Poisson Distribution (1)

A grid with 10 * 10 squares was laid over an area and the numbers of events were counted (for instance, the number of plants or a particular type of worm in the quadrat). The Poisson prediction is calculated in Example 14.8. The results were as follows.

R	***P***	
Number Found per Square	**Number of Squares**	**Poisson Prediction**
0	1	0.5
1	2	2.8
2	6	7.3
3	13	12.7
4	16	16.6
5	19	17.4
6	15	15.2
7	12	11.4
8	10	7.5
9	3	4.3
10	2	2.3
11 or more (= 11)	1	2.0
	Total	100

The significance of the differences between what was found and the Poisson estimate can be tested using the χ^2 test.

Example 14.8 The Poisson Distribution (2)

Using the data in Example 14.7 that shows the number of squares that have 0, 1, 2, 3, and so forth, plants in them:

$$\text{The total number of plants} = 1*2+2*6+3*13+\cdots+11*1 = 524$$

There are 100 squares, hence, the mean number per square = m = 5.24. The probabilities are then calculated from $P = \Sigma \frac{m^r}{r!} e^{-m}$

$$\text{The value of } (1/e^{5.24}) = e^{-m} = 0.0053$$

According to the Poisson formula, the probabilities for $P_0 = 1 / (e^{5.24}) = 0.0053$, so out of 100 squares the expected frequency of nil finds = 100 * 0.0053 = 0.53.

Thus, $P_0 = 0.53$. Hence, we can calculate:

$$P_1 = (m/1) * P_0 = 2.8$$

$$P_2 = (m/2) * P_1 = 7.3$$

$$P_3 = (m/3) * P_2 = 12.7$$

$$P_4 = (m/4) * P_3 = 16.6$$

$$P_5 = (m/5) * P_4 = 17.4$$

$$P_6 = (m/6) * P_5 = 15.2$$

and so on.

14.9 MONTE CARLO SIMULATION

The word random means an event or characteristic that is not possible to predict. Computers follow strict logical rules in the course of executing programs and are not able to come up with numbers that are random. To work around this limitation programs have been developed that can generate "pseudo" random numbers. A *pseudorandom number* is not a random number, but a sequence of pseudorandom numbers can look very much like a sequence of random numbers. Generators of pseudorandom numbers are deterministic. Designers of pseudorandom generators strive to achieve certain qualities in their generators, such as to make it not easy to tell whether the sequence is random or pseudorandom. To be suitable for use in many applications a pseudorandom number generator must produce numbers that are uniformly distributed along the range of the generator. One can envision it as the integral curve of a flat function (i.e., y = constant) with borders of $x = 0$ and 1. Each run of a pseudorandom number generator will produce completely different sets of numbers if the correct parameters are used. Therefore in many applications it is necessary to archive the pseudorandom numbers generated for troubleshooting and auditing purposes.

Generating other distributions: pseudorandom number generators are commonly used in statistical modeling and analysis. When other than a uniform distribution is required for an application, the uniform distribution is converted to the desired type of distribution, for instance, $y = f(x)$. This is done by way of discretizing the desired distribution into narrow vertical slices similar to the numerical integral methods discussed in Section 10.7. Once the pseudorandom numbers are generated, they can be made to fit within each slice of the desired distribution by scaling its value. The fraction of all the pseudorandom numbers to be put into each slice is proportional to the y value, or height, of each slice.

Monte Carlo simulation: often referred to as Stochastic Modeling, Monte Carlo simulation is named after the Monte Carlo casino in Monaco, which is one of the most famous legal gambling establishments in the world. Monte Carlo simulation is a way to compute probability via generating thousands or millions of random "trials" then attempting to simulate outcomes of events one trial at a time. In the era of fast computing, using the Monte Carlo simulation method can turn complex problems of probability computation into thousands or millions of random "trials" to be calculated on the computer.

In a Monte Carlo simulation one creates large numbers of scenarios, or "trials," in which each input variable takes on a particular value. These values are drawn from the expected distribution of each variable using a pseudorandom number generated by a pseudorandom number generator, much like randomly dealing a playing card. Once all input variables are assigned values using this method, the simulation proceeds by calculating the outcome using knowledge of real-life processes that lead to each kind of outcome. Typically there may be several or many different kinds of outcomes and each one will be predicted to occur with a specific probability. The probability of outcome X is computed by:

$$\text{Probability}(X) = \text{Number of trials with } X \,/\, \text{Total number of trials}$$

For example: we want to calculate the probability that a train we want to catch will be late by more than 30 minutes. Using Monte Carlo simulation, we first need to know what are the factors that can delay the train and what is the probability distribution that these factors would take on each value. We decide to run a Monte Carlo simulation using the following input factors:

1. The first variable is whether there is abnormal wind resistance. We have a distribution of wind speeds and wind directions for the train route. In each Monte Carlo trial, the wind speed and direction will be randomly populated yet conforming to the distribution.
2. Another variable is whether there will be emergencies that delay the train. These emergencies can range from medical emergencies of passengers, police activity, or other emergency

events that lead to the train stopping for a longer period at a station. Based on historical data, we learn that these types of events occur approximately once every N days, and there is a distribution of how much delay each kind of event can cause. We utilize this data to populate a random variable for emergency events.

3. Our last variable is the level of congestion in the train. When the train has more passengers it takes longer for passengers to on- and off- board the train at intermediate stations. We predict the number of passengers based on the date we plan to take the train, whether it's a weekday, weekend, or holiday. From historical data we can gather a distribution of crowdedness levels based on date, and this also enables us to populate each Monte Carlo trial with randomly generated data on crowdedness.

Now we can execute the Monte Carlo simulation by calculating each trial, using formulae that logically calculate if and how long the train will be late in each trial. After all trials have been run, we would then be able to calculate the probability of the train being late by 30 minutes on a specific day.

GLOSSARY

Analysis of variance Also known as ANOVA, tests the observed variance within and between samples to determine whether the differences are random or significant.

Binomial expansion An expansion of the form

$$(p+q)^n = P^n + np^{(n-1)}q + \{n(n-1)/2)\}p^{(n-2)}q^2 + \cdots +_n C_r p^{(n-r)}q^r + + \cdots + npq^{(n-1)} + q^n$$

This has a *mean* outcome of *np* with *standard deviation* $\sqrt{(npq)}$.

Central limit theorem If a sequence of independent identically distributed random variables each has a finite variance, then as the number of observed values increases, they tend to mirror a *normal distribution*.

Chi-squared test Compares the frequency of an event as observed with what might be expected in the form

$$\chi^2 = \sum (f_o - f_c)^2 / f_c$$

Degrees of freedom The minimum number of parameters necessary to describe completely the state of a system.

Descriptive statistics Sets of numbers used to summarize a set of known data in a clear and concise way, such as the mean value.

Expected value The value (E) in a set of observations that will most probably occur.

Explained variance In *analysis of variance*, the overall *variance* between columns or rows. Also known as the *between-group variability*.

First moment The *expected value* of the deviation from zero (*residual*) for all the values of a *random* variable (usually the weighted mean). The second moment is the expected value of the squares of the residuals.

***F*-test** A test that compares the variances between two samples in the form

$$F = S_A^{\ 2} / S_B^{\ 2}$$

Heuristic Using processes that are guided by experience and experiment rather than by rigorous logical argument from precisely defined axioms.

Histogram A graphical method for displaying frequencies.

Inferential statistics The use of statistical data to draw conclusions from random samples.

Mean The mean is the average value by summing all the terms x_i and then dividing by the number of terms (n). Often written as μ, its value is $\frac{\sum_{i=1}^{i=n} x_i}{n}$

Measures of central tendency Measures of the spread of a set of observations around their mean value, for instance, *variance* and *standard deviation.*

Median The median is the middle number in a frequency list of numbers.

Mode The mode is the number that most often occurs, that is, it has the highest frequency in a list of numbers.

Monte Carlo simulation Also called stochastic modeling, is a method to predict the probability of a variety of outcomes when the potential for random variables is present.

Nonparametric test Test that can be applied to nominal and ordinal data, namely data that are distinguishable only by their class (nominal) or their place in a sequence (ordinal).

Normal curve A curve that follows the normal distribution, sometimes described as being bell-shaped.

Normal distribution Also known as a *Gaussian distribution*, it is symmetrical with the mean, median, and mode coinciding and has the form

$$y = \left[\frac{1}{\left\{ \sigma\sqrt{(2\pi)} \right\}} \right] e^{-\left\{ (x-\mu)^2 / 2\sigma^2 \right\}}$$

or, using the Z statistic

$$y = \frac{1}{\sigma\sqrt{2\pi}} e^{-\left(z^2/2 \right)}$$

Null hypothesis The supposition that there is no significant difference between any particular measure and any other sample value.

One-tailed test The difference is only significant if the quantity is exceptionally greater (or less) than anticipated but not significant if it is smaller (or greater)—unlike the *two-tailed test.*

Parametric test Data that have parameters such as a mean and standard deviation.

Pascal triangle A triangular array of integers starting with 1, in which each number is the sum of the two above it.

Poisson distribution A method for predicting what might be expected at an average rate of distribution. It takes the form

$$p = \sum \frac{m^r}{r!} e^{-m}$$

Population Sometimes referred to as a universe, it is the complete set of individual components or events from which *samples* are drawn.

Probability A measure of confidence that can be had in any event, measured on a scale from 0 to 1 (which represents certainty).

Pseudorandom number A number that appears random but is actually generated by a deterministic algorithm.

Random number A number that cannot be determined from any of the previous numbers that have been selected.

Residual The difference between an observation (x_i) and the mean of all the observations (μ). The residual $= r = (x_i - \mu)$.

Residual variation The difference between the sum of the variation between rows plus the variation between columns and the variation of the whole data set.

Sample A set of individual observations of elements of a population that can be used to predict the nature of the population.

Significance level The threshold at which there is a critical level of probability of wrongly rejecting the null hypothesis.

Standard deviation The square root of the *variance.* $\sigma = \sqrt{(\Sigma(x_i - \mu)^2 / n)}$

Statistic A function of random variables that can be used as an estimator of a *population*.

Stochastic Processes that can be described by random variables.

***t*-Test** Also known as the *Student's t-test,* it tests the significance of the difference in means between two data sets, assuming that the population is normally distributed and the samples are random.

$$t = \left[(\underline{x} - \mu)\sqrt{n} \right] / S$$

Two-tailed test A test that judges the significance of whether the quantity is exceptionally greater or less than the expected value for a random sample (unlike the *one-tailed test* where the significance is only tested in one direction).

Type I error An observation that should be included is rejected.

Type II error An observation that should be rejected is included.

Unbiased estimate Using $\sigma^2 = \{1/(n-1)\}\Sigma(x_i - \underline{x})^2$ rather than $\sigma^2 = \left\{\frac{1}{n}\right\}\Sigma(x_i - \underline{x})^2$ as the estimate of the population *variance.*

Unexplained variance In *analysis of variance,* the overall *variance* within columns or rows. Also known as the *within-group variability.*

Variance The average sum of the squares of the *residuals.* Written as sigma squared, $\sigma^2 = \sum(x_i - \mu)^2 / n$.

Variance between sample means A term used in *analysis of variance.* It equals the *variance* of the means of samples and is approximately equal to σ^2 / n.

Variance within samples A term used in *analysis of variance.* It is the estimate of the overall population variance by considering the weighted average of the *variances* between columns (or rows).

Variation *n* times the *variance.*

Weight A factor by which an observation is multiplied to make it compatible with others in a data set.

Weighted mean $\Sigma w_i x_i / \Sigma w_i$ where w_i is the weight of observation *i.*

Z score The *residual* when measured in units of standard deviation.

$$z_i = (x_i - \mu) / \sigma$$

15 Correlation and Regression

15.1 CORRELATION

The statistical tests described so far apply primarily to circumstances in which there is one variable, albeit treated in different ways, giving rise to such techniques as the *t*-test and analysis of variance. The assumption is that the data come from the same population.

In practice, and especially in geomatics and geographic information system, at least two things are going on at the same time: for instance, movement in the *x*-direction and independent movement in the *y*-direction. The question then becomes, are these somehow interdependent or, put another way, are they correlated? Also, how can we find the best compromise between all the conflicting data?

Given two variables X and Y where $X = (x_1, x_2, \dots x_i \dots x_n)$ and $Y = (y_1, y_2, \dots y_i \dots y_n)$ and x_i is possibly related to y_i, then treating X and Y independently:

$$\text{Mean of } X = \bar{x} = (1/n)\sum x$$
$$\text{Mean of } Y = \bar{y} = (1/n)\sum y$$

$$\text{Variance of } X = S_x^2 = (1/n)\sum (x_i - \bar{x})^2$$
$$\text{Variance of } Y = S_y^2 = (1/n)\sum (y_i - \bar{y})^2$$

A measure of the relationship between X and Y is given by $(1/n)\Sigma\{(x_i - \bar{x})(y_i - \bar{y})\}$. This is called the *covariance* of X and Y. It is often written as "Cov *(X,Y)*." With smaller data sets the relationship is often given as $(1/(n-1))\Sigma\{(x_i - \bar{x})(y_i - \bar{y})\}$ as this gives a better estimate of the whole population. Here we will assume large samples and use:

$$\text{Cov}(X,Y) = (1/n)\sum_{i=1}^{i=n}(x_i - \bar{x})(y_i - \bar{y})$$

$$\text{Now } \sum (x_i - \bar{x})(y_i - \bar{y}) = \sum (x_i y_i - x_i\bar{y} - y_i\bar{x} + \bar{x}\,\bar{y}) \text{ for all } i = 1 \text{ to } n.$$
$$= \sum (x_i y_i) - (x_1 + x_2 + \cdots + x_n)\bar{y} - (y_1 + y_2 + \cdots + y_n)\bar{x} + n\bar{x}\,\bar{y}$$
$$= \sum (x_i y_i) - n\bar{x}\,\bar{y} - n\bar{x}\,\bar{y} + n\bar{x}\,\bar{y} = \sum (x_i y_i) - n\bar{x}\,\bar{y}$$

Hence,

$$\text{Cov }(X,Y) = (1/n)\sum (x_i y_i) - \bar{x}\;\bar{y} = \frac{n\Sigma (x_i y_i) - n\bar{x}.n\bar{y}}{n^2} = \frac{n\Sigma (x_i y_i) - \Sigma x_i \,\Sigma y_i}{n^2}$$

The covariance has a similar form to the variance but unlike the latter, which is always positive (being the sum of squares), the covariance may be either positive or negative.

If we divide Cov (X, Y) by $\sqrt{S_x{}^2 S_y{}^2}$ (which is the *geometric mean* of the two variances), we obtain a ratio r that is known as the *coefficient of correlation*. An illustration of correlation can be seen in

DOI: 10.1201/9781003464402-20

FIGURE 15.1 Correlation between *X* and *Y*.

Figure 15.1, where there is a clear trend that relates in approximately linear form the relationship between *X* and *Y*, using the data in Example 15.1. Since

$$S_x^{\,2} = \left\{\sum(x_i - \bar{x})^2\right\} / n = \left\{n\sum x^2_{\,i} - \left(\sum x_i\right)^2\right\} / n^2$$

and

$$S_y^{\,2} = \left\{\sum(y_i - \bar{y})^2\right\} / n = \left\{n\sum y^2_{\,i} - \left(\sum y_i\right)^2\right\} / n^2$$

we can express the relationship between *X* and *Y* as *r* with

$$\text{Cov}\ (X,Y) / \sqrt{S_x^{\,2} S_y^{\,2}} = r = \frac{n\sum x_i y_i - \sum x_i \sum y_i}{\sqrt{\left\{n\sum x_i^2 - \left(\sum x_i\right)^2\right\}}\sqrt{\left\{n\sum y_i^2 - \left(\sum u_i\right)^2\right\}}}$$

r is also known as the *Pearson correlation coefficient*.

Expressing *r* in this way makes it easier to program the calculation since, in tabular form, it is relatively easy to sum the raw observations without having first to calculate the means and then take the differences between the means and the observed quantities. We did this with the residuals in Chapter 14 when discussing analysis of variance. The process simplifies the way in which the residuals and hence the variance are calculated (Table 15.1).

It will be seen from the equation for *r* (Box 15.1), that if $x = y$, then the value of $r = 1$ and if $x = -y$ then $r = -1$. If we put $2x$ for *x* or any other multiple of *x*, then *r* remains the same; similarly with *y*. *r* will have a value of +1 for full agreement between *x* and *y* while $r = -1$ for full disagreement; $r = 0$ for no correlation between the two.

The significance level of *r* can be tested using the appropriate statistical tables. Alternatively, it can be shown by suitable rearrangement of the data that *t* from the *t*-test equates with $\frac{r\sqrt{(n-2)}}{\sqrt{(1-r^2)}}$ so by calculating this quantity, the *t* tables can be used.

TABLE 15.1
Framework for Calculating *r*

Observation	x	y	x^2	y^2	xy
1	x_1	y_1	x_1^2	y_1^2	x_1y_1
2	x_2	y_2	x_2^2	y_2^2	x_2y_2
…	…	…	…	…	…
n	x_n	y_n	x_n^2	y_n^2	x_ny_n
Total	Σx	Σy	Σx^2	Σy^2	Σxy

Example 15.1 Example of Correlation

In a set of 12 observations made of both x and y the following values were found:

Obs	x	y	x^2	y^2	xy
1	21	27	441	729	567
2	23	27	529	729	621
3	25	25	625	625	625
4	26	26	676	676	676
5	29	23	841	529	667
6	31	20	961	400	620
7	32	19	1024	361	608
8	35	17	1225	289	595
9	37	14	1369	196	518
10	38	15	1444	225	570
11	41	11	1681	121	451
12	43	9	1849	81	387
Total	$\Sigma x = 381$	$\Sigma y = 233$	$\Sigma x^2 = 12665$	$\Sigma y^2 = 4961$	$\Sigma xy = 6905$

$$\text{The coefficient of correlation} = r = \frac{n\Sigma x_i y_i - \Sigma x_i \Sigma y_i}{\sqrt{\left\{n\Sigma x_i^2 - (\Sigma x_i)^2\right\}}\sqrt{\left\{n\Sigma y_i^2 - (\Sigma y_i)^2\right\}}}$$

$$= (82860 - 88773)/\sqrt{\{(6819) * (5243)\}}$$

$$= -5913/5979 = -0.989$$

This shows a high level of negative correlation. Since there are two variables X and Y and 12 paired observations so there are $(n - 2) = 10$ degrees of freedom.

BOX 15.1 COVARIANCE AND COEFFICIENT OF CORRELATION

For two independent variables X and Y:

$$\text{Covariance } (X,Y) = \text{cov}(X,Y) = \frac{n\Sigma(x_i y_i) - \Sigma x_i \Sigma y_i}{n^2}$$

$$\text{Coefficient of Correlation} = r = \text{cov}(X,Y)/\sqrt{\{S^2{}_x S^2{}_y\}}$$

$$r = \frac{n\Sigma x_i y_i - \Sigma x_i \Sigma y_i}{\sqrt{\left\{n\Sigma x_i^2 - (\Sigma x_i)^2\right\}}\sqrt{\left\{n\Sigma y_i^2 - (\Sigma y_i)^2\right\}}}$$

$$-1 \le r \le +1$$

We can present the relationships that we have so far pursued in this chapter in matrix form. In Chapter 14, we introduced the idea of *expectation* or what we would most likely expect for any set of samples x. We described this as E(x). It is the mean value of all possible samples, assuming that there has been a very large, in effect infinite, number of observations. If we take the mean of different sets of samples, calling each mean m_i with an overall "mean of means" m, then the expected value E(m) also equals the mean of the infinite population.

Using "ε," the Greek letter "epsilon" or "e" for error, we can define the residuals "ε_j" for the set with mean m_i as $\varepsilon_j = m_i - x_j$. Then, $E(\varepsilon_j) = 0 = E(m - m_i)$, because the sum of the residuals is, by the definition of the mean value, equal to zero.

For the population as a whole, we can also write $E(\varepsilon^2) = n\,\sigma^2$, that is, we would expect that if we made a sufficiently large number of observations of the whole population, then the mean of the sum of the squares of the residuals would equal the population variance. If we consider the individual values as ε_1, ε_2, and so on, we can regard this as a one row matrix **V**. Its transpose is the column matrix $\mathbf{V}^\mathrm{T}$ and $\mathbf{V}\mathbf{V}^\mathrm{T} = \sum \varepsilon_i^2$ = sum of the squares of the residuals = the variation = $n\sigma^2$.

We can then write $E(\varepsilon^2) = E(\mathbf{V}^2) = E(\mathbf{V}\mathbf{V}^\mathrm{T}) = n\,\sigma^2$.

Further, the product $\left(\mathbf{V}^\mathrm{T}\mathbf{V}\right)$ is an $n*n$ matrix $\begin{bmatrix} \varepsilon_1^2 & \varepsilon_1\varepsilon_2 & \varepsilon_1\varepsilon_3\ldots\varepsilon_1\varepsilon_n \\ \varepsilon_2\varepsilon_1 & \varepsilon_2^2 & \varepsilon_2\varepsilon_3\ldots\varepsilon_2\varepsilon_n \\ \varepsilon_3\varepsilon_1 & \varepsilon_3\varepsilon_2 & \varepsilon_3^2\ldots\varepsilon_3\varepsilon_n \\ \ldots & & \\ \varepsilon_n\varepsilon_1 & \varepsilon_n\varepsilon_2 & \varepsilon_n\varepsilon_3\ldots\varepsilon_n^2 \end{bmatrix}$

where the leading diagonal is made up of the squares of the residuals and both the upper and lower triangular forms are made up of the elements of the covariance. This matrix is known as the *dispersion matrix* or *variance–covariance matrix*.

We can write this as $\begin{pmatrix} \sigma_1^2 & \sigma_{12} & \sigma_{13} & \ldots. & \sigma_{1n} \\ \sigma_{21} & \sigma_2^2 & \sigma_{23} & \ldots. & \sigma_{2n} \\ \sigma_{31} & \sigma_{32} & \sigma_3^2 & \ldots. & \sigma_{3n} \\ \ldots & & & & \\ \sigma_{n1} & \sigma_{n2} & \sigma_{n3} & \ldots. & \sigma_n^2 \end{pmatrix}$ or as $\begin{pmatrix} \sigma_1^2 & \sigma_{12} & \sigma_{13} & \ldots. & \sigma_{1n} \\ \sigma_{12} & \sigma_2^2 & \sigma_{23} & \ldots. & \sigma_{2n} \\ \sigma_{13} & \sigma_{23} & \sigma_3^2 & \ldots. & \sigma_{3n} \\ \ldots & & & & \\ \sigma_{1n} & \sigma_{2n} & \sigma_{3n} & \ldots. & \sigma_n^2 \end{pmatrix}$

since $\varepsilon_1\varepsilon_2 = \varepsilon_2\varepsilon_1$ and therefore $\sigma_{12} = \sigma_{21}$, and so forth. If all the error or residual terms are truly random and independent, the covariance will be zero and the dispersion matrix reduces to the form

$$\begin{pmatrix} \sigma_1^2 & 0 & 0 & \ldots. & 0 \\ 0 & \sigma_2^2 & 0 & \ldots. & 0 \\ 0 & 0 & \sigma_3^2 & \ldots. & 0 \\ \ldots. & & & & \\ 0 & 0 & 0 & \ldots & \sigma_n^2 \end{pmatrix}$$

The smaller the variance σ^2, the smaller the errors are likely to be, and therefore the more likely our observations are to be accurate. Hence, we can weight our observations in inverse relation to their variance. This leads to the weight matrix

$W = \begin{pmatrix} 1/\sigma_1^2 & 0 & 0 & \ldots. & 0 \\ 0 & 1/\sigma_2^2 & 0 & \ldots. & 0 \\ 0 & 0 & 1/\sigma_3^2 & \ldots. & 0 \\ \ldots. & & & & \\ 0 & 0 & 0 & \ldots & 1/\sigma_n^2 \end{pmatrix}$ that we will refer to in Section 15.5.

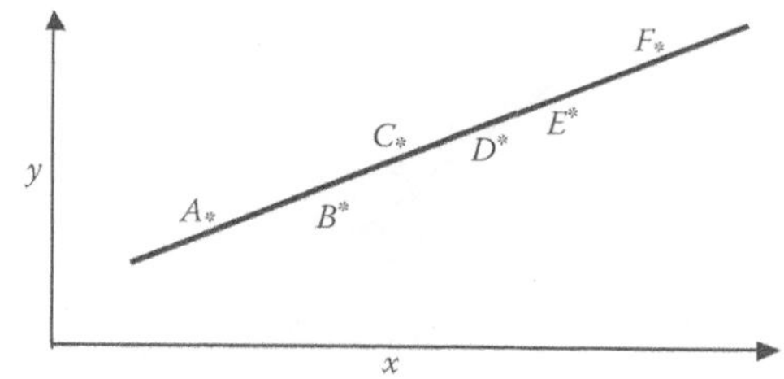

FIGURE 15.2 Regression line.

15.2 REGRESSION

For certain data sets, it is probably easier to test levels of correlation by viewing them pictorially. Consider a set of points for which there are values of X and Y in pairs with (x_A, y_A), and (x_B, y_B), and so on, as shown in Figure 15.2. They appear to fall approximately along a straight line. There appears to be some degree of correlation between them and indeed some linear function, which gives a good approximation to the relationship between the set of points X and their partners Y.

The relationship could be expressed by calculating the coefficient of correlation between them (r). Expressing the relationship in the form of $y = mx + c$ provides additional information. The process of choosing the best-fit relationship between Y and X is called *regression*, in this case *simple linear regression*.

The line that has been drawn in Figure 15.2 does not actually pass through any of the points. It could have done but then it might not. Let us exaggerate the situation as in Figure 15.3.

Each point for which there has been a pair of observations (x_A, y_A), (x_B, y_B), and so on, there will be observed ($_O$) quantities ($_Ox_i$, $_Oy_i$) where i is any point. There will also be a set of computed ($_C$) values such as ($_Cx_i$, $_Cy_i$) that are derived from the equation for the straight line.

If we assume that the Y values are dependent on the X values and that $_Ox_i = {}_Cx_i$, implying that the X values are correctly determined, then $_Oy_i$ will approximately equal $_Cy_i$. In practice, there may be small differences as shown by the short vertical lines above or below points A, B, C, and so forth, in Figure 15.3. If the line has the form

$$y = mx + c$$

then

$${}_{o}y_i = m * {}_{o}x_i + c + \varepsilon_i$$

where "ε" represents a small amount and is sometimes called the *error term.*

$$\varepsilon_i = \text{observed } y - \text{computed } y = {}_{o}y_i - {}_{C}y_i = {}_{o}y_i - c - m * {}_{o}x_i$$

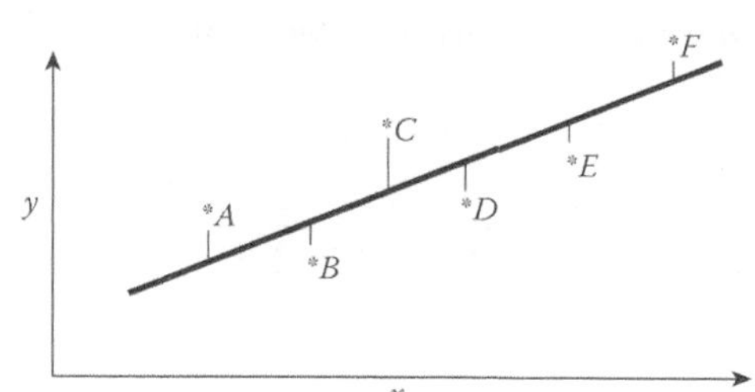

FIGURE 15.3 Residuals from the regression line.

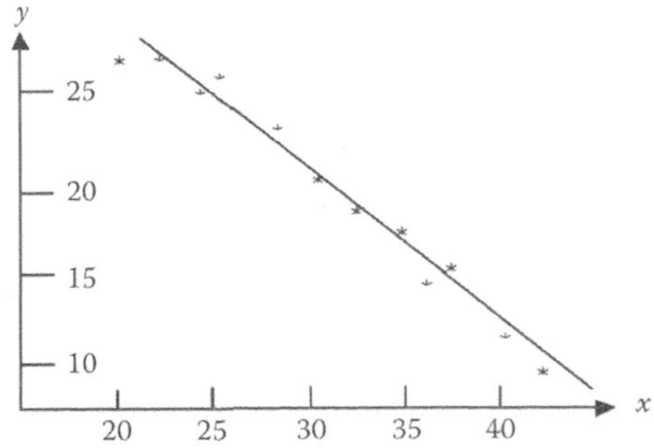

FIGURE 15.4 Example of a regression line.

As can be seen from Figure 15.3, ε will sometimes be positive and sometimes negative and in general will be small if there is good correlation between X and Y. The value of ε will depend not only upon the various values of X and Y but also upon the choice of the two constants m and c for the regression line (Figure 15.4).

If we assume that ε is normally distributed, the most likely values for m and c will be those that ensure the $\sum\varepsilon^2$ is a minimum. This is known as the *least square solution*. This fundamental assumption occurs in all sorts of problems in measurement science. In Chapter 14, we introduced the idea of residuals, variances, and standard deviations. ε is a residual and by minimizing $\sum\varepsilon^2$, we are maximizing the probability of achieving the best answer to whatever problem we may be addressing. As can be seen from the formula for the coefficient r, the maximum correlation is achieved when the denominator in the expression for r (i.e., the value of $\sqrt{S_x^2 S_y^2}$) is as small as possible. This occurs when the variance is a minimum.

In the general case, we need to choose m and c so that $\sum\varepsilon^2$ and therefore $\sum(y - c - mx)^2$ is a minimum. If there are n observations, this means that

$$\sum\left(y^2 + c^2 + m^2x^2 - 2cy - 2mxy + 2cmx\right)$$

is a minimum or if we call this relationship F, then

$$F = \sum y^2 + nc^2 + m^2\sum x^2 - 2c\sum y - 2m\sum xy + 2cm\sum x$$

is a minimum.

We can choose any values for m and c, but the minimum value for F will occur when the partial differentials with respect to m and c are zero. This will happen when:

$$(\text{Differentiating with respect to } c)\ \partial F / \partial c = 2nc - 2\sum y + 2m\sum x = 0$$

$$(\text{Differentiating with respect to } m)\ \partial F / \partial m = 2m\sum x^2 - 2\sum xy + 2c\sum x = 0$$

Hence,

$$m\sum x + cn = \sum y$$

and

$$m\sum x^2 + c\sum x = \sum xy$$

Thus, we can calculate m and c giving us

$$m = \frac{n\sum(x_i y_i) - (\sum x_i)(\sum y_i)}{n\sum x_i^2 - (\sum x_i)^2} \quad \text{and} \quad c = \frac{(\sum y_i)(\sum x_i^2) - (\sum x_i)(\sum x_i y_i)}{n\sum x_i^2 - (\sum x_i)^2}$$

Using these values, we will have the best-fit solution for a line that runs as close as possible to our observed values for x and y. See Example 15.2.

It may be that Y and X are not linear functions of each other. For example, Y may increase in the form of y^k where k is some constant (e.g., 2 or 3 so that X is related to Y^2 or Y^3). We discussed aspects of linearization in Chapter 10, Section 10.4. Here, we note a relatively simple solution.

If we consider G rather than Y as the function that is dependent on X and if we let $G = \log(y^k) = k \log y$, then we can still plot the regression line between X and G by using ($\log y_i$) instead of (y_i), the value k merely affecting the slope of the line. Plotting one set of values on a logarithmic scale does not invalidate the process as long as the relationship is correctly interpreted.

Example 15.2 Linear Regression

For two variables X and Y where Y is dependent on X, the best-fit straight line takes the form $y = mx + c$, where

$$m = \frac{n\sum(x_i y_i) - (\sum x_i)(\sum y_i)}{n\sum x_i^2 - (\sum x_i)^2}$$

and

$$c = \frac{(\sum y_i)(\sum x_i^2) - (\sum x_i)(\sum x_i y_i)}{n\sum x_i^2 - (\sum x_i)^2}$$

For example, using the data in Example 15.1,

$$\sum x = 381; \quad \sum y = 233$$

$$\sum x^2 = 12665; \quad \sum y^2 = 4961; \quad \sum xy = 6905$$

$$\begin{aligned} m &= (82860 - 88773)/(151980 - 145161) \\ &= -5913/6819 = -0.87 \end{aligned}$$

$$\begin{aligned} c &= (2950945 - 2630805)/(151980 - 145161) \\ &= 320140/6819 = 46.95 \end{aligned}$$

Thus,

$$y = -0.87x + 46.45$$

The principles enunciated above also apply in three dimensions where X and Y are freely independent variables and Z is dependent on X and Y. If Z is some function of X and Y in which $Z = f(x, y)$, then we have a surface. In particular, if the relationship is linear, then the surface is a plane. In general, the surface representing Z is known as the *trend surface*. Trend surface analysis is an extension of linear regression into three or more dimensions.

The technique of finding the best fit by minimizing the variance of the residual values can be extended to the combination of different measurements that are subject to different conditions. This is possible because as shown in Boxes 15.2 and 15.3, if

$$X = ax + by + cz + \cdots$$

BOX 15.2 THE MEAN OF LINKED INDEPENDENT VARIABLES

Consider m observations of x and n of y that are combined to form X such that

$$X = ax + by$$

There are $m * n$ possible values for X since there are m ways of choosing x and n ways of choosing y. Let

$$x_i = \bar{x} + r_i \text{ and } \quad y_i = \bar{y} + \rho_i$$

Thus, $\sum r_i = 0$ and $\sum \rho_i = 0$ (because $(1/m)\Sigma x_i = \bar{x}$ and $(1/n)\Sigma y_i = \bar{y}$). Also,

$$(1/m)\sum r^2_i = S^2_x \quad \text{and} \quad (1/n)\sum p^2_i = S^2_y$$

(since r_i and ρ_i are the residuals of x and y about their individual means). Thus,

$$\begin{aligned}
\bar{X} &= \frac{1}{mn}\sum_{i=1}^{m}\sum_{j=1}^{n}\left(ax_i + by_j\right) \\
&= \frac{1}{mn}\sum_{i=1}^{m}\sum_{j=1}^{n}\left(a\bar{x} + b\bar{y} + ar_i + b\rho_j\right) \\
&= \frac{1}{mn}\sum_{i=1}^{m}\sum_{j=1}^{n}\left(a\bar{x} + b\bar{y}\right) \text{ since both } \Sigma r_i \text{ and } \Sigma \rho_i = 0 \\
&= a\bar{x} + b\bar{y}
\end{aligned}$$

BOX 15.3 THE VARIANCE OF LINKED INDEPENDENT VARIABLES

Following on from Box 15.2, the variance of $X = S^2$

$$\begin{aligned}
&= \frac{1}{mn}\sum_{i=1}^{m}\sum_{j=1}^{n}\left(ax_i + by_j - a\bar{x} - b\bar{y}\right)^2 \\
&= \frac{1}{mn}\sum_{i=1}^{m}\sum_{j=1}^{n}\left(ar_i + b\rho_j\right)^2
\end{aligned}$$

With a little manipulation this gives us $S^2 = a^2S^2{}_x + b^2S^2{}_y$. Once again, what is true for x and y will also be true for x, y, z, and so forth. When summing independent variables, the variances are additive. Thus, if $X = ax + by + cz + \ldots$ where a, b, c are numbers and x, y, and z are related variables and subject to all possible combinations, then

$$\text{Mean of } X \text{ is } \bar{X} = a\bar{x} + b\bar{y} + c\bar{z} + \cdots$$

$$\text{The variance of } X \text{ is } S^2 = a^2S^2{}_x + b^2S^2{}_y + c^2S^2{}_z + \cdots$$

where a, b, c are any numbers, and if the means of x, y, z are $\underline{x}$, $\underline{y}$, $\underline{z}$, and so on, and the standard deviations are S_x, S_y, S_z, and so on, then:

$$\text{Mean of } X = \overline{X} = a\overline{x} + b\overline{y} + c\overline{z} + \cdots$$
$$\text{Variance of } X = \sigma^2_x = a^2S^2_x + b^2S^2_y + c^2S^2_z + \cdots$$

In the derivations given in Boxes 15.2 and 15.3, if $a = b = c$, and so on, $= 1$, then for all the possible combinations of x, y, and z, and so on,

$$\text{Mean of } X = \overline{x} + \overline{y} + \overline{z} \cdots$$
$$\text{Variance} = S^2_x + S^2_y + S^2_z + \cdots$$

15.3 WEIGHTS

In what we have explored above, the assumption has been that all the observations are equally reliable and therefore the errors in each quantity have been distributed normally with the same value for their standard deviations σ. Sometimes, the observations are of different reliability and therefore we may wish to give them different *weight*, in effect leaning more heavily toward one set of observations than another.

Thus, if we give twice as much weight to y than to x and three times as much to z, then the weighted mean will be $(x + 2y + 3z) / 6$. With weights w_1, w_2, and w_3, the weighted mean is given by

$$\frac{w_1x + w_2y + w_3z}{w_1 + w_2 + w_3}$$

Example 15.3 illustrates the use of weights. Weights are relative values, not absolute values. From a theoretical perspective, according to the normal distribution discussed in Chapter 14, the probability of an error of unit weight $(x_i - \underline{x})$ is proportional to $e^{-(x_i-\underline{x})^2/2\sigma^2}$. If the weight of observation $i = w_i$, then the probability of the error $(x_i - \underline{x})$ will be $e^{-w_i(x_i-\underline{x})^2/2\sigma^2}$. For the whole set of n values $x_1, x_2 \ldots x_i \ldots x_n$ the probability will be the sum of all these weighted values, namely,

$$e^{-\Sigma w_i(x_i-m)^2/2\sigma^2}$$

Example 15.3 Example of a Weighted Mean

A distance is measured by three different methods and the results are estimated as:

a 19.412 ± 0.005 b 19.417 ± 0.008 c 19.419 ± 0.010

The estimated standard deviation for observation a was 0.005 and the variance was 0.000025 and its inverse is 40,000 = w_a. For b the inverse = w_b = 15,625 and for c = w_c = 10,000.

Allocating weights according to these inverse variances and reducing the calculations by starting at 19.410, so that

$$a = \{19.410\} + 0.002 \pm 0.005, b = \{19.410\} + 0.007 \pm 0.008,$$
$$c = \{19.410\} + 0.009 \pm 0.010$$

The weighted mean is

$$\{19.410\} + \frac{(0.002 * w_a + 0.007 * w_b + 0.009 * w_c)}{(w_a + w_b + w_c)}$$
$$= \{19.410\} + (80 + 109.375 + 90) / (65625) = \{19.410\} + 0.004 = 19.414$$

The new residuals are −0.002, 0.003, and 0.005 while $(n - 1) = 2$. Hence,

$$\sigma^2 = \frac{1}{n-1}\sum_{i=1}^{i=n} w_i (x_i - \mu)^2$$
$$= (1/2)\frac{(0.002^2 * w_a + 0.003^2 * w_b + 0.005^2 * w_c)}{(w_a + w_b + w_c)}$$

This gives $\sigma = 0.002$. Thus, a better estimate is that the distance = 19.414 ± 0.002.

Since we may not know the true population mean, we can replace $\underline{x}$ by m. The quantity $e^{-\sum w_i(x_i - m)^2/2\sigma^2}$ will be a maximum (to give the maximum probability) when we choose m so that $\sum w_i(x_i - m)^2$ is a minimum. Differentiating with respect to m, we obtain $-2\sum w_i(x_i - m)$ and this must equal zero for a minimum value. Hence,

$$2\sum w_i x_i - 2m\sum w_i = 0 \quad \text{or} \quad m = \sum w_i x_i \,/ \sum w_i$$

Let m have the value μ as the weighted mean of x, namely, $\mu = \sum w_i x_i / \sum w_i$. This is the most probable mean value for observations, assuming that all the weights are known. Note that if all observations are of the same weight, $\sum w_i = n$ and the mean $= \sum x_i / n$.

For weighted observations the variance becomes $\sigma^2 = \frac{1}{n}\sum_{i=1}^{i=n} w_i (x_i - \mu)^2$ or, more strictly speaking for a smaller data set, $\frac{1}{n-1}\sum_{i=1}^{i=n} w_i (x_i - \mu)^2$.

15.4 SPATIAL AUTOCORRELATION

The assumption behind regression is that each error term or residual is a random number that is independent of neighboring errors or residuals. The only assumption is that the sum of the squares of the residuals should be a minimum with the errors, weighted if necessary, forming part of a normal distribution. We illustrated a linear relationship between two data sets x and y in Example 15.1 where we derived the best-fit line assuming that, although the values of x and y were correlated, the residuals were not. If successive observations do not appear to be independent, then either we have fitted the wrong type of curve or there is some sequential connection between the residuals, in which case there is said to be *autocorrelation*. The former will happen if, for instance, we try to fit a straight-line $y = mx + c$ to a relationship of the form $y = ax^2 + c$ (see Figure 15.5, which shows the apparent correlation between residuals resulting from fitting a straight line to part of an elliptical

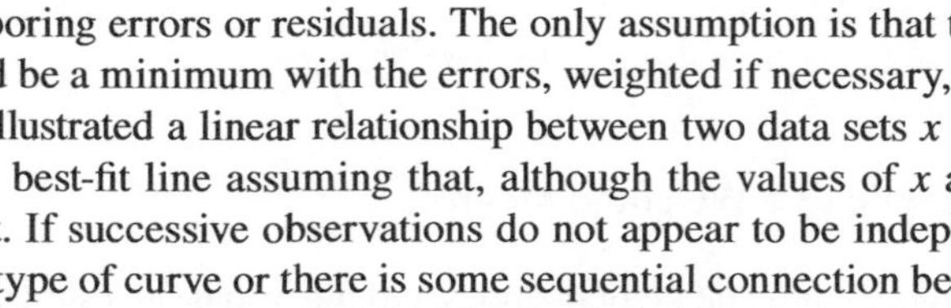

FIGURE 15.5 Residuals from fitting a straight line to an elliptical curve.

curve). If the relationship is correctly modeled, then autocorrelation will occur when the residuals (as distinct from the original observations) have a correlation that is not zero. This means that they are not independent. This will be shown when the covariance is not (approximately) zero, the condition being known as *autocovariance*. To be able to draw valid conclusions from many forms of statistical analysis, the residuals must be independent.

In the study of time series, the question can arise as to what are the chances of an event affecting a subsequent event, for example, of it being sunny tomorrow or the day after if it is sunny today. Consider two sets of observations x_s and x_t taken at time s and time t. Let their means be $E(x_s) = \underline{x_s}$ and $E(x_t) = \underline{x_t}$, where E means the "expected value" of x, which in this case is the mean value.

Similarly, the expected value of the variance $\sigma^2 = E[(x - E(x))^2] = E(x^2) - [E(x)]^2$. The autocorrelation between the two sets x_s and x_t is expressed as

$$r_{s,t} = \frac{E\left[(x_s - \bar{x}_s)(x_t - \bar{x}_t)\right]}{\sigma_s \sigma_t} = \frac{\sum_{i=1}^{n}(x_{s,i} - \bar{x}_s)(x_{t,i} - \bar{x}_t)}{(n-1)\sigma_s \sigma_t} = \frac{\sum_{i=1}^{n}(x_{s,i} - \bar{x}_s)(x_{t,i} - \bar{x}_t)}{\sqrt{\sum_{i=1}^{n}(x_{s,i} - \bar{x}_s)^2 \sum_{i=1}^{n}(x_{t,i} - \bar{x}_t)^2}}$$

If we combine the two sets into one, then $r_{s,t} = \frac{E[(x_s - \underline{x}_s)(x_t - \underline{x}_t)]}{\sigma^2}$ with, for instance, $t = s + 1$ or $s + 2$ for a one-day or two-day delay.

As we have seen in Section 15.1, the value r is often referred to as the *Pearson correlation coefficient* or more simply as the population correlation coefficient. This relationship is based on different values of x at different instances, with time here being taken as a linear or one-dimensional function.

In much geographical analysis, there is an assumption that neighbors influence what is going on around them and this means at different points in two- or three-dimensional space, as well as at different points in time. Geographers are interested in whether there is any apparent spatial correlation between quantities measured at different points. If there is, there is said to be *spatial autocorrelation*. This correlation can arise in three-dimensional space but here we will only consider two dimensions x and y.

A traditional test for spatial autocorrelation is known as *Moran's statistic I*, where

$$I = N \frac{\sum_{i=1}^{i=n}\sum_{j=1}^{j=n} W_{i,j}(x_i - \bar{x})(x_j - \bar{x})}{\left(\sum_{i=1}^{i=n}\sum_{j=1}^{j=n} W_{i,j}\right)\sum_{i=1}^{i=n}(x_i - \bar{x})^2} = \frac{N}{S} * \frac{\sum_{i=1}^{i=n}\sum_{j=1}^{j=n} W_{i,j}(x_i - \bar{x})(x_j - \bar{x})}{\sum_{i=1}^{i=n}(x_i - \bar{x})^2}$$

In this equation, N is the number of cases; x_i is the value of the variable at a particular location I, and x_j is its value at another location j; $\underline{x}$ is the mean of the N values of the variable x. W_{ij} is a weight matrix applied to the comparison between location i and location j and $S = \left(\sum_{i=1}^{i=n}\sum_{j=1}^{j=n} W_{i,j}\right)$ is the sum of all the weights so that the overall probability can remain 1.

W_{ij} is a contiguity matrix that in the simple case of a rectangular grid would be made up of 0s and 1s; if sample j is in a polygon next to sample i the weight is 1 while if it does not share a boundary with i, then the weight is zero (see Figure 15.6). In a more complex case the weight could be based on the distance between the sample points i and j—for instance, the inverse $(1 / d_{ij})$ or even the inverse square $\left(1 / d_{ij}^2\right)$.

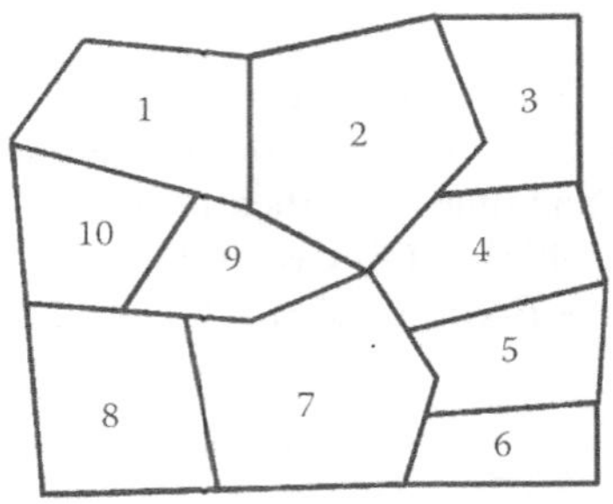

Area	1	2	3	4	5	6	7	8	9	10
1	0	1	0	0	0	0	0	0	1	1
2	1	0	1	1	0	0	0	0	1	0
3	0	1	0	1	0	0	0	0	0	0
4	0	1	1	0	1	0	1	0	0	0
5	0	0	0	1	0	1	1	0	0	0
6	0	0	0	0	1	0	1	0	0	0
7	0	0	0	1	1	1	0	1	1	0
8	0	0	0	0	0	0	1	0	1	1
9	1	1	0	0	0	0	1	1	0	1
10	1	0	0	0	0	0	0	1	1	0

FIGURE 15.6 A set of polygons and the contiguity matrix.

With a regular grid it could be based on only those squares north, south, east, or west of the chosen point (known from the game of chess as the *Rook's Case*), those on the diagonals (the *Bishop's Case*) or a combination of the two (the *Queen's* or *King's Case*)—see Figure 15.7.

Although we do not prove it here, if E(I) is the expected value of the Moran statistic I then, if there is no autocorrelation,

$$\mathrm{E}(I) = -1/(N-1)$$

If $I > \mathrm{E}(I)$ there is positive spatial autocorrelation; if $I < \mathrm{E}(I)$ there is negative spatial autocorrelation. If the value of N is large, then effectively $-1/(N-1)$ can be treated as zero.

The z statistic for the distribution of I is given by

$$\mathrm{Z}(I) = \{I - \mathrm{E}(I)\}/\sigma_{\mathrm{I}}$$

where σ_{I} is the standard deviation of I. The derivation of σ_{I} is complicated and we will not show it here. Various approximations appear in the literature. In one version, it can be shown to be:

$$\sigma_{\mathrm{I}}^2 = \text{variance of } I =$$

$$\frac{\left\{n\left[\left(n^2-3n+3\right)S_1 - nS_2 + 3S_0^2\right]\right\} - \left\{nk\left[\left(n^2-n\right)S_1 - 2nS_2 + 6S_0^2\right]\right\}}{(n-1)(n-2)(n-3)S_0^2} - \frac{1}{(n-1)^2}$$

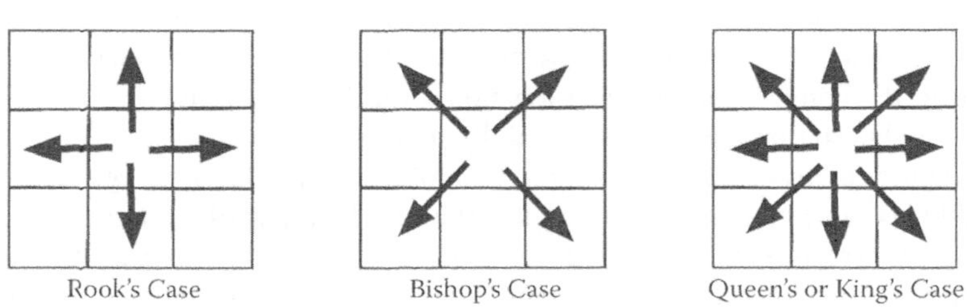

FIGURE 15.7 Contiguity cases.

where n = number of observations, $S_0 = \sum_{i=1}^{i=n} \sum_{j=1}^{j=n} W_{i,j}$ = the sum of the elements in the spatial weight matrix

$$S_1 = \frac{\sum_{i=1}^{i=n}\sum_{j=1}^{j=n}\left(W_{i,j} + W_{j,i}\right)^2}{2}$$

If the weight matrix is symmetric,

$$S_1 = 2\sum_{i=1}^{i=n}\sum_{j=1}^{j=n}\left(W_{i,j}\right)^2$$

then

$$S_2 = \sum_{i=1}^{i=n}\left(W_{i,m} + W_{m,i}\right)^2$$

W_{im} is the sum of ith column of the weight matrix $= \sum_{j=1}^{j=n} W_{ij}; W_{mi}$ is the sum of the ith row $= \sum_{j=1}^{j=n} W_{ji}$. If the matrix is symmetric, then

$$S_2 = 4\sum_{i=1}^{i=n}\left(\sum_{j=1}^{j=n} W_{i,j}\right)^2$$

Finally,

$$k = \frac{\sum_{i=1}^{n}\left(x_i - \bar{x}\right)^4}{\left(\sum_{i=1}^{n}\left(x_i - \bar{x}\right)^2\right)^2}$$

This is based on the sum of each value in the data matrix minus the mean value.

By calculating the variance, the standard deviation of I can be derived. Z(I) or $\{I - E(I)\} / \sigma_I$ then follows the normal distribution and its significance can therefore be tested using standard statistical tables.

An alternative measure of spatial autocorrelation is known as *Geary's C statistic*. It is calculated from

$$C = \frac{(n-1)}{2} * \frac{\sum_{i=1}^{i=n}\sum_{j=1}^{j=n} W_{i,j}\left(x_i - x_j\right)^2}{\left(\sum_{i=1}^{i=n}\sum_{j=1}^{j=n} W_{i,j}\right)\sum_{i=1}^{i=n}\left(x_i - \bar{x}\right)^2}$$

The theoretical mean for $C = 1$ and the value of C will lie between 0 and 2. If its value is 1 then there is no spatial autocorrelation. If $0 < C < 1$ there is positive correlation while if $1 < C < 2$, there will be negative correlation.

The two statistics—Geary's C and Moran's I—are similar and inversely related but are not the same. Moran's I depends on the difference between each value and the mean for the data set and is similar to the Pearson correlation coefficient and is a covariance test. Geary's C is more concerned with neighboring values and is in essence a variance test.

15.5 LEAST SQUARE SOLUTIONS

Returning to the ideas behind regression we have seen that, given a set of observations in which each observation or measurement may contain a small amount of error, we can obtain the most likely value of a quantity by distributing the residuals on the basis of minimum variance, which means that the sum of the squares of the residuals is a minimum. This is called the principle of *least squares*. Although the process of calculating the minimum sum of the squares can be applied to any set of residuals, the interpretation of the results assumes that the errors are normally distributed and of equal weight or that appropriate weights have been applied.

The weighted mean gives the overall most likely value of any data set while the probable errors are, in accordance with the normal distribution, of the form $e^{-w_i(x_i-\underline{x})^2/2\sigma^2}$. The maximum probability will occur when this is a maximum, which will happen when $\Sigma w_i(x_i - \underline{x})^2$ is a minimum.

We can extend this idea not just to fitting lines or surfaces, but when dealing with a set of observations, various combinations of which are subject to a set of conditions that can be determined. In the general case, if we have n observed quantities x_1, x_2, x_3,. x_n that are linearly related and m independent relationships, we can express this in the form

$$ax_1 + bx_2 + cx_3 + \cdots + nx_n + l = 0$$

where l is a numerical value. We can write such relationships in longhand as shown in Table 15.2. Alternatively, we can use the matrix form $\mathbf{MX} + \mathbf{L} = 0$ where $\mathbf{M}$ is the $(m * n)$ matrix made up from the *a*s, *b*s, and *c*s, and so forth, $\mathbf{X}$ is the column vector of unknowns, and $\mathbf{L}$ is the column vector of constants. In geomatics, the observed quantities are normally angles, distances, or time; but the technique can be used in all sorts of applications from optimizing financial balance sheets through to quality control procedures. The key point is that linear relationships can be established between the independent variables; where this is not immediately so, the relationships will need to be "linearized," an example of which is discussed later in the case of a set of angles. If $m = n$, that is, if we have the same number of independent equations as there are unknown, then we can solve for all the values of x_i precisely. If $m \neq n$, then we have to establish the best-fit solution.

Assuming that the observations are of equal weight, and that there are small errors in each of the observations x_i, then for equation j in Table 15.2, let

$$v_j = a_j x_1 + b_j x_2 + c_j x_3 + \cdots + n_j x_n + l_j$$

where j takes the values 1, 2,... m. v_j is known as the residual value. If we assume that we want to minimize the errors in each of the m equations, then we have to make the sum of the squares of the quantities v_j as small as possible by suitably amending each of the values of x_i. This will give us the most likely consistent solution.

v_j has $(n + 1)$ terms so that v_j^2 would have $(n + 1)^2$ terms and the sum of all these, Σv_j^2, would have $m * (n + 1)^2$ terms. However, we can simplify matters.

TABLE 15.2
Conditions to Be Satisfied: MX + L = 0

1:	$a_1x_1 + b_1x_2 + c_1x_3 + \cdots f_1x_i + \cdots + n_1x_n + l_1 = 0$
2:	$a_2x_1 + b_2x_2 + c_2x_3 + \cdots f_2x_i + \cdots + n_2x_n + l_2 = 0$
⋮	
j:	$a_jx_1 + b_jx_2 + c_jx_3 + \cdots f_jx_j + \cdots + n_jx_n + l_j = 0$
⋮	
m:	$a_mx_1 + b_mx_2 + c_mx_3 + \cdots f_mx_i \cdots + n_mx_n + l_m = 0$

To minimize $\sum v^2_j$ for x_1, we need to make the partial derivative $\partial(\sum v^2_j)/\partial x_1 = 0$. To minimize $\sum v^2_j$ for x_2, we need to make $\partial(\sum v^2_j)/\partial x_2 = 0$, and so forth.

$$\frac{\partial(\Sigma v_j^2)}{\partial x_1} = \frac{\partial(v_1^2 + v_2^2 + v_3^2 + \ldots)}{\partial x_1} = 2\left\{v_1 \frac{\partial v_1}{\partial x_1} + v_2 \frac{\partial v_2}{\partial x_1} + \cdots + v_m \frac{\partial v_m}{\partial x_1}\right\}$$
$$= 0 \text{ when things reach a minimum.}$$

$$\frac{\partial v_1}{\partial x_1} = a_1, \text{ and so on, hence,}$$

$$v_1 \frac{\partial v_1}{\partial x_1} = a_1\{a_1x_1 + b_1x_2 + c_1x_3 + \cdots f_1x_i + \cdots + n_1x_n + l_1\}$$
$$v_2 \frac{\partial v_2}{\partial x_1} = a_2\{a_2x_1 + b_2x_2 + c_2x_3 + \cdots f_2x_i + \cdots + n_2x_n + l_2\}, \text{ and so on,}$$

for all the m equations. If we add all these together, we obtain for x_1

$$\frac{\partial(\Sigma v_j^2)}{\partial x_1} = 2\{(\Sigma a_j^2)x_1 + (\Sigma a_jb_j)x_2 + \cdots(\Sigma a_jf_j)x_i + \cdots + (\Sigma a_jn_j)x_n + (\Sigma a_jl_j)\} = 0$$

for minimum variation and therefore maximum probability. ($\sum a_jf_j$ means "add all the m values $a_1f_1 + a_2f_2 + \cdots + a_mf_m$.") Similarly,

$$\frac{\partial(\Sigma v_j^2)}{\partial x_2} = 2\{(\Sigma a_jb_j)x_1 + (\Sigma b_j^2)x_2 + \cdots(\Sigma b_jf_j)x_i + \cdots + (\Sigma b_jn_j)x_n + (\Sigma b_jl_j)\} = 0$$

This gives us n equations for the n unknown adjusted values of x.

If we write (Σa_1^2) as $[aa]$, (Σa_jb_j) as $[ab]$, and so forth, for the sake of simplicity, then we can express these as in Table 15.3. Thus,

$$[aa] = a_1^2 + a_2^2 + a_3^2 + \cdots + a_m^2$$
$$[ab] = a_1b_1 + a_2b_2 + a_3b_3 + \cdots + a_mb_m, \text{and so forth.}$$

The equations in Table 15.3 are known as the *normal equations*. Note that in these n equations, the coefficients for x are symmetrical about the diagonal, which itself is made up from the sum of the squares of the relevant coefficients.

TABLE 15.3
The Normal Equations: $M^T(MX + L) = 0$

$[aa]x_1 + [ab]x_2 + [ac]x_3 + \cdots + [af]x_i + \cdots + [an]x_n + [al] = 0$
$[ab]x_1 + [bb]x_2 + [bc]x_3 + \cdots + [bf]x_i + \cdots + [bn]x_n + [bl] = 0$
$[ac]x_1 + [bc]x_2 + [cc]x_3 + \cdots + [cf]x_i + \cdots + [cn]x_n + [cl] = 0$
$\vdots$
$[an]x_1 + [bn]x_2 + [cn]x_3 + \cdots + [fn]x_i + \cdots + [nn]x_n + [nl] = 0$

We wrote the condition equations in Table 15.2 as $\mathbf{MX} + \mathbf{L} = 0$, where

$$\mathbf{X} = \begin{pmatrix} x_1 \\ x_2 \\ x_3 \\ \dots \\ x_n \end{pmatrix}; \quad \mathbf{L} = \begin{pmatrix} l_1 \\ l_2 \\ l_3 \\ \dots \\ l_m \end{pmatrix}$$

and

$$\mathbf{M} = \begin{pmatrix} a_1 & b_1 & c_1 & \dots\cdot & n_1 \\ a_2 & b_2 & c_2 & \dots\cdot & n_2 \\ a_3 & b_3 & c_3 & \dots\cdot & n_3 \\ \dots & \dots & \dots & \dots\cdot & \dots \\ a_m & b_m & c_m & \dots\cdot & n_m \end{pmatrix}$$

Note that

$$\mathbf{M}^\mathbf{T} = \begin{pmatrix} a_1 & a_2 & a_3 & \dots & a_m \\ b_1 & b_2 & b_3 & \dots & b_m \\ c_1 & c_2 & c_3 & \dots & c_m \\ \dots & \dots & \dots & \dots & \dots \\ n_1 & n_2 & n_3 & \dots & n_m \end{pmatrix}$$

so that

$$\mathbf{M}^\mathbf{T}\mathbf{M} = \begin{pmatrix} [aa] & [ab] & [ac] & \dots & [an] \\ [ab] & [bb] & [bc] & \dots & [bn] \\ [ac] & [bc] & [cc] & \dots & [cn] \\ \dots & \dots\cdot & \dots & \dots & \dots\cdot \\ [an] & [bn] & [cn] & \dots & [nn] \end{pmatrix}$$

or

$$\mathbf{M}^\mathbf{T}\mathbf{M} = \begin{pmatrix} a_1a_1 + a_2a_2 + a_3a_3 + \cdots + a_ma_m & a_1b_1 + a_2b_2 + a_3b_3 + \cdots + a_mb_m \ldots\ldots a_1n_1 + a_2n_2 + a_3n_3 + \cdots + a_mn_m \\ b_1a_1 + b_2a_2 + b_3a_3 + \cdots + b_ma_m & b_1b_1 + b_2b_2 + c_2c_2 + \cdots + b_mb_m \ldots\ldots b_1n_1 + b_2n_2 + c_3n_3 + \cdots + b_mn_m \\ \ldots\ldots & \\ n_1a_1 + n_2a_2 + n_3a_3 + \cdots + n_ma_m & n_1b_1 + n_2b_2 + n_3b_3 + \cdots + n_mb_m \ldots\ldots n_1n_1 + n_2n_2 + n_3n_3 + \cdots + n_mn_m \end{pmatrix}$$

and

$$\mathbf{MM}^\mathbf{T} = \begin{pmatrix} a_1a_1 + b_1b_1 + c_1c_1 + \cdots + n_1n_1 & a_1a_2 + b_1b_2 + c_1c_2 + \cdots + n_1n_2 & \ldots\ldots & a_1a_m + b_1b_m + c_1c_m + \cdots + n_1n_m \\ a_2a_1 + b_2b_1 + c_2c_1 + \cdots + n_2n_1 & a_2a_2 + b_2b_2 + c_2c_2 + \cdots + n_2n_2 & \ldots\ldots & a_2a_m + b_2b_m + c_2c_m + \cdots + n_2n_m \\ \ldots\ldots & & & \\ a_ma_1 + b_mb_1 + c_mc_1 + \cdots + n_mn_1 & a_ma_2 + b_mb_2 + c_mc_2 + \cdots + n_mn_2 & \ldots\ldots & a_m\mathrm{a}_m + b_mb_m + c_mc_m + \cdots + n_mn_m \end{pmatrix}$$

TABLE 15.4
Weighted Normal Equations: $\mathbf{M^TWMX + M^TWL = 0}$

$[waa]x_1 + [wab]x_2 + \cdots + [waf]x_i + \cdots + [wan]x_n + [wal] = 0$

$[wab]x_1 + [wbb]x_2 + \cdots + [wbf]x_i + \cdots + [wbn]x_n + [wbl] = 0$

$[wac]x_1 + [wbc]x_2 + \cdots + [wcf]x_i + \cdots + [wcn]x_n + [wcl] = 0$

$\vdots$

$[wan]x_1 + [wbn]x_2 + \cdots + [wfn]x_i + \cdots + [wnn]x_n + [wnl] = 0$

Also,

$$\mathbf{M^TL} = \begin{pmatrix} a_1 & a_2 & a_3 & \ldots & a_m \\ b_1 & b_2 & b_3 & \ldots & b_m \\ c_1 & c_2 & c_3 & \ldots & c_m \\ \ldots & \ldots & \ldots & \ldots & \ldots \\ n_1 & n_2 & n_3 & \ldots & n_m \end{pmatrix} * \begin{pmatrix} l_1 \\ l_2 \\ l_3 \\ \ldots \\ l_m \end{pmatrix} = \begin{pmatrix} [al] \\ [bl] \\ [cl] \\ \ldots \\ [nl] \end{pmatrix}$$

$\mathbf{M^T}$ has n rows and m columns, $\mathbf{M}$ has m rows and n columns so that $\mathbf{M^TM}$ is an $n * n$ matrix and $\mathbf{MM^T}$ is $m * m$. Thus, Table 15.3 can be written in shorthand as $\mathbf{M^T(MX + L)} = 0$.

Once we have formulated the necessary conditions, it becomes a matter of routine to form and transform the relevant matrices and solve for the n unknowns. An example of how this can be used is given in Example 15.4.

If the m equations are of unequal weight, then instead of minimizing $\sum v_j^2$, we need to minimize $\sum w_j v_j^2$ and the normal equations become as shown in Table 15.4 where $\mathbf{W}$ is the weight matrix. Assuming that there is no correlation between the residuals, the variance / covariance matrix will reduce to a diagonal matrix as will the weight matrix $\mathbf{W}$.

In Example 15.4, we have $m = 5$ conditions affecting $n = 3$ unknowns and we reduced these to three independent equations linking these three unknowns. We may, however, have more unknowns than conditions, that is, $m < n$. This will arise when we make redundant observations, for instance, when we measure the three angles of a triangle where the only condition is that the sum of the three observations must add up to 180°. Here we will have $x_1 + x_2 + x_3 + r = 0$ where x_1, x_2, and x_3 are the corrections to be applied to each angle and r is the amount by which our observed angles fail to add to 180°. In this case, we have:

$$\mathbf{M} = (1,1,1), \quad \mathbf{M^TM} = \begin{pmatrix} 1 & 1 & 1 \\ 1 & 1 & 1 \\ 1 & 1 & 1 \end{pmatrix}$$

and hence the three equations "$x_1 + x_2 + x_3 + r = 0$" that result from $\mathbf{M^TMX + M^TL}$ are identical and therefore offer no solution. We must therefore take a different approach, as outlined in Section 16.2.

Example 15.4 Solving for More Equations Than There Are Unknowns

Let

$$x + y + z - 9.2 = 0$$

$$x + y - z - 0.9 = 0$$

$$2x + 3y + 2z - 20.6 = 0$$

$$3x - 4y + 5z - 14.1 = 0$$

We have three values (x, y, z) with five equations, assumed to be of equal weight. In matrix form, **MX** + **L** = 0, where

$$\mathbf{M}=\begin{pmatrix}1 & 1 & 1\\ 1 & 1 & -1\\ 2 & 3 & 2\\ 3 & -4 & 5\\ 1 & 5 & -3\end{pmatrix}\quad \mathbf{X}=\begin{pmatrix}x\\ y\\ z\end{pmatrix}\quad \mathbf{L}=\begin{pmatrix}-9.2\\ -0.9\\ -20.6\\ -14.1\\ -4.9\end{pmatrix}$$

$$\mathbf{M}^{\mathrm{T}}\mathbf{M}=\begin{pmatrix}1 & 1 & 2 & 3 & 1\\ 1 & 1 & 3 & -4 & 5\\ 1 & -1 & 2 & 5 & -3\end{pmatrix}*\begin{pmatrix}1 & 1 & 1\\ 1 & 1 & -1\\ 2 & 3 & 2\\ 3 & -4 & 5\\ 1 & 5 & -3\end{pmatrix}=\begin{pmatrix}16 & 1 & 16\\ 1 & 52 & -29\\ 16 & -29 & 40\end{pmatrix}$$

$$\mathbf{M}^{\mathrm{T}}\mathbf{L}=\begin{pmatrix}1 & 1 & 2 & 3 & 1\\ 1 & 1 & 3 & -4 & 5\\ 1 & -1 & 2 & 5 & -3\end{pmatrix}*\begin{pmatrix}-9.2\\ -0.9\\ -20.6\\ -14.1\\ -4.9\end{pmatrix}=\begin{pmatrix}-98.5\\ -40\\ -105.3\end{pmatrix}$$

Thus, $\mathbf{M}^{\mathrm{T}}(\mathbf{MX}+\mathbf{L})$ or $\mathbf{M}^{\mathrm{T}}\mathbf{MX}+\mathbf{M}^{\mathrm{T}}\mathbf{L}$ gives three equations, namely,

$$16x+1y+16z-98.5=0$$
$$1x+52y-29z-40=0$$
$$16x-29y+40z-105.3=0$$

from which

$$x=2.02,\ y=2.93,\ z=3.95$$

The residuals are −0.3, 0.1, 0.13, −0.01, and −0.08. Their sum of squares = 0.1234.

GLOSSARY

Autocorrelation A condition occurring when successive items in a sequence are correlated.

Autocovariance The condition occurring when adjacent observations in a sequence are not independent.

Coefficient of correlation The extent of correspondence between two variables, expressed by the coefficient r in the form

$$r=\frac{n\sum x_i y_i-\sum x_i\sum y_i}{\sqrt{\left\{n\sum x_i^2-\left(\sum x_i\right)^2\right\}}\sqrt{\left\{n\sum y_i^2-\left(\sum y_i\right)^2\right\}}}$$

Covariance A measure of the association between two variables X and Y

$$\mathrm{Cov}(X,Y)=(1/n)\sum_{i=1}^{i=n}\left(x_i-\bar{x}\right)\left(y_i-\bar{y}\right)$$

Dispersion matrix Also known as the *variance–covariance matrix*. If **V** is a row matrix of all the error terms and its transpose is $\mathbf{V}^T$ then the dispersion matrix = $\mathbf{V}^T\mathbf{V}$.
Error term The difference between the observed and computed quantity.
Expectation E A term for the probability that an event will occur.
Geary's statistic C A measure of spatial autocorrelation in which

$$C = \frac{(n-1)}{2} * \frac{\sum_{i=1}^{i=n}\sum_{j=1}^{j=n} W_{i,j}\left(x_i - x_j\right)^2}{\left(\sum_{i=1}^{i=n}\sum_{j=1}^{j=n} W_{i,j}\right)\sum_{i=1}^{i=n}\left(x_i - \bar{x}\right)^2}$$

Geometric mean The nth root of the product of n numbers, for example, for three numbers a, b, c, it would be the cube root of $a * b * c$.
Least squares A solution that makes the sum of the squares of the *residuals* a minimum.
Moran's statistic I A measure of spatial autocorrelation in which

$$I = N\frac{\sum_{i=1}^{i=n}\sum_{j=1}^{j=n} W_{i,j}\left(x_i - \bar{x}\right)\left(x_j - \bar{x}\right)}{\left(\sum_{i=1}^{i=n}\sum_{j=1}^{j=n} W_{i,j}\right)\sum_{i=1}^{i=n}\left(x_i - \bar{x}\right)^2}$$

$$= \frac{N}{S} * \frac{\sum_{i=1}^{i=n}\sum_{j=1}^{j=n} W_{i,j}\left(x_i - \bar{x}\right)\left(x_j - \bar{x}\right)}{\sum_{i=1}^{i=n}\left(x_i - \bar{x}\right)^2}$$

Normal equations The equations that, when solved, give the best estimate of a set of values by the principle of *least squares*. They take the form $\mathbf{M}^T(\mathbf{MX} + \mathbf{L}) = 0$ where $\mathbf{MX} + \mathbf{L} = 0$ represents the *condition equations*.
Regression A measure of association between a dependent variable and one or more independent variables.
Residual The difference between an observed and a predicted value. Also referred to as the *error term*.
Simple linear regression A linear measure of association between one dependent and one independent variable.
Spatial autocorrelation The property that mapped data possess when because of their location they are not independent.
Trend surface A form of regression in which the third dimension Z is dependent on X and Y.
Variance–covariance matrix Another name for the *dispersion* matrix.
Weight A relative value of the importance of an observation.

16 Statistical Modeling and Machine Learning

16.1 MACHINE LEARNING

Machine Learning is a common term that refers to the work of drawing patterns and making predictions using data through computer algorithms. Machine Learning belongs to the broader scope of Artificial Intelligence. Many tasks of statistical data analysis can be efficiently performed with machine learning and *deep learning* algorithms. These techniques became popular as more powerful computing devices became available. As the name "machine learning" implies, a machine learning model is created through the process of learning from the input data. Goals of machine learning algorithms range from classification, prediction, exploration to inference, pattern recognition, optimization, and others, where the main idea is to train the model by showing it examples.

In previous chapters, we have discussed various statistical methods for drawing patterns and making predictions, such as regression. Regression and classification are two methods of *Supervised Learning*. Supervised learning refers to the process where the data to be learned by the machine, called *training data*, have well-defined output values, much like traditional statistical data. The machine learning algorithm is tasked with trying to predict the output values given the input data. The following are popular methods of supervised learning:

1. Linear Regression
2. Logistic Regression
3. Decision Tree
4. Random Forest
5. Naive Bayes
6. Neural Networks

Now let us consider how regression or classification problems are solved with machine learning. Suppose we have data about people's height, and we want to separate them into two classes. We can examine the data and set a threshold, T, so that people taller than T become members of one class, and all other people members of the other:

$$C = \begin{cases} c_1, & h > T \\ c_2, & h \le T \end{cases}$$

However, this requires manual labor. Setting a threshold might be a challenging task which calls for a field expert and a considerable knowledge about the area of research. It would be more convenient that the system examined the data and itself decided on an appropriate threshold. We would then only specify how many classes we want to have. Problems like this can be solved by a *clustering algorithm* (Figure 16.1).

In our example of people's heights, we prepare a set of data which becomes the training set, in which we associate the desired category for each height value. This dataset is fed to a machine learning system where the algorithm establishes a dynamic threshold T, also called a "gate" in machine learning terminology. Each input point in compared with the gate T. If the person's height

DOI: 10.1201/9781003464402-21

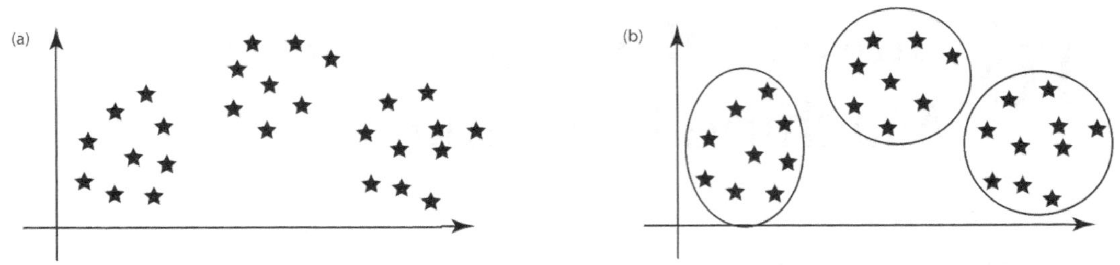

FIGURE 16.1 Goal of a clustering algorithm. (a) Samples and (b) samples categorized into clusters.

h is larger than T, and this input point is labeled, say "short", then the gate is adjusted by increasing it just enough to let this person pass under it. When a person labeled "tall" can pass under the gate, the gate is lowered so that this input point must be above it. This process repeats for each input height h_i, and in the end the gate will be set at the position so that all data points of one class will be less than the gate value, T, and the rest are greater. Therefore the machine learned its parameters (the height of the gate, T) from the input data, and from then on it is capable of classifying data points that were not part of the training set.

To judge performance of a machine learning model we need to create a metric, also known as an error function. During the training process we have the designated output D_i coupled with each data point in the training set. Therefore our error function should compare the actual output, C_i from the model with the desired output D_i. The error can be calculated in many suitable ways, for example, as an absolute value of the difference or as a square of the difference between these two values:

$$E = \left(C_i - D_i\right)^2$$

During the learning process the machine accumulates all training data points. Then during classification of an unseen data point the machine finds a training point that is the most similar to it and outputs its class. The training process is akin to finding an approximation to the ideal *transfer function* of the machine learning system. It is desirable that the machine learning system is able to learn from a limited training dataset and generalize the acquired behavior to any other data point it might receive in the future. If the system is making mistakes while trying to generalize, we call them mistakes of generalization or *generalization error. Overfitting* is one such generalization error, it occurs when the machine's output C_i matches the input data points suspiciously well, but does not perform well with additional data points.

A machine learning model that separates input data points into two classes is often called a binary classifier. The idea of a binary classifier can be extended to a classifier for any number of classes. Often such classifiers are organized as a tree structure of a graph, and they are called *decision trees*. Classification starts at the root node and proceeds to the internal nodes and then terminates at a leaf node. One application of such machine learning systems is data discovery and data mining, where, as new features are detected in the input data, a new leaf it added to the tree. Further development of tree models leads to the idea of running several models at the same time, or running a randomly selected classifier. Such algorithms are often called a *random forest*, reminiscent to having multiple trees.

So far, we have been discussing learning machines that are trained to recognize, classify, compare objects and infer information from previously unseen data. These learning machines can be characterized as *discriminative*. A different class of machine learning algorithms constructs stochastic models to generate data, and this class of machines is often called *generative*. Generative machine learning can generate data points that resemble observations, which will allow us to model the phenomenon that we study. To achieve this, the machine has to learn the probability distribution from the training dataset. The idea is to use the training dataset and create a joint probability distribution over the set of input data points and their labels or values. This probability distribution describes relationships between data points and their values, therefore it allows us to make predictions, infer trends and discover information that was not previously known.

Although discriminative and generative machine learning algorithms operate on different principles, there exist successful attempts to combine both ideas into a hybrid learning machine. For instance, document classification or language models consist of multiple modules to perform a specific task. The field of machine learning is rapidly evolving and more and more abstract ideas from pure mathematics are bridged to practical applications, and often serve as a stimulus to spawn new directions of innovation.

16.2 SURVEY ADJUSTMENTS

In much of geomatics, especially in position fixing, we deliberately make redundant observations in order to improve the accuracy of our work, both by identifying and thereby being able to eliminate any gross errors in measurement, but more particularly, by reducing the overall standard deviations of the residual errors in our measurements. This increases the probability of obtaining a better answer.

Consider then the case where we have more observations than there are conditions to satisfy. We must satisfy the basic condition while also ensuring that $\sum w_i\varepsilon_j^{\,2}$ is a minimum where w_i is the weight of each observation and ε_i is the correction to the observed quantity (o_i). i is an observation that takes the value from one to n where n is the total number of observed quantities.

If the best answer for each observation is x_i where $x_i = o_i + \varepsilon_i$, then for $\sum w_i\varepsilon_j^{\,2}$ to be a minimum, the partial differential with respect to ε must be zero. Thus,

$$2w_1\varepsilon_1\delta\varepsilon_1 + 2w_2\varepsilon_2\delta\varepsilon_2 + \cdots + 2w_n\varepsilon_n\delta\varepsilon_n = 0$$

or

$$w_1\varepsilon_1\delta\varepsilon_1 + w_2\varepsilon_2\delta\varepsilon_2 + \cdots + w_n\varepsilon_n\delta\varepsilon_n = 0$$

The weights can be expressed as a diagonal matrix **W** in the form

$$\begin{pmatrix} w1\ 0\ 0\ldots\ .0 \\ 0\ w_2\ 0\ldots\ .0 \\ \ldots\ \ldots\ \ldots\ .\ldots\ .0 \\ 0\ 0\ 0\ldots\ .w_n \end{pmatrix}$$

or, as shown at the end of Section 15.1 in Chapter 15,

$$\mathbf{W} = \begin{pmatrix} 1/\sigma_1^2 & 0 & 0 & \ldots\ldots & 0 \\ 0 & 1/\sigma_2^2 & 0 & \ldots. & 0 \\ 0 & 0 & 1/\sigma_3^2 & \ldots & 0 \\ \ldots. & & & & \\ 0 & 0 & 0 & \ldots & 1/\sigma_n^2 \end{pmatrix}$$

In the general case, we have n observations of x and m conditions to satisfy a set of equations as shown in Table 16.1. Here, $x_i = o_i + \varepsilon_i$ where x is the adjusted value and o the observed value, such

TABLE 16.1
Observations and Conditions: M(O + ε) + L = 0

$$a_1(o_1+\varepsilon_1)+b_1(o_2+\varepsilon_2)+\cdots f_1(o_i+\varepsilon_i)\cdots+n_1(o_n+\varepsilon_n)+l_1=0$$
$$a_2(o_1+\varepsilon_1)+b_2(o_2+\varepsilon_2)+\cdots f_2(o_i+\varepsilon_i)\cdots+n_2(o_n+\varepsilon_n)+l_2=0$$
$$\vdots$$
$$a_i(o_1+\varepsilon_1)+b_i(o_2+\varepsilon_2)+\cdots f_i(o_i+\varepsilon_i)\cdots+n_i(o_n+\varepsilon_n)+l_i=0$$
$$\vdots$$
$$a_m(o_1+\varepsilon_1)+b_m(o_2+\varepsilon_2)+\cdots f_m(o_i+\varepsilon_i)\cdots+n_m(o_n+\varepsilon_n)+l_m=0$$

as one of a set of angles, the difference between them being the correction terms ε. **M** is the $m * n$ matrix of conditions, **O** and ε are $n * 1$ column matrices.

The m relationships in Table 16.1 must be fulfilled while ensuring that $\sum\varepsilon^2$ is a minimum. There are n unknown errors or corrections that we must apply to satisfy the m relationships. If we consider the uncorrected observations, then we will have m equations of the form

$$a_io_1+b_io_2+c_io_3+\cdots f_io_i+\cdots+n_io_n+l_i=r_i$$

Here, r_i is the residual for condition i. The equations are of the form **MO** + **L** = **R** and the least square solution requires both $\sum\varepsilon^2$ and $\sum r^2$ to be a minimum.

We can write the m equations in Table 16.1 as Table 16.2 in which ε is the column vector of corrections to the observed quantities, and **R** is the column vector of the residuals.

If we partially differentiate the equations in Table 16.2, assuming that ε is the variable, then we derive the m equations with n unknowns shown in Table 16.3, which we call E_1, E_2, and so forth.

TABLE 16.2
The Relationships to Be Optimized: Mε + R = 0

$$a_1\varepsilon_1+b_1\varepsilon_2+c_1\varepsilon_3+\cdots f_1\varepsilon_i+\cdots+n_1\varepsilon_n+r_1=0$$
$$a_2\varepsilon_1+b_2\varepsilon_2+c_2\varepsilon_3+\cdots f_2\varepsilon_i+\cdots+n_2\varepsilon_n+r_2=0$$
$$\vdots$$
$$a_i\varepsilon_1+b_i\varepsilon_2+c_i\varepsilon_3+\cdots f_i\varepsilon_i+\cdots+n_i\varepsilon_n+r_i=0$$
$$\vdots$$
$$a_m\varepsilon_1+b_m\varepsilon_2+c_m\varepsilon_3+\cdots f_m\varepsilon_i\cdots+n_m\varepsilon_n+r_m=0$$

TABLE 16.3
The Differential Equations: M(δε) = 0

E_1	$a_1\,\delta\varepsilon_1+b_1\,\delta\varepsilon_2+c_1\,\delta\varepsilon_3+\cdots f_1\,\delta\varepsilon_i+\cdots+n_1\,\delta\varepsilon_n=0$
E_2	$a_2\,\delta\varepsilon_1+b_2\,\delta\varepsilon_2+c_2\,\delta\varepsilon_3+\cdots f_2\,\delta\varepsilon_i+\cdots+n_2\,\delta\varepsilon_n=0$
$\vdots$	
E_i	$a_i\,\delta\varepsilon_1+b_i\,\delta\varepsilon_2+c_i\,\delta\varepsilon_3+\cdots f_i\,\delta\varepsilon_i+\cdots+n_i\,\delta\varepsilon_n=0$
$\vdots$	
E_m	$a_m\,\delta\varepsilon_1+b_m\,\delta\varepsilon_2+c_m\,\delta\varepsilon_3+\cdots f_m\,\delta\varepsilon_i\cdots+n_m\,\delta\varepsilon_n=0$

Next we introduce m constants in the form of the column vector

$$\mathbf{K} = \begin{pmatrix} K_1 \\ K_2 \\ \cdots \\ K_i \\ \cdots \\ K_m \end{pmatrix}$$

Initially, the numerical values of K_1, K_2, and so forth, are unknown. We apply each of these to its equivalent differentiated equation in Table 16.3 and combine all the equations together by adding them in the form $\mathbf{M}(\delta\varepsilon)\mathbf{K}$. The equations then become $(E_1) * K_1 + (E_2) * K_2 + \cdots$, and so on, where by E_1 we mean $(a_1\ \delta\varepsilon_1 + b_1\ \delta\varepsilon_2 + c_1\ \delta\varepsilon_3 + \cdots f_1\ \delta\varepsilon_i + \cdots + n_1\ \delta\varepsilon_n)$. This gives us a single equation:

$$\begin{aligned} &K_1\left(a_1\delta\varepsilon_1 + b_1\delta\varepsilon_2 + c_1\delta\varepsilon_3 + \cdots f_1\delta\varepsilon_i + \cdots + n_1\delta\varepsilon_n\right) \\ &+K_2\left(a_2\delta\varepsilon_1 + b_2\delta\varepsilon_2 + c_2\delta\varepsilon_3 + \cdots f_2\delta\varepsilon_i + \cdots + n_2\delta\varepsilon_n\right) + \\ &+\cdots\ \cdots \\ &+K_m\left(\alpha_\mu\delta\varepsilon_1 + b_m\delta\varepsilon_2 + c_m\delta\varepsilon_3 + \cdots f_m\delta\varepsilon_i \cdots + n_m\delta\varepsilon_n\right) = 0 \end{aligned}$$

Hence, on rearranging terms

$$\begin{aligned} &\left(a_1K_1 + a_2K_2 + \cdots + a_mK_m\right)\delta\varepsilon_1 \\ &+\left(b_1K_1 + b_2K_2 + \cdots + b_mK_m\right)\delta\varepsilon_2 + \\ &+\cdots\ \cdots \\ &+\left(n_1K_1 + n_2K_2 + \cdots + n_mK_m\right)\delta\varepsilon_n = 0 \end{aligned}$$

For this to be identical with the condition that the weighted sum of the squares of the residuals is a minimum, namely, as we showed earlier:

$$w_1\varepsilon_1\delta\varepsilon_1 + w_2\varepsilon_2\delta\varepsilon_2 + \cdots + w_n\varepsilon_n\delta\varepsilon_n = 0$$

then we must have the n relationships shown in Table 16.4.

This gives us n relationships linking the m unknown quantities K with the n unknown quantities ε. Put another way, $\varepsilon_1 = (a_1K_1 + a_2K_2 + \cdots + a_mK_m) / w_1$, and so on. The quantities K are called

TABLE 16.4
The Relationships between the Correlatives: $\mathbf{M}^T\mathbf{K} = \mathbf{W}\varepsilon$

$(a_1K_1 + a_2K_2 + \cdots + a_mK_m) = w_1\varepsilon_1$
$(b_1K_1 + b_2K_2 + \cdots + b_mK_m) = w_2\varepsilon_2$
:
$(n_1K_1 + n_2K_2 + \cdots + n_mK_m) = w_n\varepsilon_n$

TABLE 16.5
The Equations for the Correlatives

1. $a_1(a_1K_1 + a_2K_2 + \cdots + a_mK_m) / w_1$
   $+ b_1(b_1K_1 + b_2K_2 + \cdots + b_mK_m) / w_2 + \cdots + r_1 = 0$
2. $a_2(a_1K_1 + a_2K_2 + \cdots + a_mK_m) / w_1$
   $+ b_2(b_1K_1 + b_2K_2 + \cdots + b_mK_m) / w_2 + \cdots + r_2 = 0$

$\vdots$

m. $a_m(a_1K_1 + a_2K_2 + \cdots + a_mK_m) / w_1$
   $+ b_m(b_1K_1 + b_2K_2 + \cdots + b_mK_m) / w_2 + \cdots + r_m = 0$

the *correlatives*. Substituting the values of ε in our original equations (Table 16.2), we obtain the m equations for the correlatives as given in Table 16.5, which we can rearrange as shown in Table 16.6. We can then solve these m linear equations to give us the values K, which we can then substitute back into the equations given in Table 16.4 to find the numerical values ε.

To summarize, if the column vector of n corrections that must be applied to meet the conditions is

$$\varepsilon = \left(\varepsilon_1, \varepsilon_2, \ldots \varepsilon_i, \ldots .\varepsilon_m\right)^{\mathrm{T}}$$

and if **K** is the column vector

$$\left(k_1, k_2, \ldots .k_i, \ldots .k_m\right)^{\mathrm{T}}$$

of correlatives and **M** is the set of m conditions that link n unknown corrections ε so that $\mathbf{M}\varepsilon + \mathbf{R} = 0$ and $\sum r^2$ and is a minimum, then assuming unit weight, if $\mathbf{MM^TK} + \mathbf{R} = 0$, we can obtain the values in **K** and thus $\varepsilon = \mathbf{M^TK}$.

Since this is very abstract, let us consider an example. If all the eight internal angles are measured in Figure 16.2 (a figure known to surveyors as a *braced quadrilateral*), then there will be redundant measurements. In fact, if A and B are known then only two measures are required to fix

TABLE 16.6
Solving for the Correlatives

1. $K_1(a_1a_1 / w_1 + b_1b_1 / w_2 + \cdots + n_1n_1 / w_n)$
   $+ K_2(a_1a_2 / w_1 + b_1b_2 / w_2 + \cdots + n_1n_2 / w_n)$
   $+ \cdots$
   $+ K_m(a_1a_m / w_1 + b_1b_m / w_2 + \cdots + n_1n_m / w_n) + r_1 = 0$
2. $K_1(a_1a_2 / w_1 + b_1b_2 / w_2 + \cdots + n_1n_2 / w_n)$
   $+ K_2(a_2a_2 / w_1 + b_2b_2 / w_2 + \cdots + n_2n_2 / w_n)$
   $+ \cdots$
   $+ K_m(a_2a_m / w_1 + b_2b_m / w_2 + \cdots + n_2n_m / w_n) + r_2 = 0$

$\vdots$

m. $K_1(a_1a_m / w_1 + b_1b_m / w_2 + \cdots + n_1n_m / w_n)$
   $+ K_2(a_2a_m / w_1 + b_2b_m / w_2 + \cdots + n_2n_m / w_n)$
   $+ \cdots$
   $+ K_m(a_ma_m / w_1 + b_mb_m / w_2 + \cdots + n_mn_m / w_n) + r_m = 0$

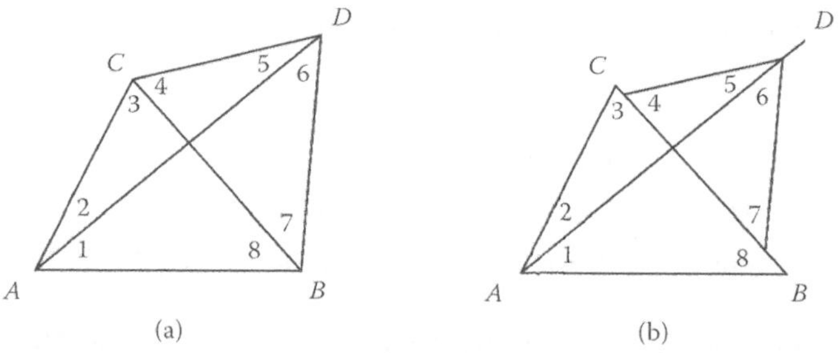

FIGURE 16.2 The braced quadrilateral: (a) consistent scale, (b) inconsistent scale

C and two to fix *D*, hence, if we have eight measurements there are four extra observations. These can be used to improve the accuracy of the determinations of *C* and *D* from *A* and *B*.

If we take the braced quadrilateral in Figure 16.2a, then obviously:

Angles

$$1 + 2 + 3 + 8 = 180$$
$$4 + 5 + 6 + 7 = 180$$
$$2 + 3 + 4 + 5 = 180$$
$$6 + 7 + 8 + 1 = 180$$
$$1 + 2 + 3 + 4 + 5 + 6 + 7 + 8 = 360$$

We have five equations and four redundant observations, since if *A* and *B* are known, then angles 1, 2, 7, and 8 are sufficient to fix *C* and *D*; angles 3, 4, 5, and 6 are redundant. However, these five equations are not independent since the fifth equation is the sum of the first and second, and also of the third and fourth. Furthermore, (4 + 5) = (1 + 8) while (2 + 3) = (6 + 7) and we cannot distinguish between the pairs. The angles in Figure 16.2b are the same as in 16.2a, but the figure is geometrically different.

The conclusion from this is that care must be taken in defining the conditions that must be satisfied. In the case of the braced quadrilateral, in order to satisfy the geometry, we need to make sure that the scale is consistent. Since

$$AB / AC = \sin 3 / \sin 8; \quad AC / CD = \sin 5 / \sin 2$$
$$CD / DB = \sin 7 / \sin 4; \quad DB / AB = \sin 1 / \sin 6$$

Then on multiplying all these together, we obtain

$$\frac{AB}{AC} * \frac{AC}{AD} * \frac{CD}{DB} * \frac{DB}{AB} = \frac{AB * AC * CD * DB}{AC * CD * DB * AB} = 1$$

Hence,

$$\frac{\sin 3 * \sin 5 * \sin 7 * \sin 1}{\sin 8 * \sin 2 * \sin 4 * \sin 6} = 1$$

or

$$\sin 3 * \sin 5 * \sin 7 * \sin 1 = \sin 8 * \sin 2 * \sin 4 * \sin 6$$

So now we have four independent equations that must be satisfied, namely,

$$(4 + 5) = (1 + 8)$$
$$(2 + 3) = (6 + 7)$$
$$(1 + 2 + 3 + 4 + 5 + 6 + 7 + 8) = 360$$
$$\sin 3 * \sin 5 * \sin 7 * \sin 1 = \sin 8 * \sin 2 * \sin 4 * \sin 6$$

Satisfying these four equations will ensure that the geometry is robust. These equations are the condition equations and must be satisfied exactly.

Three of the equations are of linear form, but at present the relationship between the sine values of the angles is not linear. We can, however, express this relationship in the form:

$$\log \sin 3 + \log \sin 5 + \log \sin 7 + \log \sin 1$$
$$= \log \sin 8 + \log \sin 2 + \log \sin 4 + \log \sin 6$$

This is a linear relationship between the log sin values. Let each observed angle have a correction ε_i seconds where $i = 1$ to 8. From the log sine tables we can find the difference for one second of arc. Let us call this $\emptyset_i$; for instance, for the angle 52° 48′ 12″, the log sine = −0.0987787 while for 52° 48′ 13″ it is −0.0987771 and hence $\emptyset_i$= 0.0000016. Provided that the corrections are small, the difference in the logarithmic values will be $\varepsilon_i\ \emptyset_i$., which in this case would be 0.0000016 ε_i. Thus,

$$\varepsilon_3\emptyset_3 + \varepsilon_5\emptyset_5 + \varepsilon_7\emptyset_7 + \varepsilon_1\emptyset_1 = \varepsilon_8\emptyset_8 + \varepsilon_2\emptyset_2 + \varepsilon_4\emptyset_4 + \varepsilon_6\emptyset_6$$

This gives us a linear equation between the error terms.

Example 16.1 establishes a relationship based on the log sine conditions by reducing these to a set of linear relationships. Example 16.2 then summarizes the corrections that must be optimized.

In Example 16.1, the logarithm of the sine of each of the eight angles has been tabulated together with the differences in each logarithm for one second of arc Δ1″.

Example 16.1 Angles in a Braced Quadrilateral

Angle	Observed	Log Sine (Odd angles)	Δ1″	Log Sine (Even angles)	Δ1″
1.	52° 48′ 12.3″	−0.0987783	16		
2.	89 33 15.8			−0.0000131	0
3.	15 13 58.9	−0.5804649	77		
4.	17 23 14.0			−0.5245787	67
5.	57 49 31.7	−0.0724090	13		
6.	68 09 09.5			−0.0323685	8
7.	36 38 19.6	−0.2241944	28		
8.	22 24 31.0			−0.4188365	51
Σ =	360° 00′ 12.8″	−0.9758466		−0.9757968	

If we adjust each angle by an amount to make the sine values consistent, then

$$16\varepsilon_1 + 77\varepsilon_3 + 13\varepsilon_5 + 28\varepsilon_7 - 9758466 = 67\varepsilon_4 + 8\varepsilon_6 + 51\varepsilon_8 - 9757968$$

or

$$16\varepsilon_1 + 77\varepsilon_3 - 67\varepsilon_4 + 13\varepsilon_5 - 8\varepsilon_6 + 28\varepsilon_7 - 51\varepsilon_8 - 498 = 0$$

This allows us to express the errors in terms of linear combinations derived from a relationship that was expressed as a series of sine values that were multiplied together.

Example 16.2 shows how the process is carried out, leading to a simple linear relationship between the corrections to be applied to each of the angles. The box also shows the simple relationships between the sums of the angles. There are four independent equations relating to eight unknowns. In addition, we want $\sum\varepsilon^2$ to be a minimum. Using the formulae derived above, $n = 8$, the number of observed angles and $m = 4$, the number of conditions. We can tabulate the four condition equations (Eqn1), (Eqn2), (Eqn3), and (Eqn4), and eight unknowns as in Example 16.3.

In Example 16.3, the rows are the condition equations and the columns the corrections. Here, the row called Eqn1 shows that from Eqn1 in Example 16.2,

$$1*\varepsilon_1 - 1*\varepsilon_4 - 1*\varepsilon_5 + 1*\varepsilon_8 - 2.4 = 0$$

Using the equations in Example 16.3 and assuming the weights all = 1, we can substitute numbers for all the expressions such as

$$K_1\left(a_1a_1 / w_1 + b_1b_1 / w_2 + \cdots + n_1n_1 / w_n\right) + K_2\left(a_1a_2 / w_1 + b_1b_2 / w_2 + \cdots + n_1n_2 / w_n\right),$$ and so forth.

Example 16.2 An Example of a Survey Adjustment

Using the data in Example 16.1:

$$\text{Angles } 1+8 = 75°12'43.3'' \text{ and } 4+5 = 75°12'45.7''$$

Hence,

$$\begin{aligned}\varepsilon_1 + \varepsilon_8 + 75°12'43.3'' &= \varepsilon_4 + \varepsilon_5 + 75°12'45.7'' \\ \varepsilon_1 - \varepsilon_4 - \varepsilon_5 + \varepsilon_8 - 2.4 &= 0\end{aligned} \quad (Eqn1)$$

Also, angles 2 + 3 = 104° 47′ 14.7″ and 6 + 7 = 104° 47′ 29.1″

$$\varepsilon_2 + \varepsilon_3 - \varepsilon_6 - \varepsilon_7 - 14.4 = 0 \quad (Eqn2)$$

$$\varepsilon_1 + \varepsilon_2 + \varepsilon_3 + \varepsilon_4 + \varepsilon_5 + \varepsilon_6 + \varepsilon_7 + \varepsilon_8 + 12.8 = 0 \quad (Eqn3)$$

Also, sin1 * sin3 * sin5 * sin7 = sin2 * sin4 * sin6 * sin8. Hence,

$$\text{logsin1} + \text{logsin3} + \text{logsin5} + \text{logsin7} = \text{logsin2} + \text{logsin4} + \text{logsin6} + \text{logsin8}$$

The log (sine) values of the four odd numbered angles add up to −0.9758466 and the evens to −0.9757968 (see Example 16.1). The change in the log sine that arises from a change of 1″ can be obtained from tables or using a pocket calculator.

This gives us, as shown in Example 16.1

$$16\varepsilon_1 + 77\varepsilon_3 - 67\varepsilon_4 + 13\varepsilon_5 - 8\varepsilon_6 + 28\varepsilon_7 - 51\varepsilon_8 - 498 = 0 \; (Eqn4)$$

Hence, we have four linear equations, Eqn1, Eqn2, Eqn3, and Eqn4 above linking the eight unknowns ε_i each of which represents a correction to one of the angles.

Example 16.3 Forming the Correlatives

	$a = \varepsilon_1$	$b = \varepsilon_2$	$c = \varepsilon_3$	$d = \varepsilon_4$	$e = \varepsilon_5$	$f = \varepsilon_6$	$g = \varepsilon_7$	$h = \varepsilon_8$	r
Eqn1	1	0	0	−1	−1	0	0	1	−2.4
Eqn2	0	1	1	0	0	−1	−1	0	−14.4
Eqn3	1	1	1	1	1	1	1	1	12.8
Eqn4	16	0	77	−67	13	−8	28	−51	−498

Thus, in the first equation the coefficient of

$$K_1 \text{ from } Eqn1^2 = 1^2 + 0^2 + 0^2 + (-1)^2 + (-1)^2 + 0^2 + 0^2 + 1^2 = 4$$
$$K_2 \text{ from } Eqn1 * Eqn2 = 1*0 + 0*1 + 0*1 - 1*0 - 1*0 + 0*(-1) + 0*(-1) + 1*0 = 0$$
$$K_3 \text{ from } Eqn1 * Eqn3 = 1 + 0 + 0 - 1 - 1 + 0 + 0 + 1 = 0$$
$$K_4 \text{ from } Eqn1 * Eqn4 = 16 + 0 + 0 + 67 - 13 + 0 + 0 - 51 = 19$$

This gives us

$$4 * K_1 + 0 * K_2 + 0 * K_3 + 19 * K_4 - 2.4 = 0$$

Similarly, for the second equation the coefficients are derived from

$$(Eqn1 * Eqn2), (Eqn2)^2, (Eqn2 * Eqn3), (Eqn2 * Eqn4)$$

For the third,

$$(Eqn1 * Eqn3), (Eqn2 * Eqn3), (Eqn3)^2, (Eqn3 * Eqn4)$$

For the fourth,

$$(Eqn1 * Eqn4), (Eqn2 * Eqn4), (Eqn3 * Eqn4), (Eqn4)^2$$

Hence, using the numbers in Example 16.3 and substituting them in the equations given in Table 16.6 we obtain the following:

1. $4 * K_1 + 0 * K_2 + 0 * K_3 + 19 * K_4 - 2.4 = 0$
2. $0 * K_1 + 4 * K_2 + 0 * K_3 + 57 * K_4 - 14.4 = 0$
3. $0 * K_1 + 0 * K_2 + 8 * K_3 + 8 * K_4 + 12.8 = 0$
4. $19 * K_1 + 57 * K_2 + 8 * K_3 + 14292 * K_4 - 498 = 0$

From 1,

$$K_1 = 0.6 - 4.75K_4$$

From 2,

$$K_2 = 3.6 - 14.25K_4$$

From 3,

$$K_3 = -1.6 - K_4$$

From 4,

$$K_4 = (294.2)/(13381.5)$$
$$= 0.022$$

So,

$$K_1 = 0.496 \quad K_2 = 3.287 \quad K_3 = 1.622 \quad K_4 = 0.022$$

Substituting back, using Example 16.3 and reading the columns downward so that under the column headed "$a = \varepsilon_1$" we have:

$$\varepsilon_1 = 1 * Eqn1 + 0 * Eqn2 + 1 * Eqn3 + 16 * Eqn4$$
$$\varepsilon_2 = 0 * Eqn1 + 1 * Eqn2 + 1 * Eqn3 + 0 * Eqn4$$
$$\varepsilon_3 = 0 * Eqn1 + 1 * Eqn2 + 1 * Eqn3 + 77 * Eqn4, \text{ and so on.}$$

The results are shown in Example 16.4.

Example 16.4 The Finished Solution

$$\varepsilon_1 = 1 * K_1 + 0 * K_2 + 1 * K_3 + 16 * K_4 = -0.774 = -0.8$$

Angle 1 = 52° 48′ 11.5″

$$\varepsilon_2 = 0 * K_1 + 1 * K_2 + 1 * K_3 + 0 * K_4 = 1.665 = +1.7$$

Angle 2 = 89° 33′ 17.5″

$$\varepsilon_3 = 0 * K_1 + 1 * K_2 + 1 * K_3 + 77 * K_4 = 3.358 = +3.4$$

Angle 3 = 15° 14′ 02.3″

$$\varepsilon_4 = -1 * K_1 + 0 * K_2 + 1 * K_3 - 67 * K_4 = -3.591 = -3.6$$

Angle 4 = 17° 23′ 10.4″

$$\varepsilon_5 = -1 * K_1 + 0 * K_2 + 1 * K_3 + 13 * K_4 = -1.832 = -1.8$$

Angle 5 = 57° 49′ 29.9″

$$\varepsilon_6 = 0 * K_1 + -1 * K_2 + 1 * K_3 - 8 * K_4 = -5.085 = -5.1$$

Angle 6 = 68° 09′ 04.4″

$$\varepsilon_7 = 0 * K_1 + -1 * K_2 + 1 * K_3 + 28 * K_4 = -4.293 = -4.3$$

Angle 7 = 36° 38′ 15.3″

$$\varepsilon_8 = 1 * K_1 + 0 * K_2 + 1 * K_3 - 51 * K_4 = -2.248 = -2.2$$

Angle 8 = 22° 24′ 28.8″

In matrix form this can be more elegantly expressed. To summarize:

$$\text{If } \mathbf{M}=\begin{pmatrix} 1 & 0 & 0 & -1 & -1 & 0 & 0 & 1 \\ 0 & 1 & 1 & 0 & 0 & -1 & -1 & 0 \\ 1 & 1 & 1 & 1 & 1 & 1 & 1 & 1 \\ 16 & 0 & 77 & 67 & 13 & -8 & 28 & -51 \end{pmatrix}; \mathbf{R}=\begin{pmatrix} -2.4 \\ -14.4 \\ 12.8 \\ -498 \end{pmatrix}; \mathbf{M}^{\mathrm{T}}=\begin{pmatrix} 1 & 0 & 1 & 16 \\ 0 & 1 & 1 & 0 \\ 0 & 1 & 1 & 77 \\ -1 & 0 & 1 & -67 \\ -1 & 0 & 1 & 13 \\ 0 & -1 & 1 & -8 \\ 0 & -1 & 1 & 28 \\ 1 & 0 & 1 & -51 \end{pmatrix} \text{ then}$$

$$\mathbf{M}*\mathbf{M}^{\mathrm{T}}*\mathbf{K}=\begin{pmatrix} 4 & 0 & 0 & 19 \\ 0 & 4 & 0 & 57 \\ 0 & 0 & 8 & 8 \\ 19 & 57 & 8 & 14292 \end{pmatrix}*\mathbf{K}=-\mathbf{R}=-\begin{pmatrix} -2.4 \\ -14.4 \\ 12.8 \\ -498 \end{pmatrix} \text{ where } \mathbf{K}=\begin{pmatrix} K_1 \\ K_2 \\ K_3 \\ K_4 \end{pmatrix}$$

This gives

$$\mathbf{K}=-\begin{pmatrix} 4 & 0 & 0 & 19 \\ 0 & 4 & 0 & 57 \\ 0 & 0 & 8 & 8 \\ 19 & 57 & 8 & 14292 \end{pmatrix}^{-1}*\begin{pmatrix} -2.4 \\ -14.4 \\ 12.8 \\ -498 \end{pmatrix}=-\begin{pmatrix} 0.2517 & 0.0051 & 0.0004 & -0.0004 \\ 0.0051 & 0.2652 & 0.0011 & -0.0011 \\ 0.0004 & 0.0011 & 0.1251 & -0.0001 \\ -0.0004 & -0.0011 & -0.0001 & 0.0001 \end{pmatrix}$$

$$*\begin{pmatrix} -2.4 \\ -14.4 \\ 12.8 \\ -498 \end{pmatrix}=\begin{pmatrix} 0.496 \\ 3.287 \\ -1.622 \\ 0.022 \end{pmatrix}$$

Hence,

$$\varepsilon=\mathbf{M}^{\mathrm{T}}*\mathbf{K}=\begin{pmatrix} 1 & 0 & 1 & 16 \\ 0 & 1 & 1 & 0 \\ 0 & 1 & 1 & 77 \\ -1 & 0 & 1 & -67 \\ -1 & 0 & 1 & 13 \\ 0 & -1 & 1 & -8 \\ 0 & -1 & 1 & 28 \\ 1 & 0 & 1 & -51 \end{pmatrix}*\begin{pmatrix} 0.496 \\ 3.287 \\ -1.622 \\ 0.022 \end{pmatrix}=\begin{pmatrix} -0.774 \\ 1.665 \\ 3.358 \\ -3.591 \\ -1.832 \\ -5.085 \\ -4.293 \\ -2.248 \end{pmatrix}$$

The somewhat tortuous calculations that have led to Example 16.4 provide a solution in which, rounding off to the nearest 0.1″, gives $\sum\varepsilon^2 = 80.6$. This is the minimum least square value.

Returning to the problem posed at the end of Section 15.5 where we had three angles of a triangle with corrections $x_1 + x_2 + x_3 + r = 0$, and $\mathbf{M} = (1, 1, 1)$ then $\mathbf{M} * \mathbf{M}^T = 3$ and $\mathbf{M} * \mathbf{M}^T * \mathbf{K} = 3 * \mathbf{K} = -r$. Thus, $\mathbf{K} = (-r / 3)$ and

$$X = \begin{pmatrix} -r/3 \\ -r/3 \\ -r/3 \end{pmatrix} \text{ and } x_1 = x_2 = x_3 = -r/3$$

meaning that we apply equal corrections to each angle as the optimum least square solution.

Very many types of observation can be processed in this way, provided that we know the conditions that must be fulfilled and the relative weighting of the different observations. Although tedious to go through by longhand, once the equations are formed, the electronic computer can take over and provide a most probable value for each of the observations.

The technique of least square adjustments has many applications in surveying and mapping, and indeed other areas where a best-fit solution is required. The algorithms are ideally suited to electronic data processing but the quality of the results will depend on the quality of the observations and the assumptions underlying the adjustment procedures. If, for example, each observation is not truly independent and is somehow correlated with other observations, then the underlying assumptions of the least square process are not fully valid. Techniques are available to test for autocorrelation using variance and covariance analysis and were touched on in Chapter 15.

16.3 SATELLITE POSITION FIXING

Global positioning systems (GPS) make use of several mathematical processes to determine the coordinates of a point on the Earth's surface. A series of satellites circling the Earth transmit data about their position and time, the signals being picked up by a receiver. In essence, if satellite i has position (x_i, y_i, z_i) at time t_i and if its signal is received at a ground station R in position (x, y, z) at time t_r according to the local ground station clock, then the distance between the satellite and point R will be $\sqrt{\{(x - x_i)^2 + (y - y_i)^2 + (z - z_i)^2\}}$. The coordinates for this system are rectangular Cartesian and geocentric in that they are expressed relative to the notional center of the Earth, the z-axis lying through the North and South Poles, the direction of the x-axis being determined by the meridian through Greenwich (G in Figure 16.3) and the y-axis being at right angles thereto.

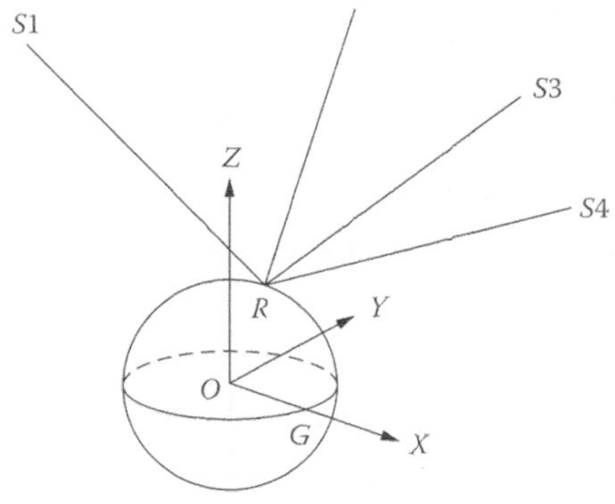

FIGURE 16.3 Four satellites and a receiver.

The time taken for the signal to travel at the speed of light will be $(t_r - e - t_i)$ where e is the error between the two clocks. The distance between the two is therefore $(t_r - e - t_i) * c$ where c = the speed of light. This is known precisely and = 299,792,458 meters per second.

We therefore have a distance (derived from a measurement of time) from a known point in space. We can imagine a sphere of radius $(t_r - e - t_i)$ around each satellite with the receiver R being the point where these spheres all intersect with the Earth's surface.

Using Pythagoras, we have $\sqrt{\left\{(x-x_i)^2+(y-y_i)^2+(z-z_i)^2\right\}}=(t_r-e-t_i)*c.$

In this equation, there are four unknowns; namely x, y, z, and e. We therefore need four sets of observations to solve for the unknowns so that we must take observations to at least four satellites. This assumes that the clock error e is the same for each set of observations. In practice, the signals sent out by the satellites are based on identically the same time reference system. e is solely dependent on the receiver station and will be the same for all observations that are taken at the same time from the same place with the same equipment. The assumption that all we need is a minimum of four satellites is therefore valid.

We can express these four relationships as

$$\sqrt{\left\{(x-x_i)^2+(y-y_i)^2+(z-z_i)^2\right\}}+ec=(t_r-t_i)*c$$

where $i = 1, 2, 3, 4$. The value $(t_r - t_i) * c$ is known as the *pseudo-range* and is often designated ρ_i.

The four equations are not linear but can be solved iteratively using Newton's method, as discussed later. Alternatively, if we write ec as l then the equation above can be expressed in the form

$$(x-x_i)^2+(y-y_i)^2+(z-z_i)^2=(\rho_i-l)^2=\rho_i^2-2l\rho_i+l^2$$

or

$$F_i=\left(x_i^2+y_i^2+z_i^2-\rho_i^2\right)-2\left(x_ix+y_iy+z_iz-l\rho_i\right)+\left(x^2+y^2+z^2-l^2\right)=0$$

Let us call this relationship F_i. The term $(x^2 + y^2 + z^2 - l^2)$ is the same for all observations when the receiver is at a given point. If we have five or more satellite observations then we can eliminate the term by working with pairs of observations giving us four or more linear relationships between x, y, z, and l. We can also use the method of least squares to determine the optimum solution by treating F_i not as zero but as the residual for which we must minimize the sum of its squares. The determination of our position should become more accurate the more observations that we make.

The equations F_i are quadratic implying that there will be two possible solutions, although only one of these will be point R where the receiver is located. Let us assume that there are n satellites.

Let

$$\mathbf{S}_i=\begin{bmatrix} x_i \\ y_i \\ z_i \\ \rho_i \end{bmatrix}$$

where (x_i, y_i, z_i) are the geocentric coordinates for each satellite, $i = 1, 2, \ldots, n$. ρ_i is the pseudo-range to satellite i.

Let

$$\mathbf{X}=\begin{bmatrix} x \\ y \\ z \\ l \end{bmatrix}$$

represent the ground receiver coordinates (x, y, z) and l is the distance represented by the clock error multiplied by the speed of light.

$$\textbf{If } \mathbf{N} = \begin{bmatrix} 1 & 0 & 0 & 0 \\ 0 & 1 & 0 & 0 \\ 0 & 0 & 1 & 0 \\ 0 & 0 & 0 & -1 \end{bmatrix} \text{ then } \mathbf{X}^{\mathrm{T}} * \mathbf{N} = \begin{bmatrix} x & y & z-l \end{bmatrix}$$

$(\mathbf{X}^{\mathrm{T}}\mathbf{N})\mathbf{X} = (x^2 + y^2 + z^2 - l^2)$. This is a constant for all observations from the ground station. Let us call it k so that $k = (\mathbf{X}^{\mathrm{T}}\mathbf{N})\mathbf{X} = \mathbf{K}$.

Let $\mathbf{A}$ = column matrix of terms $(x_i^2 + y_i^2 + z_i^2 - p_i^2)$ of size n rows and one column; also let $\mathbf{B}$ = matrix with n rows and four columns made of the elements of $[x_j\ y_j\ z_j,\ -\rho_j]$ where $j = 1, 2, \ldots, n$

$$\mathbf{A} = \begin{bmatrix} x_1^2 + y_1^2 + z_1^2 - \rho_1^2 \\ x_2^2 + y_2^2 + z_2^2 - \rho_2^2 \\ x_3^2 + y_3^2 + z_3^2 - \rho_3^2 \\ x_4^2 + y_4^2 + z_4^2 - \rho_4^2 \\ \cdots \\ x_n^2 + y_n^2 + z_n^2 - \rho_n^2 \end{bmatrix} \text{ and } \mathbf{B} = \begin{bmatrix} x_1 & y_1 z_1 - \rho_1 \\ x_2 & y_2 z_2 - \rho_2 \\ x_3 & y_3 z_3 - \rho_3 \\ x_4 & y_4 z_4 - \rho_4 \\ \cdots & \cdots \\ x_n & y_n z_n - \rho_n \end{bmatrix} \quad \mathbf{BX} = \begin{bmatrix} x_1 x + y_1 y + z_1 z - \rho_1 l \\ x_2 x + y_2 y + z_2 z - \rho_2 l \\ x_3 x + y_3 y + z_3 z - \rho_3 l \\ x_4 x + y_4 y + z_4 z - \rho_4 l \\ \cdots\ \cdots \\ x_n x + y_n y + z_n z - \rho_n l \end{bmatrix}$$

The equations F_i show that we need to satisfy the condition that $\mathbf{A} - 2\mathbf{BX} + \mathbf{K} = 0$. Because of potential for error in all the measurements, slightly different answers will be obtained for each set of four satellites observed. For now, let us assume that we have four satellites, hence $n = 4$ and $\mathbf{B}$ is a square matrix. Let $\mathbf{B}^{-1}$ be the inverse of $\mathbf{B}$, and finally let $\mathbf{K}$ be the column vector with $n = 4$ rows, each row being equal to k where, as before, $k = (x^2 + y^2 + z^2 - l^2)$.

Then the equations F_i can be written in the form

$$\mathbf{A} - 2\mathbf{BX} + \mathbf{K} = 0 \text{ or}$$
$$\mathbf{BX} = 0.5(\mathbf{A} + \mathbf{K}) \text{ or}$$
$$\mathbf{X} = 0.5\mathbf{B}^{-1}(\mathbf{A} + \mathbf{K})$$

If, for convenience, we write $\mathbf{C}$ for $0.5\mathbf{B}^{-1}$ where $\mathbf{C}$ is a 4 * 4 matrix, then

$$\mathbf{X} = \mathbf{C}(\mathbf{A} + \mathbf{K})$$

and

$$\mathbf{X}^{\mathrm{T}} = (\mathbf{A} + \mathbf{K})^{\mathrm{T}} \mathbf{C}^{\mathrm{T}}$$

The elements of $\mathbf{C}$ and $\mathbf{A}$ are all observed quantities and hence are all known.

Each element of $\mathbf{K}$, namely, k, is a scalar constant made up from x, y, z, and l. If we knew k we would have a series of at least four linear equations linking x, y, z, and l.

Now $k = (\mathbf{X}^{\mathrm{T}} * \mathbf{N}) * \mathbf{X} = \{(\mathbf{A} + \mathbf{K})^{\mathrm{T}} * \mathbf{C}^{\mathrm{T}} * \mathbf{N} * \mathbf{C} * (\mathbf{A} + \mathbf{K})\}$. If in $\mathbf{A}$ we write $a_j = x_j^2 + y_j^2 + z_j^2 - p_j^2$, all of which are observed quantities, then

$$\mathbf{A} + \mathbf{K} = \begin{bmatrix} a_1 + k \\ a_2 + k \\ a_3 + k \\ a_4 + k \end{bmatrix} \text{ and } (\mathbf{A} + \mathbf{K})^{\mathrm{T}} = \begin{bmatrix} a_1 + k, & a_2 + k, & a_3 + k, & a_4 + k \end{bmatrix}$$

Let the elements of **C** be such that

$$\mathbf{C} = \begin{bmatrix} c_{11} & c_{12} & c_{13} & c_{14} \\ c_{21} & c_{22} & c_{23} & c_{24} \\ c_{31} & c_{32} & c_{33} & c_{34} \\ c_{41} & c_{42} & c_{43} & c_{44} \end{bmatrix}$$

Therefore,

$$(\mathbf{A}+\mathbf{K})^{\mathrm{T}} * \mathbf{C}^{\mathrm{T}} = \left[a_1+k,\ a_2+k,\ a_3+k,\ a_4+k\right] * \begin{bmatrix} c_{11} & c_{21} & c_{31} & c_{41} \\ c_{12} & c_{22} & c_{32} & c_{42} \\ c_{13} & c_{23} & c_{33} & c_{43} \\ c_{14} & c_{24} & c_{34} & c_{44} \end{bmatrix}$$

Hence,

$$(\mathbf{A}+\mathbf{K})^{\mathrm{T}} * \mathbf{C}^{\mathrm{T}} * \mathbf{N} \text{ is a } 1*4 \text{ row vector } \left[v_1, v_2, v_3, v_4\right]$$

where

$$\begin{aligned} v_1 &= c_{11}(a_1+k)+c_{12}(a_2+k)+c_{13}(a_3+k)+c_{14}(a_4+k) \\ v_2 &= c_{21}(a_1+k)+c_{22}(a_2+k)+c_{23}(a_3+k)+c_{24}(a_4+k) \\ v_3 &= c_{31}(a_1+k)+c_{32}(a_2+k)+c_{33}(a_3+k)+c_{34}(a_4+k) \\ v_4 &= -\left[c_{41}(a_1+k)+c_{42}(a_2+k)+c_{43}(a_3+k)+c_{44}(a_4+k)\right] \end{aligned}$$

Also,

$$\mathbf{C}*(\mathbf{A}+\mathbf{K}) = \begin{bmatrix} c_{11} & c_{12} & c_{13} & c_{14} \\ c_{21} & c_{22} & c_{23} & c_{24} \\ c_{31} & c_{32} & c_{33} & c_{34} \\ c_{41} & c_{42} & c_{43} & c_{44} \end{bmatrix} * \begin{bmatrix} a_1+k \\ a_2+k \\ a_3+k \\ a_4+k \end{bmatrix} \text{ which is a } 4*1 \text{ column vector } \begin{bmatrix} v_1 \\ v_2 \\ v_3 \\ v_4 \end{bmatrix}$$

If we call the row vector $[v_1, v_2, v_3, v_4]$ as $[p_1 + q_1k, p_2 + q_2k, p_3 + q_3k, p_4 + q_4k]$, then the elements of the column vector are the same, except for a change of sign at the end. We therefore have

$$(\mathbf{X}^{\mathrm{T}} * \mathbf{N}) * \mathbf{X} = k = (p_1+q_1k)^2 + (p_2+q_2k)^2 + (p_3+q_3k)^2 - (p_4+q_4k)^2$$

or

$$(q_1^2+q_2^2+q_3^2-q_4^2)k^2 + (2p_1q_1+2p_2q_2+2p_3q_3-2p_4q_4-1)k + (p_1^2+p_2^2+p_3^2-p_4^2) = 0$$

This is a quadratic in k, which will give two possible values for k only one of which will be the ground receiver, the other being somewhere out in space.

We can now write $F_i = \left(x_i^2 + y_i^2 + z_i^2 - p_i^2\right) - 2\left(x_i x + y_i y + z_i z - l\rho_i\right) + k$, giving us four linear equations relating x, y, z, and l. With four satellites we can solve the equations directly, as we do in Example 16.5. If there are more than four then we will need to derive a least squares solution using the principles outlined in Sections 16.2 and 16.3 to obtain a more accurate solution for the coordinates of point R.

Example 16.5 Satellite Positioning

Consider four satellites where their (x, y, z) is known, along with the pseudo-ranges:

Satellite x	Satellite y	Satellite z	$t_r - t_s$	$(t_r - t_s) * c$	A
15629930.79	17229314.77	12797138.77	71,410,508.00	21,408,331.72	2.46594E + 14
1198579.24	15672887.77	21381561.43	76,622,350.95	22,970,802.93	1.76589E + 14
4927877.54	24742040.32	8352009.51	80,056,010.16	24,000,188.06	1.302E + 14
18740047.66	−4918293.33	18270380.04	70,918,574.34	21,260,853.72	2.57162E + 14

Column A is made up of terms $x_i^2 + y_i^2 + z_i^2 - \rho_i^2$. The notation E + 14 indicates an exponential value or scientific notation showing that the number is large with 14 terms before the decimal point. Here we have rounded off all the numbers.

The matrix **B** with elements $[x_j\ y_j\ z_j, -\rho_j]$ =

$$\begin{bmatrix} 15629930.79 & 17229314.77 & 12797138.77 & -21{,}408{,}331.72 \\ 1198579.24 & 15672887.77 & 21381561.43 & -22{,}970{,}802.92 \\ 4927877.54 & 24742040.32 & 8352009.51 & -24{,}000{,}188.06 \\ 18740047.66 & -4918293.33 & 18270380.04 & -21{,}260{,}853.72 \end{bmatrix}$$

$$0.5\mathbf{B}^{-1} =$$

$$\begin{bmatrix} 0.000000044454 & -0.00000001484 & -0.00000002227 & -0.00000000359 \\ 0.000000037673 & 0.000000007393 & -0.00000001620 & -0.00000002763 \\ 0.000000034215 & 0.000000037870 & -0.00000005129 & -0.00000001747 \\ 0.000000059872 & 0.000000017754 & -0.00000005996 & -0.00000003530 \end{bmatrix}$$

We defined p_1 as $c_{11}\,a_1 + c_{12}a_2 + c_{13}a_3 + c_{14}a_4$ and $q_1 = c_{11} + c_{12} + c_{13} + c_{14}$, and so on. This gives values for p_1, q_1, and so on, as

	1	2	3	4
p	4,518,807.6085	1,379,841.7715	3,953,880.9441	−1,013,796.077
q	0.00000000375	0.00000000124	0.00000000333	0.00000001763

This in turn gives k = 4.0865E + 13. The coordinates and error term for the receiver then become

$$\left(x_i^2 + y_i^2 + z_i^2 - \rho_i^2\right) - 2\left(x_i x + y_i y + z_i z - l\rho_i\right) + k = 0, \text{ or}$$

$$2x_i x + 2y_i y + 2z_i z - 2l\rho_i = \left(x_i^2 + y_i^2 + z_i^2 - \rho_i^2\right) + k \text{ as shown below}$$

x	y	z	l	Equals
31259861.58	34458629.54	25594277.54	−42,816,663.44	2.87081E + 14
2397158.48	31345775.54	42763122.86	−45,941,605.86	2.17076E + 14
9855755.08	49484080.64	16704019.02	−48,000,376.13	1.70686E + 14
37480095.32	−9836586.66	36540760.08	−42,521,707.44	2.97648E + 14

On solving these four quadratics:

$x = 4670813.04; y = 1429701.03; z = 4088501.02$, and the time error $l = 299792.477$

(l suggests that the receiver clock was almost exactly one second out.)

A word of caution should, however, be given. The orbit of the satellites is around 26.5 million meters from the center of the Earth. The determinant for **B** in the above equations is a number in excess of 1,000,000,000,000,000,000,000,000,000. Even when working in kilometers rather than meters it is still in excess of 1,000,000,000,000,000. Any computation system must be capable of handling such large numbers. In Example 16.5, we show part of the calculations but with rounding up of some of the numbers in the printout.

It should also be noted that the description above is a simplification as, for example, it assumes perfect measurement conditions and ignores the theory of relativity.

An alternative approach makes use of a series of iterations to come up with a solution where there are four satellites and the approximate position of the receiver station and the time delay is known. If the "o" for "old" coordinates are

$$\mathbf{F} = \begin{bmatrix} f_1 \\ f_2 \\ f_3 \\ f_4 \end{bmatrix}$$

where

$$f_i = \sqrt{\left\{(x_o - x_i)^2 + (y_o - y_i)^2 + (z_o - z_i)^2\right\}} - \rho_i + l \text{ and } i = 1,2,3,4$$

The partial derivatives of these equations are:

$$\partial f_i / \partial x = (x - x_i) / \left(\sqrt{\left\{(x - x_i)^2 + (y - y_i)^2 + (z - z_i)^2\right\}}\right)$$

$$\partial f_i / \partial y = (y - y_i) / \left(\sqrt{\left\{(x - x_i)^2 + (y - y_i)^2 + (z - z_i)^2\right\}}\right)$$

$$\partial f_i / \partial z = (z - z_i) / \left(\sqrt{\left\{(x - x_i)^2 + (y - y_i)^2 + (z - z_i)^2\right\}}\right)$$

$$\partial f_i / \partial l = 1$$

We then use a version of Newton's method (sometimes referred to as Newton–Raphson), which is an extension to what was discussed in Chapter 10. We form the matrix **J** from these partial derivatives so that

$$\mathbf{J} = \begin{bmatrix} \partial f_1 / \partial x \partial f_1 / \partial y \partial f_1 / \partial z \partial f_1 / \partial l \\ \partial f_2 / \partial x \partial f_2 / \partial y \partial f_2 / \partial z \partial f_2 / \partial l \\ \partial f_3 / \partial x \partial f_3 / \partial y \partial f_3 / \partial z \partial f_3 / \partial l \\ \partial f_4 / \partial x \partial f_4 / \partial y \partial f_4 / \partial z \partial f_4 / \partial l \end{bmatrix}$$

This is known as the *Jacobian* of **F**. It can be evaluated from the known coordinates of the satellites and the assumed value for the receiver station, for instance,

$$\partial f_1 / \partial x = (x_o - x_1) / \left(\sqrt{\{(x_o - x_1)^2 + (y_o - y_1)^2 + (z_o - z_1)^2\}}\right)$$

We then calculate the inverse of $\mathbf{J} = \mathbf{J}^{-1}$. Then we derive a new set of coordinates by calculating

$$X_{new} = X_{old} - J^{-1} F_{old} \text{ where } \mathbf{J}^{-1} \text{ is a } 4 * 4 \text{ square matrix}$$

or

$$\begin{bmatrix} x_n \\ y_n \\ z_n \\ l_n \end{bmatrix} = \begin{bmatrix} x_o \\ y_o \\ z_o \\ l_o \end{bmatrix} - \mathbf{J}^{-1} \begin{bmatrix} f_1 \\ f_2 \\ f_3 \\ f_4 \end{bmatrix}$$

We can then replace x_o by x_n, and so forth, and repeat the process. If the original estimate is reasonably close to the true value, the iteration should converge quite rapidly. See Example 16.6.

Example 16.6 Positioning From Satellites Using Iteration

Consider four satellites with geocentric coordinates (*x*, *y*, *z*), an apparent time difference $t_r - t_s$

Satellite *x*	Satellite *y*	Satellite *z*	$t_r - t_s$	$(t_r - t_s) * c$
15629930.79	17229314.77	12797138.77	71,410,508.00	21,408,331.72
1198579.24	15672887.77	21381561.43	76,622,350.95	22,970,802.93
4927877.54	24742040.32	8352009.51	80,056,010.16	24,000,188.06
18740047.66	−4918293.33	18270380.04	70,918,574.34	21,260,853.72

Let us assume an approximate position for the receiver station as

$$(4{,}671{,}000;\ 1{,}430{,}000;\ 4{,}089{,}000) \text{ with zero clock error.}$$

We can calculate the distances to each satellite by Pythagoras. These are 21,108,012.509, 22,670,470.636, 23,700,009.711, and 20,960,688.688 from the trial point. The pseudo-ranges are 21,408,331.720; 22,970,802.929; 24,000,188.064; and 21,260,853.719 but these distances include the clock error.

Using the partial derivative equations based on the estimated position

$$\partial f_i / \partial x = (x - x_i) / \left(\sqrt{\{(x - x_i)^2 + (y - y_i)^2 + (z - z_i)^2\}}\right), \text{ and so on,}$$

we can obtain the Jacobian with elements as shown in the following tabular form:

diff *x* / old dist	diff *y* / old dist	diff *z* / old dist	$\partial f_i / \partial l$
−0.519183451549186	−0.748498455879073	−0.412551336422655	1
0.1531693283319	−0.628257260285913	−0.762779110723671	1
−0.010838710329384	−0.983629990443701	−0.188042391302602	1
−0.671211135864479	0.302866638866031	−0.676570329051984	1

The inverse of the Jacobian becomes:

$$\begin{bmatrix} -2.023620428365 & 0.6221991481610 & 1.203438354395 & 0.1979829258089 \\ -1.644050579201 & -0.357954526276 & 0.8253402169906 & 1.176664888486 \\ -1.594018146643 & -1.783706488009 & 2.593669274997 & 0.7840553596546 \\ -1.938813875222 & -0.6807634043134 & 2.312592882192 & 1.306984397343 \end{bmatrix}$$

Applying $\mathbf{X}_{new} = \mathbf{X}_{old} - \mathbf{J}^{-1}\mathbf{F}_{old}$ the new approximate coordinates become:

$$\left(4{,}670{,}808.109;\ 1{,}429{,}697.652;\ 4{,}088{,}490.454\right) \text{ with } l = 299783.064$$

The next iteration gives

$$\left(4{,}670{,}813.115;\ 1{,}429{,}701.081;\ 4{,}088{,}501.240\right) \text{ with } l = 299792.677$$

A further iteration gives

$$\left(4{,}670{,}813.009;\ 1{,}429{,}701.009;\ 4{,}088{,}501.012\right) \text{ with } l = 299{,}792.473$$

And once more gives

$$\left(4{,}670{,}813.011;\ 1{,}429{,}701.010;\ 4{,}088{,}501.016\right) \text{ with } l = 299{,}792.477$$

Thus, quite quickly, we have honed in on the probable position and clock error using four satellites. Any difference with Example 16.5 is due to rounding errors in the computation.

GLOSSARY

Clustering algorithm An unsupervised learning technique used to group similar data points based on their characteristics.

Correlatives Unknown constants introduced in the solution by least squares.

Decision tree A supervised learning algorithm utilized for both classification and regression.

Deep learning A type of machine learning that uses artificial neural networks inspired by the human brain.

Discriminative A type of machine learning that learns the boundaries between different classes in a dataset.

Dispersion matrix Also known as the *variance–covariance matrix*. If $\mathbf{V}$ is a row matrix of all the error terms and its transpose is $\mathbf{V}^T$ then the dispersion matrix = $\mathbf{V}^T\mathbf{V}$.

Error term The difference between the observed and computed quantity.

Generalization error The difference between a model's performance on training data and its performance on unseen data.

Generative A machine learning model designed to create new data that is similar to its training data.

Jacobian A matrix made from the partial derivatives of given functions.

Least squares A solution that makes the sum of the squares of the *residuals* a minimum.

Machine learning A type of artificial intelligence that teaches machines to learn and perform tasks.

Normal equations The equations that, when solved, give the best estimate of a set of values by the principle of *least squares*. They take the form $\mathbf{M}^{\mathrm{T}}(\mathbf{MX} + \mathbf{L}) = 0$ where $\mathbf{MX} + \mathbf{L} = 0$ represents the *condition equations*.

Overfitting A machine learning problem that occurs when a model is too closely tuned to training data, making it unable to generalize to new data.

Pseudo-range The distance to a satellite based on the time for light to travel from the satellite to the receiver, assuming the clocks are synchronized. The speed of light equals 299,792,458 meters per second.

Residual The difference between an observed and a predicted value. Also referred to as the *error term*.

Supervised learning A machine learning technique that uses labeled data to train algorithms to predict outcomes.

Training data A collection of labeled information used to train a machine learning model.

Transfer function A mathematical function that models a system's output for each possible input.

Variance–covariance matrix Another name for the *dispersion* matrix.

Further Reading

There are many excellent books covering basic mathematics and more specific techniques in topics such as algebra, calculus, geometry, and statistical analysis. For the more advanced, CRC Press/Taylor & Francis publishes a wide range of texts on both mathematics and geoscience—see www.crcpress.com. The following is a small selection of some others that are available.

GIS

Heywood I., Cornelius C., and Carver S. (2012) *An Introduction to Geographical Information Systems*, 3rd ed. New York, NY: Prentice Hall.

Longley P.A., Goodchild M.F., Maguire D.J., and Rhind D.W. (2015) *Geographical Information Science and Systems*, 4th ed. Hoboken, NJ: Wiley.

MATHEMATICS

Allan A.L. (2019) *Maths for Map Makers*, 2nd ed. Scotland, UK: Whittles Publishing.

Borowski E.J. and Borwein J.M. (2005) *Collins Internet-Linked Dictionary of Mathematics*. New York, NY: Collins.

de Smith M.J., Goodchild M.F., and Longley P.A. (2018) *Geospatial Analysis: A Comprehensive Guide*, 6th ed. London, UK: Winchelsea Press.

Ghilani G.D. (2017) *Adjustment Computations—Spatial Data Analysis*, 6th ed. Hoboken, NJ: John Wiley & Sons.

SPECIFIC TOPICS

Ghilani G.D. and Wolf P.R. (2017) *Elementary Surveying. An Introduction to Geomatics*. New York, NY: Pearson.

Iliffe J.C. and Lott R. (2008) *Datums and Map Projections: For Remote Sensing, GIS, and Surveying*, 2nd ed. Scotland, UK: Whittles Publishing.

Mikhail E., Bethel J., and McGlone J.C. (2001) *Introduction to Modern Photogrammetry*. Hoboken, NJ: John Wiley and Sons.

Salomon D. (2022) *Curves and Surfaces for Computer Graphics*. New York, NY: Springer.

Vince J. (2010) *Mathematics for Computer Graphics*. New York, NY: Springer.

Wolf P.R., Dewitt B.A. and Wilkinson Benjamin E. (2014) *Elements of Photogrammetry with Applications in GIS*. New York, NY: McGraw Hill.

Index

D

E

F

G

O

P

Q

R

S

W

X

Y

Z